AI in Banking

Liyu Shao • Qin Chen • Min He

AI in Banking

Practical Applications and Case Studies

 Springer

Liyu Shao
Beijing, China

Qin Chen
Chongqing, China

Min He
Chongqing, China

ISBN 978-981-96-3836-9   ISBN 978-981-96-3837-6 (eBook)
https://doi.org/10.1007/978-981-96-3837-6

Jointly published with China Machine Press Co., Ltd.
The print edition is not for sale in China (Mainland). Customers from China (Mainland) please order the print book from: China Machine Press Co., Ltd.

Jointly published with China Machine Press Co., Ltd.

© China Machine Press Co., Ltd 2025

This work is subject to copyright. All rights are solely and exclusively licensed by the Publisher, whether the whole or part of the material is concerned, specifically the rights of reprinting, reuse of illustrations, recitation, broadcasting, reproduction on microfilms or in any other physical way, and transmission or information storage and retrieval, electronic adaptation, computer software, or by similar or dissimilar methodology now known or hereafter developed.

The use of general descriptive names, registered names, trademarks, service marks, etc. in this publication does not imply, even in the absence of a specific statement, that such names are exempt from the relevant protective laws and regulations and therefore free for general use.

The publishers, the authors, and the editors are safe to assume that the advice and information in this book are believed to be true and accurate at the date of publication. Neither the publishers nor the authors or the editors give a warranty, express or implied, with respect to the material contained herein or for any errors or omissions that may have been made. The publishers remain neutral with regard to jurisdictional claims in published maps and institutional affiliations.

This Springer imprint is published by the registered company Springer Nature Singapore Pte Ltd.
The registered company address is: 152 Beach Road, #21-01/04 Gateway East, Singapore 189721, Singapore

If disposing of this product, please recycle the paper.

Foreword 1

This book provides a comprehensive look at the practical applications of AI in banking, recording the genuine experiences of innovators and practitioners.

When the author sent me the manuscript, I hesitated to accept it immediately, with a reverence for new technologies. I agreed to read it, despite not being a technical expert. The digital transformation of banks is a major contemporary challenge, tempting me to explore this book. It includes numerous specific AI technologies and concepts, along with successful applications, which captivated my curiosity like a refreshing scent. I consulted Mr. Liang Lifang, who served as the General Manager of the Software Center at the Industrial and Commercial Bank of China from 2001 to 2010, and is one of the most renowned technical experts in the industry. His support encouraged me.

In the context of banking digitalization, two key questions must be understood to grasp the significance of this book. First, what is a digital bank? How do we transition to digitalization? Second, what is AI? Once we understand the meaning of a digital bank, the applications of AI become clearer. Since the turn of the twenty-first century, with the rapid development of China's financial industry and the rise of new information technologies, computer applications in banking have been flourishing, propelling the banking industry into the digital age and transforming traditional operations. Reflecting on the entire development process of financial technology, our understanding of banking digitalization is no longer superficial.

First, what is digital banking? Throughout the emergence and development of any significant new technology, people continuously reveal, explore, and understand its crucial functions and significance from different perspectives, witnessing its vast potential. This is the charm of new technology and thinking. There are various definitions of digital banking:

- Digital banking increasingly moves financial services to digital devices, providing products and services to customers without relying on time, place, tellers, customer managers, or risk control personnel. Banks can better understand customer needs, continuously optimizing business processes.

- Digital banking involves the complete digitization of management activities, enabling clients to benefit from digital services and products, and facilitating a seamless digital connection between front-end and back-end processes, leading to fully automated bank operations. It represents an upgrade of online or mobile banking.
- Digital banking utilizes the latest information technologies, including mobile communication, big data, artificial intelligence, biometrics, and blockchain, to achieve automated business processes and intelligent operations, breaking the constraints of time and space to greatly improve efficiency, optimize services, and reduce costs while enhancing effectiveness.

The discourse on various expressions often focuses on the technical aspects of banking digitization, emphasizing business and service operations. However, when considering a higher, more profound perspective to unveil its core principles, the significance of digital banks in achieving modern society becomes clearer. General Secretary Xi Jinping pointed out, "Digital technology and the digital economy are the forefront of the world technological revolution and industrial transformation, key areas of the new round of international competition. We must seize the initiative and capture the commanding heights of future development." In this context, digitalization represents the call of the future, an era-defining marker of technological application level, and the development model of modern banks.

Historically, banking technology has generally remained at the level of technological application, evolving from electronification, networking, and informatization to the current stage of datafication. From the initial promotion and application of various individual technologies such as the Internet, cloud computing, big data, blockchain, and the mobile Internet to artificial intelligence, these technologies have become standard in banking operations. These foundational applications have laid the groundwork for digital banks by integrating various technologies through digitization, ultimately aiming to fundamentally transform the business management model.

The overarching direction for banking technology is to advance financial digital transformation, driven by digital technology and leveraging the development of artificial intelligence as a new opportunity. This involves understanding the new trends in digital economic development, embedding digital elements in the entire service process, and infusing digital thinking throughout the business operations chain. The goal is to build a modern financial service system centered around users and scenarios, rebuilding intelligent services and achieving comprehensive intelligent financial services for public benefit. Digitalized operations will also create new momentum, promoting innovation across the entire process and chain and establishing "boundaryless" all-channel financial service capabilities. Digitalization thus becomes the most crucial tool for realizing the modern bank.

The focus of FinTech development lies in the process of digitalization and popularization within banks. Purely technology-driven approaches only solve application issues at the management level. Digital transformation must ultimately lead to innovative business models; technology serves merely as a bridge, reflecting the profound changes in banking operations and societal service. This is the essence of "digitalization"! It is not simply an extension of electronic systems but an advanced stage of banking informatization. While earlier technological applications represent quantitative changes, digitalization is a qualitative leap, innovating business models and products. The future modern bank is a digital bank, and accelerating digital transformation is the blueprint and mission of FinTech development.

Digital banking should meet two demands: first, to serve the digital society, as highlighted in the 14th Five-Year Plan, necessitating synchronized progress; second, to build digital banks, with both aspects complementing each other. This entails fully leveraging the latest technologies to analyze internal management data and external sources, yielding valuable insights. This process opens new markets, expands customer bases, innovates business models, and develops new products, creating value for customers, aiding in scientific pricing, precise marketing, and stringent risk control. It supports the sustainable development of the banking sector, achieving transformation, serving modern society, and enhancing operational capability and financial service efficiency.

Second, what is AI? Artificial intelligence (AI) is the abbreviation. It studies and develops theories, methods, technologies, and systems to simulate, extend, and enhance human intelligence. It is a branch of computer science that involves areas such as robotics, speech recognition, image recognition, natural language processing, and expert systems.

In banking technology, AI's application typically includes big data, biometrics, cloud platforms, and blockchain. AI is particularly crucial. Its applications integrate closely with big data platforms and usually incorporate intelligent customer service, smart investment advisory, intelligent voice and image analysis, online loans, targeted marketing, risk control, and more.

This book excels in two aspects: First, it provides valuable case studies in precise marketing, client acquisition, risk control, intelligent voice and image analysis, and intelligent customer service—projects proven to be successful and effective. Second, given AI's multidisciplinary nature in banking applications, this book offers an unprecedented, detailed examination of background analysis, concept introductions, design schemes, implementation processes, and project summaries. This is rare and innovative.

In summary, the author provides an outstanding guide for banks seeking digital transformation. This practical experience manual is highly recommended for IT decision-makers and tech professionals in the banking industry.

Industrial and Commercial Bank of China, Beijing, China

Zhang Qu

Foreword 2

China's banking industry has evolved through distinct phases: the branch era (Bank 1.0), online banking era (Bank 2.0), mobile Internet era (Bank 3.0), and the technology finance era (Bank 4.0). These phases mirror the stages of digitalization within the financial sector. Analysis indicates that a 10% increase in digitalization correlates with a 0.5–0.62% rise in per capita GDP. In 2021, China's digital economy reached 45.5 trillion yuan, accounting for 39.8% of its GDP. By 2025, it is expected that half of the global economic value will be derived from the digital economy. This economy depends on digital technology, which now deeply integrates with all sectors, driving socio-economic development. Digital technology is crucial for industrial transformation, economic and social transitions, fostering new economic drivers, and enhancing commercial banks' competitive advantages in the post-financial crisis era. The integration of digital technology in commercial banking has led to technology finance. Innovative digital technologies like big data and AI have reshaped banking operations fundamentally. Numerous Internet banks and Internet of Things banks, such as WeBank, Xinnet Bank, E-Business Bank, Baixin Bank, and Xishang Bank, have gained regulatory approval, embedding financial services into various online non-financial scenarios. The emergence of open banking is gradually eroding the boundaries of traditional banking operations. Commercial banks are now exploring new cross-border cooperation models. New digital technologies—data-driven intelligent decision-making, blockchain, 5G, intelligent wearables, biometrics, and edge computing—are widely adopted. Consequently, digital technology has become the most significant productivity factor for banks.

In recent years, the banking industry has coalesced around the notion of "digital transformation," with a significant focus on "scene digitization." Scene digitization, a digital representation of scene finance, involves quantifying various financial scenarios (clients, products, settings, objectives, etc.) and utilizing mathematical models to derive optimal decisions. By harnessing data, algorithms, and computational power, it enables comprehensive operation of scene finance, providing critical data for high-level decisions and ongoing business iteration. Intelligent data systems manage business operations, customer analytics and maintenance, personalized long-tail services, and fund management. The concept of "scene" is expansive, encompassing aspects of production and life, such as rural revitalization, online lending, healthcare education, government-enterprise alliances, matrimonial and real estate sectors, gaming, and entertainment. Commercial banks integrate financial services into these scenarios via cooperative channels, smartphones, and mobile devices, thus significantly broadening their business scope. In these scenarios, commercial banks find avenues for customer acquisition, battlefield engagement, conversion, and retention. This approach is essential for fostering comprehensive social and economic advancement.

In the era of Bank 4.0, digital technology penetrates deeply into every aspect of commercial bank operations. This includes technology strategy, technology planning, architecture, capabilities, platform construction, and innovation culture. Scenario digitalization largely dictates a bank's technological and financial capabilities and is crucial for future competition.

The book "Bank AI Project Practice: AI Solutions and Case Implementation for Typical Business Scenarios" exemplifies digitalization in commercial bank scenarios, summarizing the author's extensive practical experience. It leverages various digital technologies to enable computers to listen, speak, read, write, and think. The book covers intelligent marketing, risk control, and operations, providing comprehensive theoretical and technical explanations suitable for banking professionals.

The core of this book is to share practical experiences that can be immediately applied, promoting quick iteration from scratch. It encourages more banking professionals to engage in scenario digitalization, collectively shaping the future of bank digital transformation.

Bank 4.0 begins with scenario digitalization! Start this journey by reading this book!

Zhongguancun Internet Finance
Research Institute, Beijing, China

Liu Yong

Zhongguancun Financial Technology
Industry Development Alliance, Beijing,
China

Preface

Artificial intelligence (AI) stands as one of the foremost promising technological fields today and in the future. In recent years, with the explosion of terminal data and cloud data and the maturity of algorithms and computing power, more and more difficult problems have been solved by AI, which makes research talents, venture capital, and application scenarios pour into the field of AI continuously. AI has become a technology that is widely applied in many fields, such as education, medical care, food, industry, finance, tourism, and government affairs. In terms of academic research, new research achievements have been made, resulting in many new interdisciplinary disciplines and subdivisions. In terms of business, technology enterprises with AI background have easy financing, and AI enterprises and AI scenarios are blooming everywhere. In terms of talents, enterprises have strong demand, positions are well paid, and talents are in short supply. Phenomena such as "a paper opening up a new field," "a startup becoming a unicorn," and "new technology disrupting an industry" have become commonplace. AI has thus ushered in profound societal changes and new development opportunities for countless startups.

For commercial banks, AI represents the high ground today and in the future, being crucial for their digital transformation. China is in the period of data explosion, information explosion, and knowledge explosion. Commercial banks, as the core elements of the national economy, must adapt to the needs of economic and social development and explore a new model to lead economic development. This requires a transformation from the development model of network building, investment fees, and networking to the development model of scenario-based, intelligent, and digital development. At present, the vast majority of newly registered private commercial banks are almost pure AI technology companies, while traditional commercial banks have also set up technology subsidiaries or independent AI data departments to full-time digital operations. It can be said that the application ability of AI technology of commercial banks determines its business model, market share, customer experience, and profitability to a large extent, all of which are key elements of market competitiveness. In the future, the bank's business will surely be seamlessly implanted into various life scenarios and production scenarios through the Internet,

AR glasses, embedded devices, third-party channels, and other forms. It can be said that the future competition of commercial banks is, to a large extent, the competition of AI capabilities.

This book shares the author's AI practical projects in the scientific and technological innovation work of commercial banks. The application scenarios include intelligent marketing, intelligent risk control, and intelligent operation, covering four major sectors: retail marketing, electronic banking, credit business, and science and technology operation and maintenance. The content of this book involves the AI thinking brain (left brain) and the AI perception brain (right brain), which gives the computer the ability to listen, speak, read, write, and think. It provides scenario applications of AutoML, ensemble learning, graph computing, recommendation system, causal inference, generative adversarial network, supervised learning, unsupervised learning, computer vision, reinforcement learning, fuzzy control, automatic control, speech semantics, Bayesian networks, edge computing, and other technologies. Each chapter is a project, respectively, from the project background significance, project implementation, algorithm principle, code explanation, project effect, and other aspects of the elaboration. The practical projects introduced in this book have the nature of research and exploration, avoiding obscure mathematical formulas and principle analysis and explaining how to apply AI technology in the operational scene of commercial banks in simple and understandable popular language.

This book shares the author's practical AI projects in the domain of commercial banking technology innovation. The applications cover smart marketing, intelligent risk control, and intelligent operations, encompassing retail marketing, electronic banking, credit services, and technology operations. The content involves AI thinking brain (left brain) and AI perception brain (right brain), which provide computers with abilities, such as listening, speaking, reading, writing, and thinking. It includes the application of various technologies like automatic machine learning, ensemble learning, graph computing, recommendation systems, causal inference, generative adversarial networks, supervised learning, unsupervised learning, computer vision, reinforcement learning, fuzzy control, automatic control, speech semantics, Bayesian networks, and edge computing. Each chapter represents a project, detailing the project's background significance, implementation, algorithm principles, code explanation, and project effects. The practical projects described in this book are exploratory in nature and avoid complex mathematical formulas and principles, using easy-to-understand language to illustrate how AI technology can be applied in commercial banking.

Sharing real case practical experience is a major feature of this book. For the reader's convenience, all projects provide steps for environment setup, data files, and project source codes. Readers can follow the WeChat official account "Artificial Intelligence and Metaverse Industry Application" (WeChat ID: AI7Meta), register, and download all data files and codes of this book.

This book aims to provide practical experience to accelerate readers' self-built project development speed. It has a certain level of professionalism, and readers need to have basic knowledge of mathematics and machine learning. It is recommended to first understand the project design and algorithm concepts before running the projects to deepen the understanding.

Looking ahead, AI will continue to have a profound impact on commercial banking in several ways: data-driven banking decision-making will shift from partial intelligence to comprehensive intelligence; AI customer service will replace manual labor, significantly boosting productivity; blockchain combined with supply chain will achieve the integration of information flow, capital flow, and logistics across multiple institutions; the rise of AI personalized private banking will fully transform high-end services in commercial banking; leveraging AI technology, banking business will be extensively embedded into various online non-banking business scenarios, collaborating with cross-industry partners to achieve open banking.

Undoubtedly, AI has become a strategic development priority for commercial banks. The cultivation and introduction of AI talents, the establishment of AI thinking, and the construction of an AI ecosystem greatly influence the future success of commercial banks.

Due to the author's limited capabilities, this book may have shortcomings. Readers are welcome to provide criticism and corrections. The author's email address is chenchenqin_88@qq.com. Thanks to our family and friends for their understanding and support during our writing and to the Machinery Industry Press for their strong support.

Chongqing, China
May 15, 2022

Chen Qin

Contents

Part I Smart Marketing

1 Mobile Banking Potential Monthly Active Customer Mining: Automated Machine Learning Techniques 3
   1.1 Introduction to AutoML 6
   1.2 Developing Frameworks and Libraries 8
      1.2.1 An Important Feature Selection Library: Feature_selector 8
      1.2.2 An Important Feature Selection Library: Boruta 13
      1.2.3 An AutoML Modeling Framework: Flaml 15
      1.2.4 AutoML Framework: Autogluon 20
      1.2.5 Bayesian Optimization Library: Bayesian Optimization 21
   1.3 Case Practice 27
      1.3.1 Operation Environment Construction 27
      1.3.2 Data Set Preparation 29
      1.3.3 Feature Selection Code Practice 34
      1.3.4 Actual Automation Modeling Code 39
      1.3.5 Automated Reference Code Practice 42
   1.4 Case Summary 44

2 Retail Potential High-value Customer Identification: Graph Neural Network Technology 47
   2.1 Introduction to Graph Neural Network 48
      2.1.1 The Concept of Graph Neural Network 48
      2.1.2 The Advantages of Graph Neural Network 51
      2.1.3 The Development of Graph Neural Networks 54
      2.1.4 Graph Neural Network as the Product of Big Data Era 56
   2.2 Scheme Design 58
   2.3 Graph Convolutional Neural Network Algorithm 60

	2.4	Development Framework	62
		2.4.1 Neo4j Graph Database	62
		2.4.2 DGL Graphical Neural Network Framework	63
	2.5	Case Practice	65
		2.5.1 Environment Preparation	66
		2.5.2 Code Practice	74
	2.6	Case Summary	85
3	**Accurate Recommendation for Banking: Recommender System**		**89**
	3.1	Introduction to the Recommendation System	90
	3.2	Recommendation Algorithm	92
		3.2.1 Collaborative Filtering Algorithm	92
		3.2.2 PersonalRank Diagram Recommendation	96
		3.2.3 Text Convolutional Neural Network	99
		3.2.4 Two-Tower Model	99
	3.3	Development Framework	103
		3.3.1 Computing Framework: PySpark	103
		3.3.2 Word Segmentation Framework: Pkuseg	104
		3.3.3 Deep Learning Framework: TensorFlow and Keras	105
	3.4	Case Practice	105
		3.4.1 Data Preparation	105
		3.4.2 Environment Preparation	107
		3.4.3 Code Practice	107
	3.5	Case Summary	126
4	**Assessing the Value of Bank Online Marketing Posts: Reinforcement Learning Techniques**		**127**
	4.1	Introduction to Reinforcement Learning	128
		4.1.1 Development of Artificial Intelligence and Reinforcement Learning	128
		4.1.2 Basic Concepts of Reinforcement Learning	131
		4.1.3 Q-Learning Algorithm	132
	4.2	Case Practice	134
	4.3	Case Summary	139
5	**Modeling Binary Causal Effects of Related Repayments: Causal Inference Techniques**		**141**
	5.1	Introduction to Causal Science	142
	5.2	Causal Forest Algorithm	145
	5.3	Developing the Library	148
	5.4	Case Practice	149
		5.4.1 Data Preparation	149
		5.4.2 Environment Setup	150
		5.4.3 Code Practices	150
	5.5	Case Summary	156

Part II Intelligent Risk Control

6 Telecom Fraud Money Laundering Account Recognition Case: Multiple Machine Learning Techniques 159
   6.1 Case Pain Point: Limitations of Anti-Telecom Fraud Risk Control Rules in the Banking Industry 160
   6.2 Modeling Techniques and Scenario Analysis 162
       6.2.1 A Solution for "Real-Time Dynamic Adjustment of Risk Control Rules": Continuous Real Depth Feature Synthesis Technology 162
       6.2.2 A Solution for "Non-objective and Incomplete Risk Control Rules": Unsupervised Adversarial Machine Learning Techniques 167
       6.2.3 A Solution for "Unclear Expression of Fuzzy Risk Control Rules": Fuzzy Control Techniques 180
   6.3 Case Practice 182
       6.3.1 Environment Setup 183
       6.3.2 Code Practice 184
   6.4 Case Summary 196

7 Developing a Dialectal Speech Phone Collection Bimodal Robot from Scratch: Intelligent Voice Q&A Technology 199
   7.1 Scheme Design 202
   7.2 Intelligent Q&A Technology 205
       7.2.1 Basic Tasks of Intelligent Voice Q&A System 205
       7.2.2 ASR Automatic Speech Recognition Technology 208
       7.2.3 QuartzNet Model 211
       7.2.4 Q&A Technology Based on Free Text Reading Comprehension 214
       7.2.5 Text-to-Speech Synthesis Techniques 217
       7.2.6 Transfer Learning 217
   7.3 Development Framework 218
       7.3.1 Nvidia NEMO Conversational AI Framework 218
       7.3.2 ESPnet End-to-End Voice Processing Framework 219
       7.3.3 Transformers Model Library 219
       7.3.4 PyQt5 Cross-Platform GUI Framework 221
       7.3.5 SIP Protocol and PJSIP Framework 222
   7.4 Case Practice 224
       7.4.1 Hardware and Software Environment Setup and Case Runs 225
       7.4.2 Code Practice 234
   7.5 Case Summary 255

8 Chattel Collateral Warehouse Visual Monitoring Project: Image Understanding Technology 257
    8.1   Scheme Design 258
    8.2   Development Libraries and Frameworks 260
           8.2.1   Computer Vision Processing Library: OpenCV 262
           8.2.2   Open-Source Library for Face Recognition: Face_Recognition 263
           8.2.3   Instance Segmentation Open-Source Library: Yolact 266
           8.2.4   Deep Learning Image Processing Library with Target Detection Migration Learning: ImageAI 270
           8.2.5   Framework and Pyecharts Data Visualization Library: Django 272
    8.3   Case Practice 274
           8.3.1   Hardware and Software Environment Setup and Case Runs 274
           8.3.2   Code Practice 280
    8.4   Case Summary 297

9 Personal Loan Delinquency Prediction Project: Bayesian Network Techniques 299
    9.1   Introduction to Bayesian Networks 300
           9.1.1   Bayesian Learning Concepts 300
           9.1.2   From Bayesian Learning to Bayesian Networks 302
    9.2   Probability Graph Calculation Library: Pgmpy 305
    9.3   Case Practice 305
           9.3.1   Environment Setup and Case Runs 305
           9.3.2   Code Practice 306
    9.4   Case Summary 312

Part III Intelligent Operation

10 Enterprise WeChat Private Traffic Customer Cold Start Program: Automated Control Technology 315
    10.1   Program Design 318
    10.2   Development Library 319
           10.2.1   Underlying Interface Library: Pywin32 320
           10.2.2   Image Processing Library: Pillow 322
           10.2.3   Computer Vision Processing Library: OpenCV 322
           10.2.4   Data Processing Library: Pandas 322
           10.2.5   Pynput Library 323
    10.3   Case Practice 323
           10.3.1   Environment Setup and Case Runs 323
           10.3.2   Code practice 325
    10.4   Case Summary 334

11 Intelligent Inspection Robot for Commercial Bank Data Centers: Computer Vision Technology 335
   11.1 Scheme Design 336
   11.2 Computer Vision Technology 337
      11.2.1 Raspberry PI 338
      11.2.2 HSV Color Space 340
      11.2.3 Median Filtering 343
      11.2.4 Edge Computing 343
   11.3 Developing Libraries 345
      11.3.1 Computer Vision Processing Library: OpenCV 345
      11.3.2 Scientific Computing Library: Numpy 345
   11.4 Case Practice 345
      11.4.1 Environment Setup and Case Runs 346
      11.4.2 Code Practice 347
   11.5 Case Summary 354

About the Authors

Liyu Shao is a senior banking technology expert with over 26 years of experience. He has extensive expertise in both managing large-scale banking IT projects and architectural planning of large projects. He has made significant contributions to the fields of big data assets, data element markets, and artificial intelligence. Mr. Shao has led numerous major IT projects for commercial banks and has received multiple prestigious awards. He is the author of *Research and Practice of Big Data Governance in Commercial Banks* and has published several papers in authoritative journals, including "Construction and Practice of Bank Big Data Risk Control Capability" and "Analysis of Core Data Capabilities in Commercial Bank Data Governance."

Qin Chen He is a banking technology expert with over 23 years of industry experience, currently serving as Deputy General Manager of the Information Technology Department at a commercial bank branch. He was honored as one of the bank's inaugural "Top 10 Technology Stars." He serves as a researcher at the Chongqing Branch of the National New-Type Crime Research Center and is a member of the Financial Technology Working Group within Chongqing's Anti-Money Laundering Talent Pool. With a specialization in data intelligence, computer vision, recommendation systems, natural language understanding, and knowledge graphs, he brings 10 years of experience in developing AI applications for a major commercial bank. Her independently developed banking AI projects include "End-to-End AI Applications in Financial Consumer Complaint Management," "Intelligent Conference Behavior Management System," "AI-Powered Telecom Fraud Account Detection Model," "AR-Based Interactive Financial Scenarios," "High-Value Customer Mining Based on Social Network Analysis," and "Intelligent Financial Scene Text Recognition." These projects have earned her the bank's First Prize in Software Development, First Prize in Big Data Innovation, Second Prize at the 2021 Chongqing Banking Association Outstanding Research Project, Chongqing Financial Data Comprehensive Pilot Project, and Third Prize in

Chongqing's 2019 Financial Technology Research. He has published multiple academic papers, including "Graph Neural Networks in Banking: Marketing and Risk Control Applications," "The Middle Way to Resolve Banking Technology Practical Contradictions," and "Analysis of the Disconnect and Integration Between Bank IT and Business Operations."

Min He a Senior Banking Architect with a decade of experience in core banking project development. He specializes in banking application architecture planning and has conducted extensive research in blockchain, artificial intelligence, and big data domains. He has led multiple digital innovation projects in the financial scenarios and participated in numerous provincial-level key research initiatives. His notable achievements include receiving the third prize from the Banking and Insurance Regulatory Commission for the research project "Research and Practice of Traditional and Internet Core Dual Integration Architecture," the third prize from the People's Bank of China for "Robotic Process Automation and AI Applications in Bank Operations Data Management," and an excellence award in the National Financial Standardization Research program for "Research on National Cryptographic Standards Promoting Financial Information Security." His paper, "Technical Innovation and Optimization Practices in Core Banking Systems," was published in "Financial Technology Time" magazine.

Part I
Smart Marketing

Chapter 1
Mobile Banking Potential Monthly Active Customer Mining: Automated Machine Learning Techniques

Mobile banking refers to the use of smartphones to complete various banking services. As the most important portal for retail banking, it functions to attract, engage, retain, and convert customers. It plays a vital role in competing for long-tail customers, collecting merchant entry commissions, reducing bank operating costs, enhancing customer experience, and boosting financial product sales. The "2021 China Digital Finance Survey Report" highlights four observations that underscore the market position of mobile banking.

Observation 1: Mobile banking growth outpaces personal online banking. Since 2015, mobile banking has maintained double-digit growth rates for 6 consecutive years, increasingly becoming the main output product of financial technology and a key tool for digital transformation. Online banking is gradually becoming marginalized, with its customer growth rate slowing down. In 2021, the customer penetration rate for mobile banking was 81%, with a growth rate of 15%, while personal online banking had a penetration rate of 63% and a growth rate of 7%, as shown in Fig. 1.1.

Observation 2: Mobile banking has become the fastest growing e-banking channel. From 2017 to 2021, the growth rate of mobile banking is more than 6%, higher than that of online banking, WeChat banking and telephone banking channels, as shown in Fig. 1.2.

Observation 3: Mobile banking has become the most used e-banking channel. The results of the online survey show that the usage rate of mobile banking customers will be 85% in 2021, which is higher than other electronic channels, as shown in Fig. 1.3.

Observation 4: Mobile banking has become the focus of commercial bank strategists. Banks frequently introduce hardcore tricks, constantly enrich mobile banking functions, and accelerate innovation through artificial intelligence and scenarioization. Mobile banking has ushered in the post-app era, promoted the "mobile banking +" mobile terminal layout, and opened up the three-party traffic entrance. The financial attribute of mobile banking has gradually weakened, replaced by the "scene + social" attribute, that is, with basic financial services as the support,

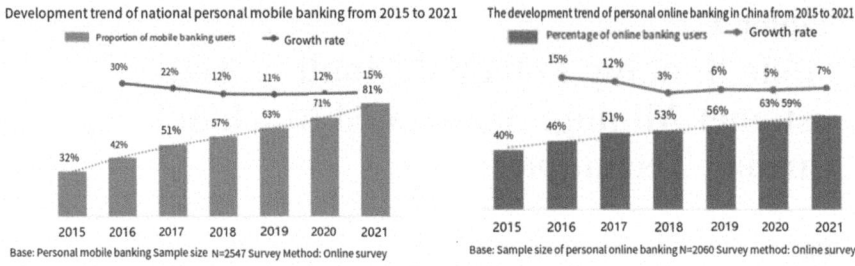

Fig. 1.1 Mobile banking is growing faster than personal online banking

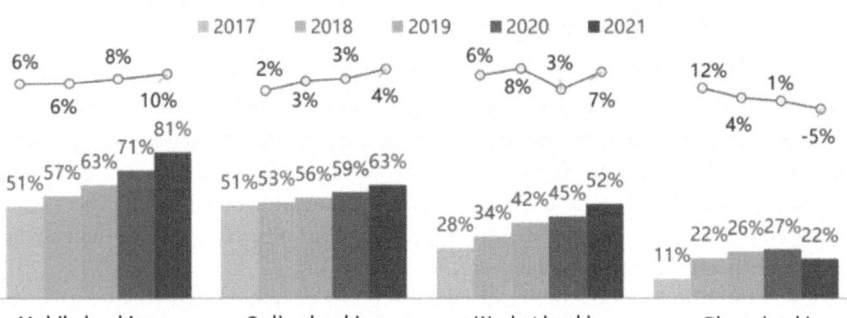

Fig. 1.2 Growth trend of retail e-banking channels

Fig. 1.3 Customer usage rate of retail e-banking channels

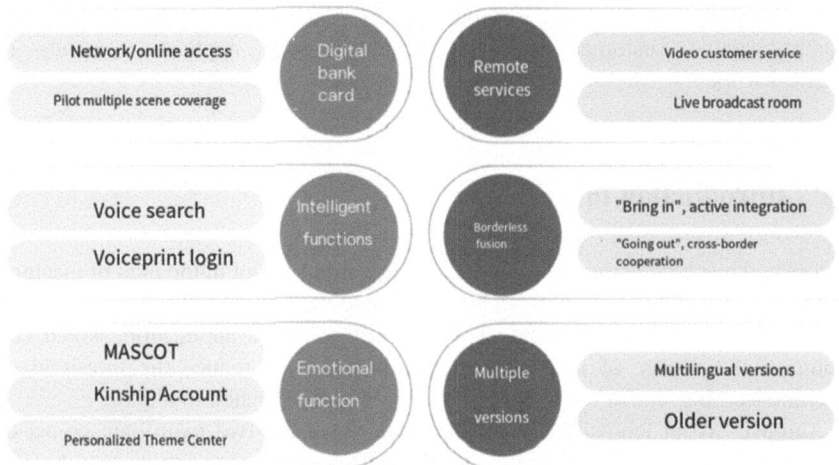

Fig. 1.4 New business form of mobile banking in the post-APP era

mobile App as the carrier, and high-frequency life scenes as the driver, to reshape the new business form of mobile banking. At the same time, various commercial banks have begun to explore the use of WeChat, Alipay, and other Internet to divert external customers and achieve effective transformation. Commercial banks position mobile banking as the main starting point and landing platform of digital transformation, as well as the strategic core of retail business, as shown in Fig. 1.4.

Due to the infrequency of financial transactions, customer loyalty to banks is low, leading to easy attrition, difficulty in uncovering customer value, and challenges in establishing an open banking ecosystem. Banks must focus on operating mobile banking by integrating financial functions with open internet channels and various online lifestyle scenarios. This includes onboarding merchants, offering diverse marketing activities, enhancing usability and customer experience, and increasing the number of active monthly mobile banking customers. Transforming low-frequency scenarios into high-frequency ones can help acquire richer customer data, maintain and uncover customer value, and boost customer loyalty and retention. This leads to the development of a new financial ecosystem adapted to the internet era.

Monthly active mobile banking customers refer to those who log in and use mobile banking within a given month, reflecting the service utilization efficiency and maturity of the financial ecosystem. Potential monthly active customers are those who did not log in during the month but have characteristics making them likely to convert to active status under targeted marketing strategies. Maximizing the number of mobile banking monthly active customers is the goal of mobile banking operations. Accurately identifying potential active customers is an effective approach to achieving this goal, holding significant strategic importance for bank operations.

This case proposes a set of methods based on AutoML to predict the potential monthly active customers and has achieved good application results in practical work.

1.1 Introduction to AutoML

Automated machine learning (AutoML) is a research hotspot in the field of machine learning in recent years. It refers to the theory and method of automating the whole process of machine learning model from construction to application, which can minimize the degree of manual participation and thus reduce the threshold of machine learning. It can realize fast and convenient automated modeling.

AutoML, as an innovative technology, addresses the two main pain points of traditional machine learning: being cumbersome and time-consuming, and having high entry barriers. Traditional machine learning modeling tasks involve several steps, such as problem definition, data collection, data cleaning, feature generation, feature selection, algorithm selection, model training, hyperparameter optimization, model evaluation, and model deployment. The process from data collection to model evaluation is iterative and reliant on personal experience, requiring significant manual input and often taking months to complete. Traditional machine learning usually necessitates modelers to possess mathematical knowledge, such as calculus, probability theory, linear algebra, statistics, and graph theory, as well as an understanding of various machine learning algorithms, such as classification, clustering, regression, dimensionality reduction, and graph algorithms, along with experience in hyperparameter tuning, making it difficult and with high development barriers. Automated machine learning views this iterative process as an optimal solution search process, creating a mathematical implementation within predefined model and parameter spaces. It automatically completes tasks such as automated feature engineering, automated algorithm selection, automated model training, automated hyperparameter search, and automated pipeline matching without manual intervention. This approach reduces time and labor input while minimizing the skill requirements for modelers, making machine learning modeling faster and easier. The comparison between traditional machine learning and automated machine learning is shown in Fig. 1.5.

AutoML technology was first proposed by Google in late 2017. By January 2018, they launched their first AutoML product, AutoML Vision. Currently, AutoML is rapidly advancing. Major tech companies like Baidu, Ali, Tencent, and 4Paradigm have developed their own AutoML products. Meanwhile, mainstream development frameworks such as TPOT, Auto-sklearn, Auto_ml, and HyperOpt have emerged. AutoML has found applications across various industries, including finance, education, and government affairs.

Figure 1.6 illustrates the common machine learning frameworks at the top and AutoML frameworks at the bottom. Machine learning frameworks encompass three primary sections: feature engineering, model creation, and parameter tuning, each

1.1 Introduction to AutoML

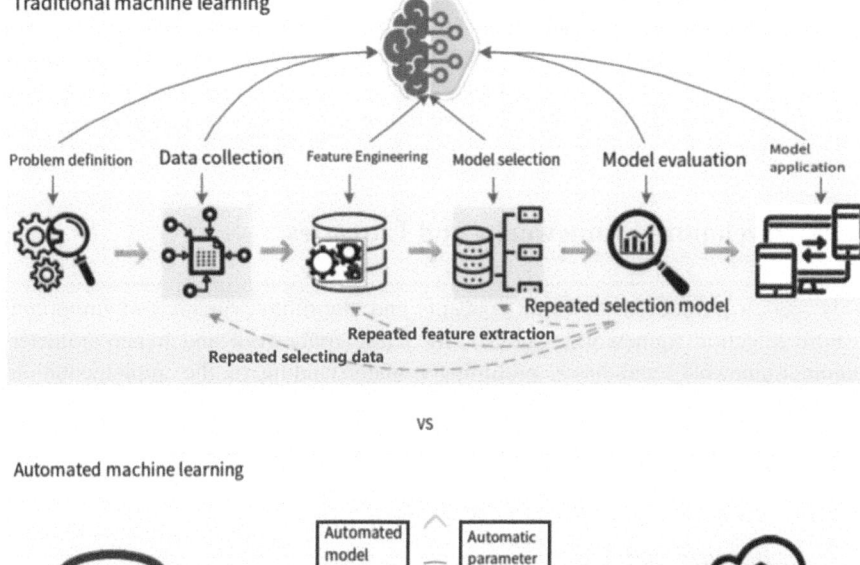

Fig. 1.5 Comparison between traditional machine learning and AutoML

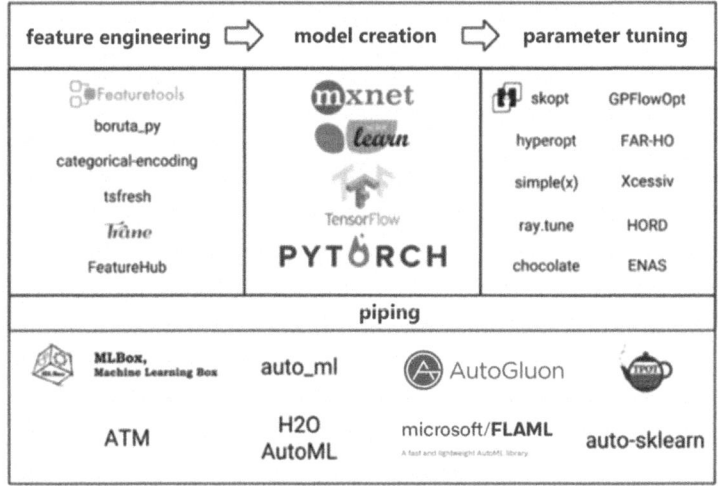

Fig. 1.6 Machine learning and AutoML frameworks

with its own dedicated development framework. AutoML frameworks often use pipelining to automate the entire machine learning process. Recently, new techniques in AutoML have emerged, such as automated ensemble learning, multi-learner learning, and automatic hyperparameter search. This book highlights several cutting-edge AutoML frameworks that we utilize in our financial practices.

1.2 Developing Frameworks and Libraries

This section introduces the basic concepts and algorithm principles of important feature selection framework, AutoML modeling framework and hyperparameter tuning framework, and has a preliminary understanding of the implementation principle and development framework of AutoML.

1.2.1 An Important Feature Selection Library: Feature_selector

In the realm of machine learning, feature engineering represents the preliminary phase of any modeling task. This critical process addresses two fundamental issues: feature generation and feature selection. Feature generation encompasses the creation of new features necessary for modeling by transforming several initial features into derived variants. Concurrently, feature selection involves rigorous evaluation to identify the most fitting features for effective modeling. A widely adopted method during this phase involves computing the information value (IV), which quantifies the significance of a specific feature's contribution to the model's predicted outcome, denoted as the y value. IV helps to gauge how well a feature informs the predictive accuracy of the model. Practitioners can conduct feature selection by either crafting bespoke code or leveraging existing machine learning tools and libraries designed to streamline this process.

Feature_selector is a feature selection library developed by Williamkoehrsen, a data scientist at Feature Labs. The library implements the following five types of feature selection:

- Features with a high proportion of missing values (identify_missing function)
- Collinear features with high correlation (identify_collinear function)
- Features with zero importance to the predicted value (identify_zero_importance function)
- Features that contribute little to model predictions (identify_low_importance function)
- Features with a single value (identify_single_unique function)

1.2 Developing Frameworks and Libraries

Feature_selector's open-source repository is simple, efficient, and open source based on the GPL-3.0 license, see https://github.com/WillKoehrsen/feature-selector. The following is an example of a credit card customer transaction prediction model developed by the authors to illustrate the main process of feature_ selector's main use process.

1. Instantiate the class: To use the FeatureSelector() function to create an instance for feature selection. Specify the training feature matrix and the training label vector to obtain a feature selection instance: fs = FeatureSelector (data=train_features, labels=train_labels).
2. Analyze the absence rate: Utilize the fs member function identify_missing() to select features with a missing rate exceeding a specified threshold. For example, the following code sets the missing rate threshold at 60%: fs.identify_missing (missing_threshold=0.6). The results of fs's operations are recorded in the dictionary member ops. To view features with a missing proportion above 60%, inspect the missing key in ops: print(fs.ops['missing']). Given that feature missingness is only related to the features themselves and not the type of task, the identify_missing() function is suitable for both supervised and unsupervised learning. Feature_selector provides visualization functions for each type of feature selection. Simply call the corresponding plot_XXXX() function to use them. To observe the missing feature situation, invoke fs.plot_missing() to generate a bar chart with the missing ratio on the x-axis and the number of features on the y-axis. Figure 1.7 demonstrates that one feature has a missing rate of 90–100%, while 50 features have a missing rate of 0–10%.
3. Analyze collinear features: In geometry, a set of points is considered collinear if they lie on a single straight line. In feature engineering, features with high correlation are termed collinear features. To identify pairs of features with correlations exceeding a specified threshold, utilize the member function "identify_collinear()" from the Feature Selector module. For instance, to detect all input features with a correlation above 30%, without applying one-hot encoding to any feature, use the following command: "fs.identify_collinear (correlation_threshold=0.3, one_hot=False)." To examine the features with correlations higher than 30%, simply access the "ops" member field with the key "collinear": "print(fs.ops['collinear'])." Given that feature correlation depends solely on the features themselves and is independent of the target task, the "identify_collinear()" method is applicable in both supervised and unsupervised learning scenarios. Moreover, the function "fs.plot_collinear()" provides a visual representation of the collinear features, where color changes indicate the strength of their correlation. Collinear features indicate the presence of inter-variable relationships, which can complicate modeling and are commonly removed. The collinear feature correlation analysis within Feature Selector is presented as a visual heatmap, as illustrated in Fig. 1.8.
4. Analyze zero-importance features: The "identify_zero_importance()" member function of the "fs" module identifies features that do not contribute to the predictive target, termed as zero-importance features. The "Feature_selector"

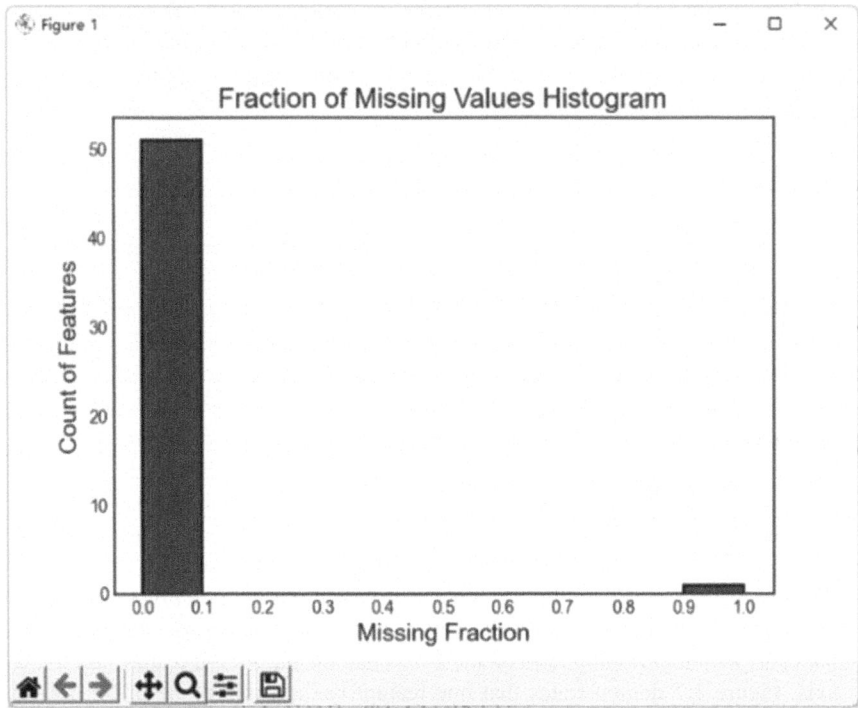

Fig. 1.7 The feature loss rate analysis of feature_selector

employs a gradient boosting machine (GBM) to assess the importance score for each feature. It normalizes these scores to identify features with an importance score of zero. To ensure minimal variance in the importance scores, "identify_zero_importance" conducts multiple GBM training iterations and averages the results to derive the final importance scores. Additionally, it extracts a subset of the data as a validation set. During GBM training, it calculates metrics on this validation set, terminating the training once predefined conditions are met. For instance, in classification tasks, the training concludes early if the validation set accuracy attains a pre-set threshold. The process involves a total of ten training iterations, as illustrated in the following code:

```
fs.identify_zero_importance(task='classification',
eval_metric='auc',n_iteration=10,early_stopping=True)
```

To check the selected zero-importance feature, simply query the ops dictionary for the zero_importance key value:

```
fs.ops['zero_importance']
```

1.2 Developing Frameworks and Libraries

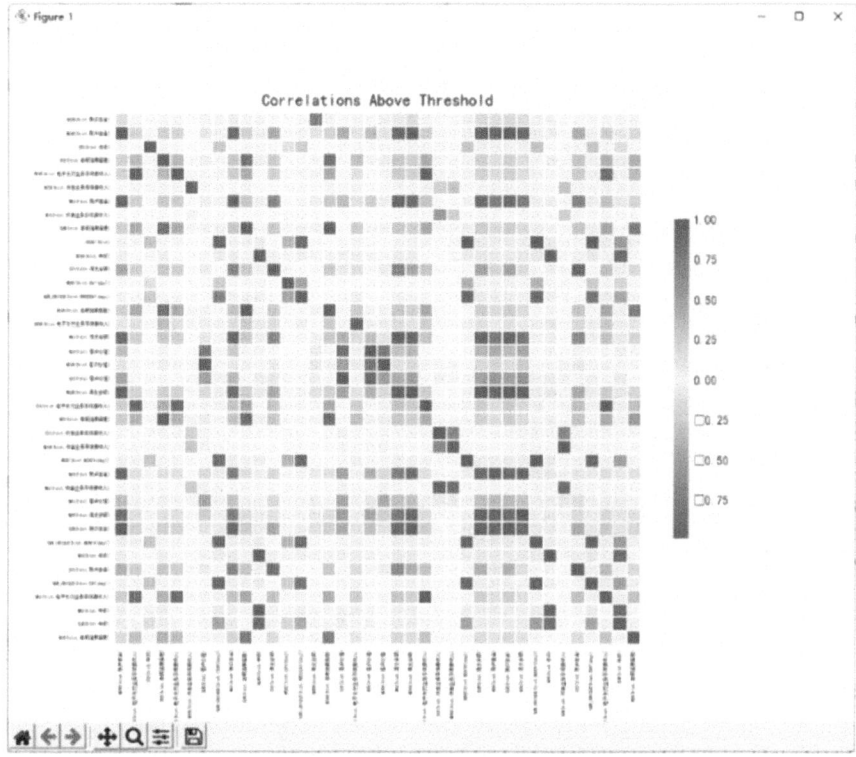

Fig. 1.8 The collinear feature correlation analysis of feature_selector

When training GBM, we get an importance score for each feature, and after normalizing the importance, we can draw an importance bar. For example, we draw the normalized bar chart of the first 51 important features (plot_n parameter), and calculate the number of features required when the cumulative importance of all features reaches 90% (threshold parameter):

```
fs.plot_feature_importances(threshold=0.9, plot_n=51)
```

Figure 1.9 shows the visual output of Feature_selector for feature importance analysis. Note that the feature importance scores on the horizontal axis are normalized, so the sum of all feature importance scores must be equal to 1.

Feature_selector then draws a dotted line. It shows about 25 features that are needed if the cumulative importance score of all features reaches 90% in the current dataset, as shown in Fig. 1.10.

Because of the randomness of GBM training, there will be some differences in the feature importance scores obtained from each training. Because the target field needs to be specified for GBM training, labels must be passed to

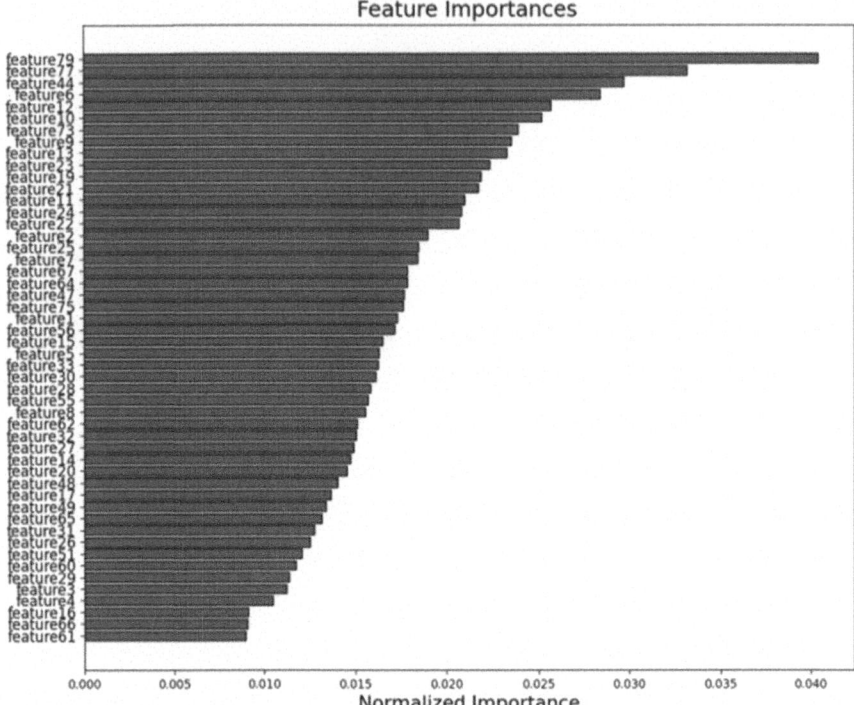

Fig. 1.9 The feature importance analysis of feature_selector

FeatureSelector() function. identify_zero_importance() applies only to supervised learning, not unsupervised learning.

5. Analysis of low importance characteristics: Using the fs member function identify_low_importance(), you can obtain the unimportant features to the right of the dotted line in Fig. 1.10. Since it is executed on a GBM basis, it can only be used for supervised learning. Execute the following code to get a list of low importance features:

```
Fs. Identify_low_importance (cumulative_importance = 0.9)

print('fs.ops[low_importance'])
```

6. Analyze the single value feature: Use fs member function identify_single_unique () to get features with only a single value. This feature has no meaning for model training and can be directly deleted. It is suitable for supervised learning and unsupervised learning. Execute the following code to get a list of single value features:

1.2 Developing Frameworks and Libraries

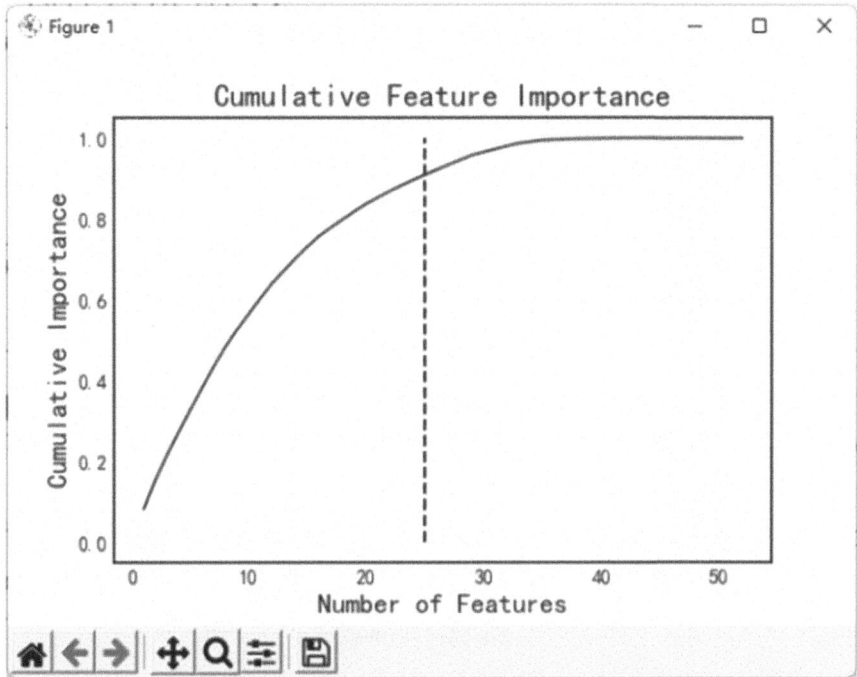

Fig. 1.10 The cumulative feature importance analysis of feature_selector

```
fs.identify_single_unique()

fs.ops['single_unique']
```

Feature_selector provides the plot_unique() function to draw a histogram of unique values for each feature, as shown in Fig. 1.11. That is, counting the number of unique values for each feature and then plotting the frequency ratio.

1.2.2 An Important Feature Selection Library: Boruta

In the realm of machine learning, the terms "cost function" and "loss function" both pertain to the model's prediction bias. However, there is a distinction between their applications. The loss function typically applies to an individual data sample, while the cost function is computed for the entire dataset and represents the mean value of all individual loss functions. In feature selection, the primary objective is to identify the subset of features that minimizes the cost function for the current model. This entails selecting only the essential features, thereby reducing redundancy. Optimal

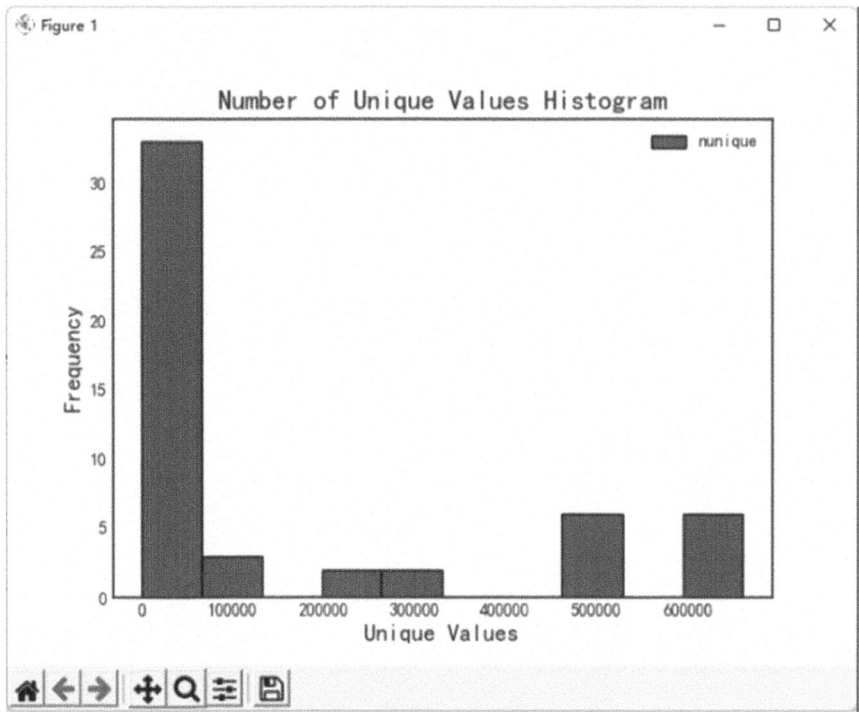

Fig. 1.11 The feature unique value analysis of feature_selector

and minimal feature selection enhances the model's fit to the dataset and mitigates unnecessary computational overhead.

Boruta stands as a significant feature selection library offering an alternative approach to the conventional methods described. Utilizing Boruta, the goal is to identify all feature sets that exhibit a correlation with the dependent variable. Through this method, we can streamline the selection process to highlight features that directly influence the prediction of the dependent variable. The Boruta algorithm is pivotal because it enables a comprehensive understanding of the factors influencing the dependent variable, facilitating more effective and efficient feature selection.

Boruta's algorithm entails an intricate procedure involving several steps to assess feature importance within a dataset. Initially, the algorithm conducts a Shuffle operation on the original real features dataset to create shadow features. Subsequently, it integrates the original and shadow features into a synthetic feature matrix for training purposes. The training phase employs a tree-based classifier, specifically a random forest, wherein the target features are substituted by synthetic features. The algorithm then evaluates the performance differences among all features to determine their relative significance. In a cyclical manner, the importance of each feature is assessed, iterating through multiple rounds of comparison between the original feature and its corresponding shadow feature. When the original features' importance markedly exceeds that of the shadow feature, the original feature is deemed significant. Contrarily, if there is no substantial difference, the original feature is categorized as unimportant. Ultimately, the importance scores derived from the shadow features establish a reference baseline. Through this comparison, the algorithm discerns the subset of features genuinely correlated with the dependent variable, selectively emphasizing those from the original dataset that hold substantial predictive value. The comprehensive process of the Boruta algorithm is thus meticulously structured to ensure the identification of truly impactful features. The algorithm process is as follows:

1. Each feature value of the feature matrix X is randomly shuffled, and the shadow feature after Shuffle is spliced with the original feature to form a new feature matrix.
2. Using the new feature matrix as input, the model that can output the feature importance is trained.
3. Calculate the Z_score of the original feature and the shadow feature. The formula for calculating Z_score is: mean of feature importance/standard deviation of feature importance.
4. Find the largest Z_score_{max} in the shadow feature and mark the original feature with Z_socre greater than Z_{max} as "important," mark the original feature with Z_score significantly less than Z_{max} as "unimportant," and permanently remove it from the feature set.
5. Delete all shadow features.
6. Repeat (1) through (5) until all features are marked as "important" or "not important."

1.2.3 An AutoML Modeling Framework: Flaml

Before introducing Flaml framework, let's first look at the core problem of AutoML to solve: optimal model search. This refers to how to find the best performing model algorithm and the best hyperparameters through algorithms when the data set and modeling task are determined. Simply put, it is to find the best solution among a set of candidate solutions. Figure 1.12 shows an optimal model search path from the highest point to the lowest point.

To optimize model performance, treat different models and parameters (Param1, Param2... ParamN) as independent variables across two dimensions, and use the cost function value or other performance metrics as dependent variables. This relationship forms a high-dimensional hypersurface. Begin searching for the optimal model by starting at an initial point, adjusting the values across model and parameter dimensions, and identifying the path that leads to the lowest Cost (illustrated as a dark line). The corresponding model and hyperparameters at this point are optimal. This search process typically considers four key factors impacting model performance.

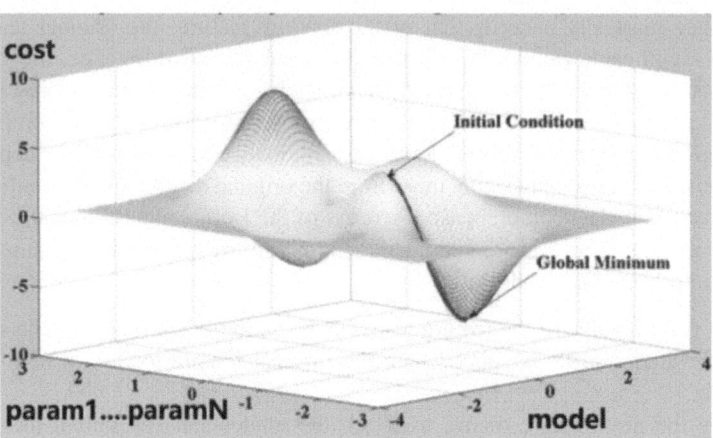

Fig. 1.12 Schematic of the optimal model search path

1. Sampling strategy: refers to how a model collects data samples, usually employing K-fold cross-validation and holdout strategies. K-fold cross-validation involves dividing the dataset into K similar-sized, mutually exclusive subsets. Each time, the union of K-1 subsets serves as the training set, and the remaining subset acts as the test set. This process yields K training/testing sets, leading to K test results, whose average represents the final result. The advantage is balanced data use and strong model generalization, but it is computationally costly for large datasets. The holdout strategy divides the dataset into two disjoint sets: training set S and test set T. After training the model on S, it is tested on T to estimate generalization error. Due to potential feature distribution differences between the test set and the original dataset, multiple random splits and repeated evaluations are needed to obtain a reliable average estimate of the model's generalization error. K random holdouts may lead to overlapping between training and validation sets, unlike K-fold cross-validation, which avoids overlap, potentially giving K-fold a stability advantage. However, with small datasets, K-fold takes longer than a single holdout, hence preferring the latter for faster model evaluation, enabling rapid iteration from low to high-cost regions in model search, thus enhancing efficiency. We must dynamically choose between these strategies based on the situation. Optimal model search should dynamically consider the following four factors to identify the best model and hyperparameters:
2. Learner: refers to which algorithm or model is used, such as XGBoost, LightGBM, and Random Forest.
3. Hyperparameters: refers to a variety of parameters that define the machine learning model, and different hyperparameters have a greater impact on the model performance.
4. Sampling sample size: refers to the number of samples sent to the machine learning model. A small sample size means less information can be learned, which may affect the generalization ability of the model; a large sample size

1.2 Developing Frameworks and Libraries

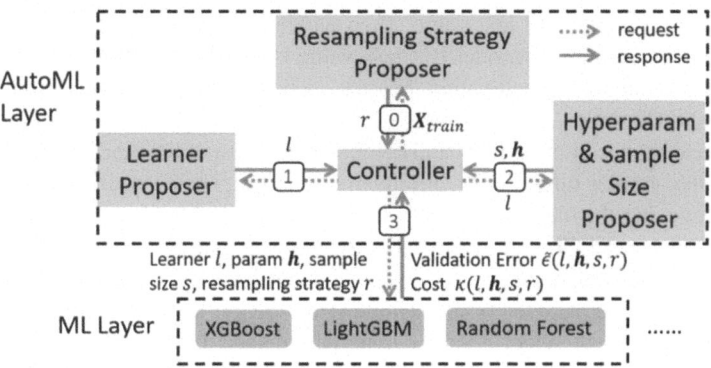

Fig. 1.13 Flaml automatic model search framework

will result in a large amount of computation and slow model training. Therefore, it is also necessary to conduct dynamic selection according to the actual situation.

Flaml is a lightweight automated machine learning framework developed by Microsoft, from a Microsoft paper "FLAML: A FAST AND LIGHTWEIGHT AUTOML LIBRARY." The full name of the framework is Fast and lightweight AutoML library. Due to its ease of use, strong readability, excellent performance, and other features, it became popular in the community once it was open-sourced. It has now been integrated into Microsoft's official Automated Tuning Library and has become a heavily promoted AutoML framework by Microsoft. The official URL of the framework is https://www.microsoft.com/en-us/research/publication/flaml-a-fast-and-lightweight-automl-library/ and the code repository is https://github.com/microsoft/FLAML, open source under the MIT license, the current star number is 1.9k, the highest version is 1.0.8.

In Flaml framework, the optimal model search not only considers the model performance, but also the time cost and computational complexity cost of model training. The working process is shown in Fig. 1.13.

In the Flaml framework, a single controller dictates the search for the optimal model. This search algorithm, once initiated, loops through feedback from the AutoML layer and the machine learning layer. It continually refines the AutoML layer's data, inching closer to the optimal model and hyperparameters. The AutoML layer sends data such as learner, model parameters, sample size, and resampling strategy to the machine learning layer. The machine learning layer then uses this data to train the model on the training set, returning verification loss and computation cost to the AutoML layer. The AutoML layer's mechanism involves three key steps:

1. Determine the resampling strategy: determine the size of the data set. If it is a small data set, K-fold cross-validation is adopted; otherwise, reservation strategy is adopted.
2. Determine the learner strategy: In a group of candidate models (algorithms or learners), start with an initial model A and perform a round of training on a small

sample set to obtain the loss value. Then switch to model B and calculate the loss value after training. If model B's loss value is higher than model A's, the ECI index is calculated by estimating the time needed for model B to reach model A's loss value based on model B's rate of loss reduction. ECI measures the potential for performance improvement and computational cost of the model. A higher ECI indicates greater difficulty in improvement and higher computational expense, thus lowering the likelihood of selecting the model. In Flaml, a model's selection probability is 1/ECI. Note that ECI is dynamic as training progresses, so model selection is also dynamic.

3. Determine hyperparameters and sample size strategy: With the sampling strategy and learner fixed, select a set of hyperparameters. If the loss value does not decrease, choose another set in the opposite direction and retrain. During training, the sample size gradually increases. Early on, training with small samples helps judge the quality of the learner and parameters, pushing model search quickly from low computational cost to high computational cost regions. Increasing sample size raises computational costs but also approaches the model's true performance, aiding in a rapid and efficient search. The search path follows set objectives (e.g., validation accuracy, loss, R^2). Notably, computational cost depends on both sample size and model parameters. More complex parameters increase costs. For instance, the cost for a random forest of 100 trees is less than that for 200 trees. Thus, FLAML's search strategy starts with simpler, fewer parameters, increasing complexity gradually.

Figure 1.14 shows FLAML's search path where darker regions indicate higher values. The left figure is a loss value heatmap; the right is a computation overhead heatmap. The controller search path points sequentially from 1 to 6. In the right figure, the color transition from light to dark indicates efficient traversal from low to high computational cost regions. When reaching point 5, computational cost significantly increases, but the loss value difference with point 4 is negligible. The algorithm thus shifts direction and moves to point 6 instead of continuing from point 5. Upon completing the iteration at point 6 and hitting the pre-set search

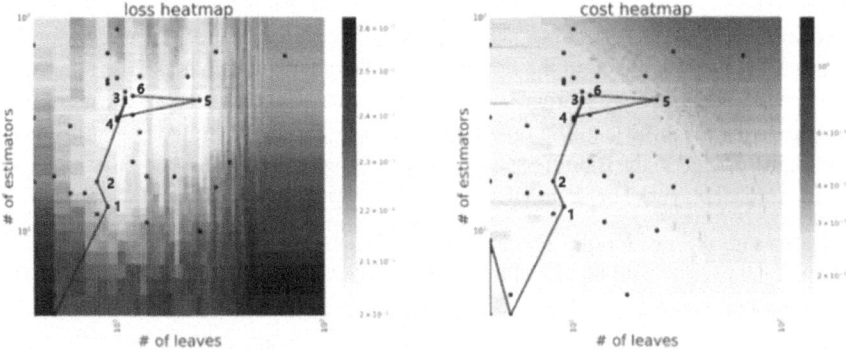

Fig. 1.14 Flaml automated model search process

duration, the search ends. This process, termed cost-frugal optimization (CFO) by FLAML, starts from a low-cost initial point (specified by low_cost_init_value in the search space) and executes local updates based on its random local search strategy. This allows CFO to quickly converge toward lower computational cost regions, demonstrating robust performance, enabling prompt initial explorations, and swiftly transitioning to higher cost regions for continued search.

The Flaml framework also provides a new global search algorithm: BlendSearch, which optimizes local hyperparameter search by taking into account global search. Blendsearch is an extended algorithm for the CFO, which adds Bayes-like exploration capabilities to the CFO's computational cost control. Like the CFO, BlendSearch starts with a model search from an initial point of low computational cost. Unlike the CFO, BlendSearch does not wait until the local search has fully converged before attempting a new starting point. BlendSearch works well to improve search results in cases where the global search space contains multiple disjoint, discontinuous subspaces.

Table 1.1 provides the model search results of our classification task on the airline dataset. It can be seen that Flaml chose the LightGBM learner at the beginning, when the sample size was only 10,000, the model parameters were very small, there were only four trees, and the training time was very short. With the deepening of the search, the model complexity increased, the training time and sample size also increased, and the validation set loss decreased. Moreover, Xgboost was selected by ECI, and the final XGBoost model complexity was 23,825 trees, the training time and sample size were significantly improved, while the validation set loss was very weak. This shows that the search strategy only uses less computational cost to find a better model and its parameters, Flaml is very fast and efficient.

Table 1.1 Flaml model search parameters observation

Execution time	Validation set loss	Parameters	Learner	Sample size
0.08	0.38	n_estimators: 4	lgbm	10,000
0.12	0.37	n_estimators: 12	lgbm	10,000
0.12	0.36	n_estimators: 14	lgbm	10,000
0.15	0.36	n_estimators: 53	lgbm	10,000
0.28	0.36	n_estimators: 53	lgbm	40,000
0.32	0.35	n_estimators: 89	lgbm	40,000
0.33	0.35	n_estimators: 53	lgbm	40,000
1.75	0.35	n_estimators: 53	lgbm	364,083
2.58	0.35	n_estimators: 112	lgbm	364,083
3.53	0.34	n_estimators: 224	lgbm	364,083
2.82	0.34	n_estimators:159	lgbm	364,083
4.38	0.33	n_estimators: 100	lgbm	364,083
7.86	0.33	n_estimators: 220	lgbm	364,083
35.61	0.33	n_estimators: 1002	lgbm	364,083
188.93	0.33	n_estimators: 5805	xgboost	364,083
189.93	0.33	n_estimators: 11,051	xgboost	364,083
259.24	0.32	n_estimators: 23,825	xgboost	364,083

1.2.4 AutoML Framework: Autogluon

AutoGluon, an open-source automated machine learning framework from Amazon, employs ensemble learning techniques. Similar to Flaml, it gained rapid popularity in the community for its promise of generating high-performance models with just three lines of code, automating processes such as algorithm selection, model training, hyperparameter tuning, and neural architecture search. AutoGluon's performance in neural architecture search outperformed manually tuned Faster Rcnn models. Amazon's Chief Scientist and MXNet co-creator Mu Li stated that this framework marks a new era in machine learning where manual hyperparameter tuning becomes obsolete. AutoGluon is an out-of-the-box product capable of quickly accomplishing tasks in tabular prediction, image classification, text classification, and object detection. Its official website is https://auto.gluon.ai/stable/index.html, documentation is available at https://auto.gluon.ai/, and the code repository is hosted at https://github.com/awslabs/autogluon under the Apache-2.0 license.

In supervised learning tasks, the goal of machine learning is to develop a robust model with consistent performance across various aspects. However, in practice, we often get multiple weak models that perform well in specific areas. We hope that even if one weak model makes a wrong prediction, other weak models can correct it. This is the purpose of ensemble learning, which achieves this by combining multiple weak models in parallel or series. In the automated machine learning framework, Flaml relies on hyperparameter search, aiming to find the best hyperparameters from a set of candidates. AutoGluon, on the other hand, avoids hyperparameter search by training multiple weak models of single algorithms and combining them through ensemble learning to produce a more robust and accurate model. AutoGluon believes that ensemble learning can try more model algorithms in the same time frame, achieving better performance and computational efficiency. The ensemble learning methods used by AutoGluon include stacking, k-fold cross-bagging, and multi-layer stacking, as shown in Fig. 1.15.

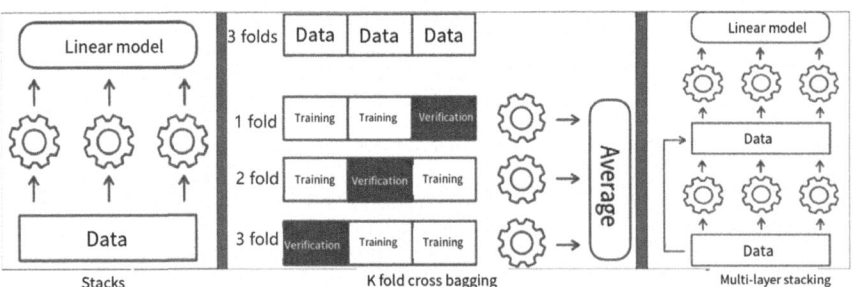

Fig. 1.15 AutoGluon's model combination technique

1.2 Developing Frameworks and Libraries

Fig. 1.16 Performance comparison of AutoGluon with common AutoML frameworks

	Champion
AutoGluon	30
TPOT	5
GCP	7
Auto-sklearn	4
H20	2
Auto-WEKA	1

Stacking: Multiple single algorithm models are trained on the same data using different algorithms (such as KNN and tree algorithm), and the output of these models is used as the input of another linear model, that is, the final output is obtained through weighted summation, and the weight of the linear model is obtained through training.

K-fold cross-bagging: The data set is cut into K data blocks, each time using a different data block as the training set and verification set. Different initial weights or different data blocks are then used to train the model of multiple algorithms of the same class. For classification problems, take the prediction results of these models in the form of voting to get the final result; For regression problems, the mean of the predictions of these models is used as the final result. AutoGluon sets the K value by specifying the num_bag_folds parameter in the fit() function.

Stacking Multiple layers: Train multiple models on the same piece of data, combine the output of these models, and stacking again. Train multiple models on top, and finally use a linear model as the output. Each model here is a Bagging of K models, and its output to the next stacking layer is a merging of the output on the corresponding verification set for each bagging model. AutoGluon sets the additional number of stacking layers by specifying the num_stake_levels parameter in the fit() function and the num_bag_folds parameter to set the K value.

In terms of performance, Amazon used 50 different types of data sets with a limited 4-h runtime to compare the computational performance of AutoGluon to that of the common AutoML framework. The results showed that AutoGluon calculated 30 data sets, far ahead of other frameworks. This shows that its integrated learning strategy is highly efficient, as shown in Fig. 1.16.

1.2.5 Bayesian Optimization Library: Bayesian Optimization

In automated hyperparameter optimization, traditional methods are grid search and random search. Grid search fundamentally is exhaustive; for example, with model parameters A, B, and C, each having ten choices, it exhausts 1000 possibilities. Random search, on the other hand, randomly seeks the optimal parameters. Both

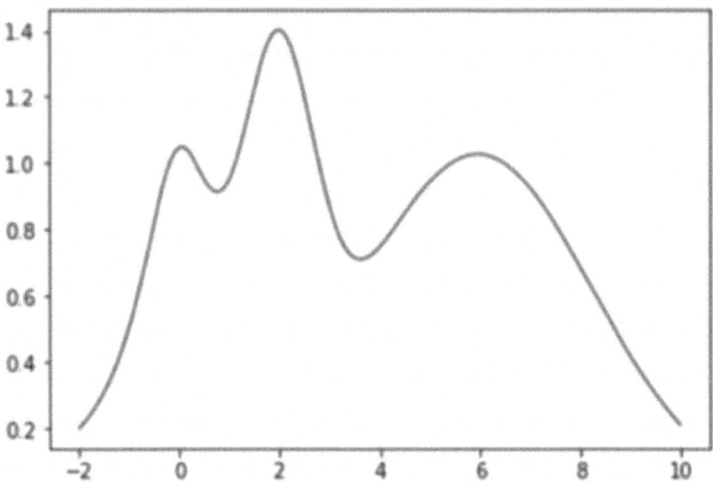

Fig. 1.17 Model parameters and model performance curves

have the common problems of high computational cost and low efficiency. Viewing model parameters X as independent variables and model performance (such as accuracy, R^2, etc.) as dependent variables, the values of the independent variables determine the dependent variables' values under the same dataset and learning task, forming an objective function to be solved. Typically, the model parameters X are a set of vectors. We aim to solve for the suitable X that maximizes the model performance dependent variable. Two issues arise: first, we lack a mathematical expression of the objective function, hence we cannot determine if the function is differentiable and consequently, cannot use gradient-based methods directly; second, the computational complexity of this function is unknown, differing greatly with different model parameters X as their computational costs vary significantly, like the number and depth of trees in a random forest. These challenges complicate the implementation of grid and random searches. Fortunately, Bayesian optimization provides a new solution approach as a mature method. Figure 1.17 represents an objective function, with the x-axis representing model parameters X and the y-axis representing model performance.

Randomly find some initial points on the graph and calculate the y value corresponding to x to get Fig. 1.18.

A random process involves sampling within a sample space, where each sample results in a process like a sequence or function. The Gaussian process (GP) generates multiple surrogate functions that fit these initial samples. Values at each surrogate function's data points follow Gaussian distributions, and the values corresponding to any k data points form a joint Gaussian distribution. Since Gaussian distributions can assume different parameter values, the GP returns a set of surrogate functions, each with varying probabilities based on prior distributions. The range distribution of these surrogate functions appears in the light gray area of Fig. 1.20. Another

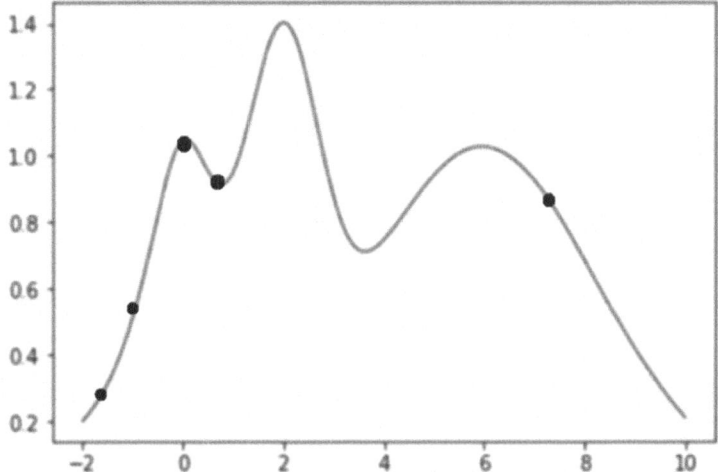

Fig. 1.18 Initial points for Bayesian optimization

acquisition function determines the location for the next sample calculation, either "exploit" (sample the known region near the current best point) or "explore" (sample the unknown region far from the known points). The acquisition function uses a specific strategy to select whether to "exploit" or "explore" the next sample point.

Figure 1.19 illustrates the Bayesian optimization process. The first figure displays the proxy function distribution fitted by a Gaussian process with a limited number of sampled points. The larger y-value range on the right side indicates a higher likelihood of the global optimal solution residing in that region, whereas the left side shows denser points, indicating previous sampling focused on "exploitation" strategies. The second figure's more uniform sampling points suggest the use of "exploration" strategies during algorithm iteration. After each new sample calculation, the Gaussian process is updated using Bayesian methods, generating a new posterior probability for the proxy function. This process iteratively updates the light blue region, forming a revised set of proxy functions. The iteration continues until the termination condition is met, yielding the x value of the sample point with the maximal y value as the best model parameter.

Bayesian optimization fundamentally revolves around approximating the most favorable data points by analyzing the distribution of a surrogate function. This is achieved by employing either an exploitation or exploration strategy to identify new sample points, thereby gathering new observational data. Consequently, the estimation of the Y-value range for the surrogate function is updated continuously. Through repeated iterations, the clarity of the surrogate function improves progressively, marked by the gradual reduction of the light gray uncertainty region. Importantly, this iterative calculation remains unaffected by the intricacies of the objective function and whether it is differentiable, thereby addressing significant issues related to derivation and computational complexity. At its core, Bayesian optimization focuses on constructing the posterior probability distribution that most closely

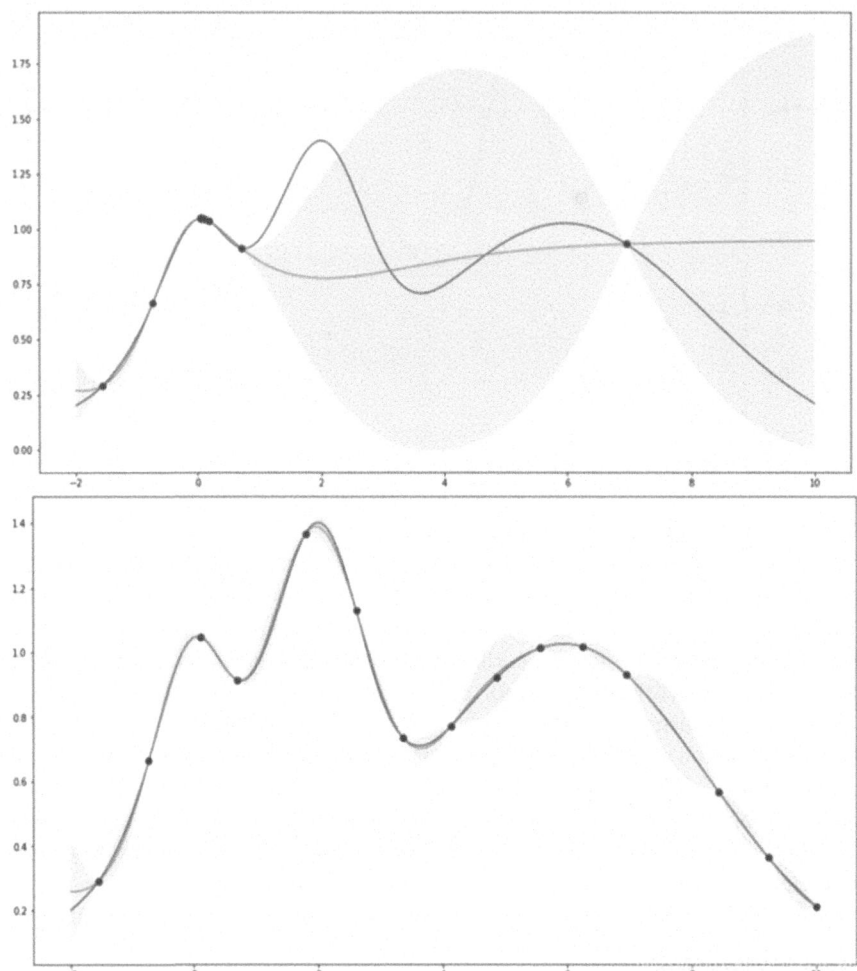

Fig. 1.19 Bayesian optimization process

approximates the function being optimized, utilizing Gaussian processes. As the quantity of observed samples increases, the accuracy of the posterior distribution enhances, allowing the algorithm to confidently discriminate between regions in the parameter space that warrant further exploration and those that do not. Figure 1.20 illustrates the Gaussian process, displaying the upper limit contour of the Gaussian surrogate function, with the algorithm capable of estimating the optimal location, denoted by the pentagram. With an increasing number of samples, the posterior distribution of the surrogate function becomes more defined, ultimately leading to a close approximation of the optimal solution.

Bayesian optimization is a Python implementation library for Bayesian global optimization using Gaussian processes. Its code repository is hosted at https://github.com/fmfn/BayesianOptimization, released under the MIT license. Currently, it has a

1.2 Developing Frameworks and Libraries

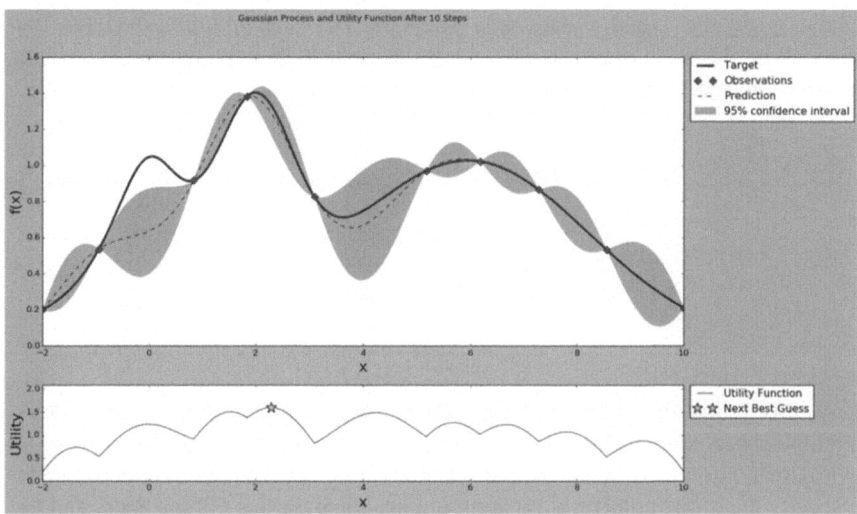

Fig. 1.20 Gaussian process

popularity rating of 5.7K stars, with the highest version being 1.2.0. The library is completely written in Python, providing out-of-the-box concept, the user can follow the following steps to call the wrapper function to achieve the whole process of Bayesian optimization, easy to use, efficient, and beautiful. The steps of the library are as follows: problem definition, exploration iteration, parameter solving. In the problem definition stage, we need to define the function to be optimized and the exploration range of each parameter. If the algorithm to be optimized is a random forest, the example step code of Bayesian optimization is as follows:

```
from sklearn.ensemble import RandomForestClassifier
def rf_cv(n_estimators, min_samples_split, max_features,
max_depth):
    val = cross_val_score(
      RandomForestClassifier(n_estimators=int(n_estimators),
        min_samples_split=int(min_samples_split),
        max_features=min(max_features, 0.999),
        max_depth=int(max_depth),
        random_state=2
      ),
      x, y, scoring='roc_auc', cv=5
    ).mean()
    return val
```

Here we define a function to be optimized, rf_cv(), which passes in each parameter of the random forest and returns the average accuracy of the fivefold cross-validation on the *x* and *y* data sets. Then the BayesianOptimization() function provided by Bayesian optimization defines the range of each parameter, and passes the function to be optimized into Bayesian optimization:

```
rf_bo = BayesianOptimization(
        rf_cv,
        {'n_estimators': (10, 250),
        'min_samples_split': (2, 25),
        'max_features': (0.1, 0.999),
        'max_depth': (5, 15)}
    )
```

At this point, the problem definition phase is complete. In the exploration iteration phase, use the maximize() function of the instantiated object rf_bo to approximate the maximum objective function value, use the init_points parameter to specify the initial number of random samples for Bayesian optimization, and use the n_iter parameter to specify how many random exploration steps to go through:

```
rf_bo.maximize(init_points=2, n_iter=5)
```

After each iteration of Bayesian optimization, the explored sample point parameters are stored in the rf_bo object. We can use the ergodic method to observe the model parameters obtained from each sample:

```
for i, res in enumerate(rf_bo.res):
    print(i, res)
```

Run the above code and you will get the output shown in Fig. 1.21. The model parameters of the random forest from each sample are saved, Where max_depth represents the maximum depth of the tree, max_features represents the number of features to be considered during optimal segmentation, min_sample_split represents the minimum number of samples required for splitting internal nodes, and n_estimators represents the number of decision trees.

If the object of Bayesian optimization is another algorithm, the resulting sample parameters are the parameters of the other algorithm. In the parameter solving stage,

```
0 {'target': 0.9590090179017903, 'params': {'max_depth': 5.456431528339038, 'max_features': 0.683937162572351, 'min_samples_split': 15.006534448317478, 'n_estimators': 128.1638689773331}}
1 {'target': 0.961479111911191, 'params': {'max_depth': 11.878092970292787, 'max_features': 0.6594194899287734, 'min_samples_split': 9.3184991825318, 'n_estimators': 185.17107521205858}}
2 {'target': 0.9620490769076907, 'params': {'max_depth': 12.847489529725674, 'max_features': 0.647615336678432, 'min_samples_split': 8.682077042676077, 'n_estimators': 185.48203604183368}}
3 {'target': 0.9623191049104911, 'params': {'max_depth': 14.564993659724847, 'max_features': 0.492923000764626, 'min_samples_split': 4.684182789244805, 'n_estimators': 184.69873846643927}}
4 {'target': 0.9613190539053905, 'params': {'max_depth': 13.152661624351412, 'max_features': 0.6079068202243004, 'min_samples_split': 5.912834991600972, 'n_estimators': 190.15241886057217}}
5 {'target': 0.9603090599059907, 'params': {'max_depth': 14.765921929968883, 'max_features': 0.8468699651624888, 'min_samples_split': 6.418766363389819, 'n_estimators': 178.53535160571826}}
6 {'target': 0.9479388858885889, 'params': {'max_depth': 11.27892852596105, 'max_features': 0.16455217225871382, 'min_samples_split': 2.177090824575627, 'n_estimators': 185.64559964355624}}
```

Fig. 1.21 Running output of the Bayesian optimization library

directly take out the sample point parameter of the rf_bo object about the maximum objective function value, and call the code:

```
params = rf_bo.max['params']
```

rf_bo.max will return a dictionary of the objective function value target and the model parameter, which is the point with the largest objective function value in the 7 sample point calculations:

```
{'target': 0.9623191049104911, 'params': {'max_depth':
14.564993659724847, 'max_features': 0.492923000764626,
'min_samples_split': 4.684182789244805, 'n_estimators':
184.69873846643927}}
```

Next, feed the best parameters into the random forest model, and you can use the model directly to make predictions:

```
rf1 = RandomForestClassifier(n_estimators=round_up(str(params
['n_estimators'])),
 max_depth=round_up(str(params['max_depth'])),
 min_samples_split=round_up(str(params['min_samples_split'])),
 max_features=params['max_features'])
```

1.3 Case Practice

This section introduces the development and operation environment construction, data preparation, feature engineering, modeling code writing, project operation, and model evaluation of this case.

1.3.1 Operation Environment Construction

Install machine learning suite Anaconda first. Anaconda is an open-source Python distribution that contains over 180 science packages and their dependencies for Conda, Python, and more. Anaconda comes with a virtual environment manager called Conda, which can install different versions of machine learning virtual environments on the same machine and be able to switch between them. Anaconda is also cross-platform and supports Windows, Linux, and MacOS operating systems. The official website of Anaconda is https://www.anaconda.com/products/individual#Downloads, and the download interface is shown in Fig. 1.22.

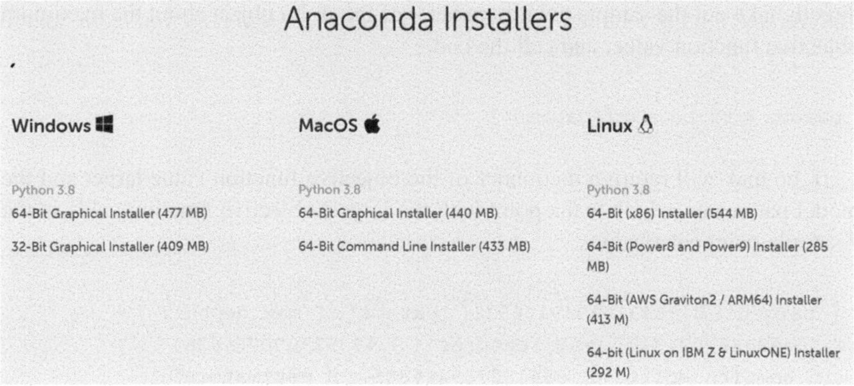

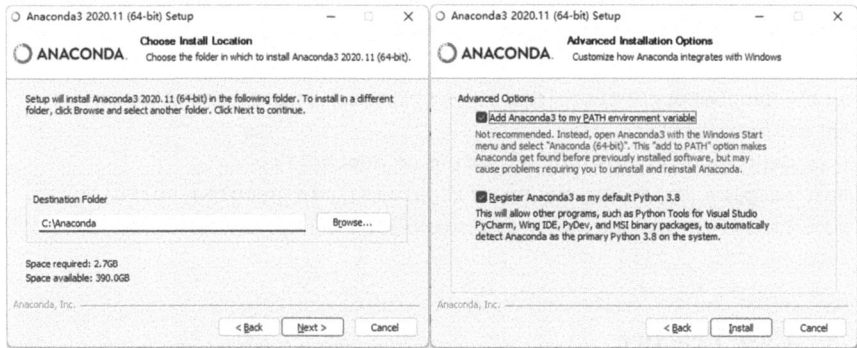

Fig. 1.22 Downloading Anaconda machine learning suite

Fig. 1.23 Anaconda installation process

During installation, the dialog box shown in Fig. 1.23 will be displayed. Select environment variables and specify the installation path to C:\Anaconda. In particular, the installation path of the author of this book is C:\Anaconda, readers can also install to other paths, just modify the corresponding path in the subsequent test.

After the installation is complete, create a virtual environment named Automl and install Python3.7 by executing the following statement from the command line:

```
conda create -n automl python=3.7 -y
```

To activate the virtual environment, execute the following statement:

```
conda activate automl
```

At this point, an "(automl)" prompt appears on the command line, indicating that you have entered the virtual environment. Run the following command to install the relevant framework without displaying an error:

1.3 Case Practice

```
pip install bayesian-optimization ray optuna flaml feature-selector boruta
pip install autogluon
```

It is worth noting that the above two install commands cannot be installed in a virtual environment due to the version conflict of the dependency package. At this point, the runtime environment is set up.

1.3.2 Data Set Preparation

In this instance, we identify and utilize the pertinent attributes based on prior business intelligence to forecast the likelihood of each customer's mobile banking activity status. It is essential to highlight that the characteristic indicators of the current month serve as predictors for the subsequent month's activity status. Consequently, each dataset entry comprises three segments: "Customer identification number, current month's characteristics, and the mobile banking activity status for the following month." The comprehensive data fields include all necessary variables for precise prediction and analysis.

Customer number, monthly average of customer's nine assets, age, number of months since account opening, gender, marital status, occupation, educational level, days since the most recent transaction, risk rating, activity sleep indicator, asset rating, number of products held, loan holdings, credit card holdings, whether signed up for electronic channels, whether signed up for express payment, total transaction amount this month, total transaction count this month, total cash transaction amount this month, total cash transaction count this month, total disbursement amount this month, total disbursement count this month, total online banking amount this month, total online banking count this month, total counter amount this month, total counter count this month, whether active on mobile banking next month.

According to the data observation, the missing (or inaccurate) situation of the two characteristics of risk rating and occupation is relatively serious, we need to abandon these two fields, and the transaction proportion can better reflect the customer's transaction preference. We need to convert the number of transactions and the amount into the proportion, and fill the remaining few missing fields with the number 0. The data preprocessing process is as follows:

```
import pandas as pd
df = pd.read_csv('data.txt',encoding='utf-8',dtype={'customer number':'str'})
del df['date of statistics']
del df['risk rating']
del df['occupation']
df['Proportion of total cash transaction count this month'] = round(df
```

```
['Total cash transaction count this month']/df['Total transaction
count this month'],2)
    df['Proportion of total cash transaction amount this month'] = round
(df['Total cash transaction amount this month']/df['Total transaction
amount this month'],2)
    df['Proportion of total disbursement count this month'] = round(df
['Total disbursement count this month']/df['Total transaction count
this month'],2)
    df['Proportion of total disbursement amount this month'] = round(df
['Total disbursement amount this month']/df[''Total cash transaction
amount this month '],2)
    df['Proportion of total online banking count this month'] = round(df
['Total online banking count this month']/df['Total transaction count
this month'],2)
    df['Proportion of total online banking amount this month'] = round(df
['Total online banking amount this month']/df[''Total cash transaction
amount this month '],2)
    df['Proportion of total counter count this month'] = round(df['Total
counter count this month']/df['Total transaction count this month'],2)
    df['Proportion of total counter amount this month'] = round(df['total
counter amount this month']/df[''Total cash transaction amount this
month '],2)
    df = df.fillna(0)
```

For numerical-type features, larger values do not inherently signify greater significance. This observation necessitates implementing a binning operation. This operation entails the algorithm's determination of optimal bin boundaries, which are based on the data distribution characteristics of each feature column. Consequently, raw features are categorized into specific groups. These grouped data points are subsequently incorporated into the model to mitigate inconsistencies in value ranges across different features. In this context, we employ the decision tree binning algorithm to ascertain the appropriate number of bins and their respective boundaries, ensuring a systematic and accurate categorization process.

As depicted in Fig. 1.24, the objective of decision tree bin division is intricately connected to the optimal segmentation of the independent variable x vector into features, relative to the dependent variable y vector. This segmentation aims to minimize the entropy of y. Here, entropy quantifies the disorder or uncertainty within the dataset. High entropy indicates numerous categories and a significant degree of confusion, while low entropy reflects fewer categories and greater order. The principle of increasing entropy illustrates that natural states tend toward greater disarray and disorder. In data modeling, the core objective is to transform raw data into a standardized, well-structured form, effectively reducing disorder and thus minimizing entropy. Decision tree bin division leverages a tree-based model to discretize the independent variable by sequentially selecting feature splits that adhere

1.3 Case Practice

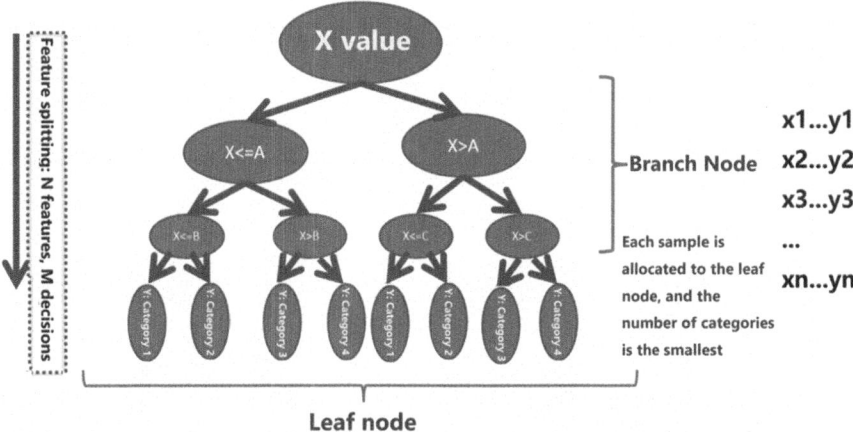

Fig. 1.24 Schematic diagram of decision tree box division

to the principle of minimizing entropy increase at every level. This process maximizes information gain and reduces data chaos, thereby achieving optimal classification and ensuring a systematic, organized representation of the data.

In the decision tree bin partitioning algorithm, each node performs a feature splitting, which is equivalent to cutting a knife in the corresponding data block, as shown in Fig. 1.25. For each cut, the information gain is calculated once. For example, the entropy of the original data set y is S_1, the x feature has five values, the probability ratio of five values is L1, L2, L3, L4, and L5, and the entropy of the result set calculated with five values is R1, R2, R3, R4, and R5, Then the entropy value of x feature selected as the branch node is S_2=L1*R1+L2*R2+L3*R3+L4*R4+L5*R5, and the information gain is S_1–S_2. The decision tree bin partitioning algorithm will determine the feature splitting boundary values of nodes in each layer from top to bottom according to the information gain from large to small, so that the y data falling on the leaf node has the minimum entropy.

The box splitting code is as follows:

```
def optimal_binning_boundary(x, y):  # Obtaining optimal bins using decision trees
    boundary = []  # List of bounding values of the bins to be returned
    x = x.values  # Filling in missing values
    y = y.values
    clf = DecisionTreeClassifier(criterion='entropy',  # Entropy minimization criterion division
                    max_leaf_nodes=5,    # Maximum number of leaf nodes
                    min_samples_leaf=0.05) # Minimum percentage of leaf node sample size
    clf.fit(x.reshape(-1, 1), y)  # Training decision trees
    n_nodes = clf.tree_.node_count
```

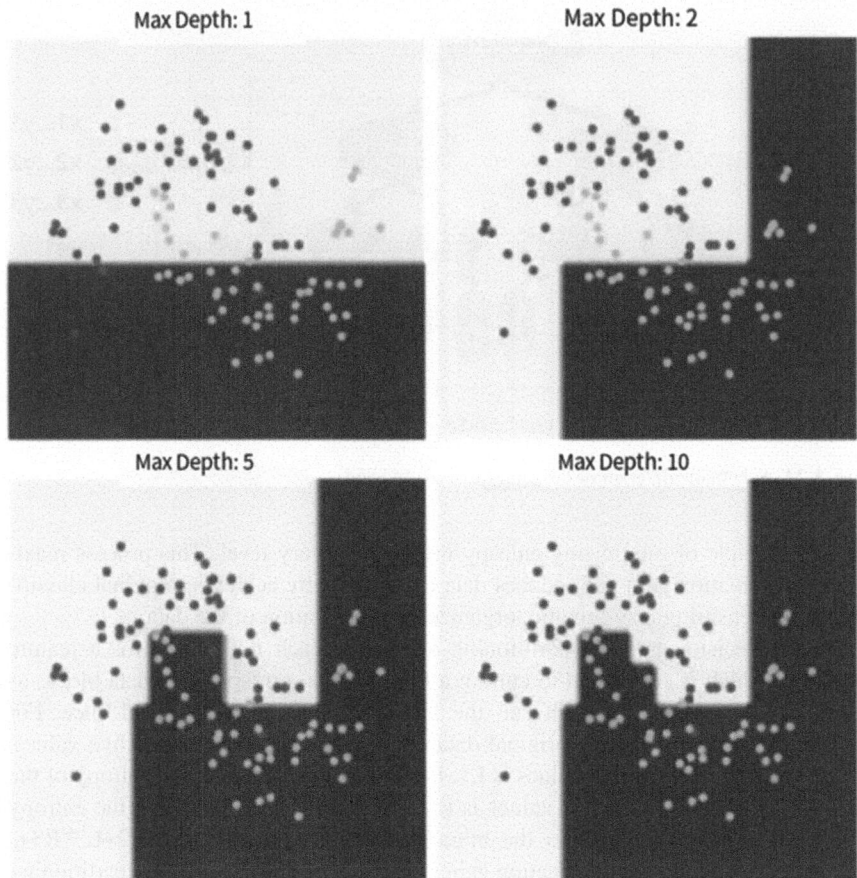

Fig. 1.25 Feature splitting process of decision tree bin splitting

```
children_left = clf.tree_.children_left
children_right = clf.tree_.children_right
threshold = clf.tree_.threshold
for i in range(n_nodes):
    if children_left[i] != children_right[i]:  # Obtain division boundary values on decision tree nodes
        boundary.append(threshold[i])
boundary.sort()
min_x = x.min()
max_x = x.max() + 0.1  # +0.1 is to account for subsequent groupby operations that can contain samples with the largest values of the feature
boundary = [min_x] + boundary + [max_x]
return boundary
```

1.3 Case Practice

Next, each column feature in the data set is traversed, and the numerical field is divided into decision tree boxes, that is, one column feature is split into multiple column features (that is, into several boxes). In this case, a maximum of five columns are split, and then the original feature column is replaced with the fissile onehot coded bin category. For example, if the algorithm decides to divide a column into three boxes, the column is split into 3 columns with the values of "0,0,1," "0,1,0," and "1,0,0."

```
df2 = DataFrame()
for a in list(df):
    if a=='whether active on mobile banking next month':
        continue
    if a=='customer number' or a=='gender' or a=='marital status' or a=='educational level' or a=='activity sleep indicator' \
    or a=='asset rating' or a=='loan holdings' or a=='credit card holdings' \
    or a=='whether signed up for electronic channels' or a=='whether signed up for express payment':
        df2[a] = df[a]
        continue
    df[a][np.isinf(df[a])] = 999 # Assign to infinite data
    ps = optimal_binning_boundary(x=df[a], y=df['whether active on mobile banking next month'])
    # The following boundary-value transformed features are taken by decision tree bins (one to many)
    if len(ps) == 6:
        Df2 [a] = pd. The cut (x = df [a], bins = ps, right = False, labels = ['0,0,0,0,1', '0,0,0,1,0', '0,0,1,0,0', '0,1,0,0,0', '1,0,0,0,0'])
    if len(ps) == 5:
        Df2 [a] = pd. The cut (x = df [a], bins = ps, right = False, labels = ['0,0,0,1', '0,0,1,0', '0,1,0,0', '1,0,0,0'])
    if len(ps) == 4:
        Df2 [a] = pd. The cut (x = df [a], bins = ps, right = False, labels = ['0, 1', '0, 0', '0, 1])
    if len(ps) == 3:
        Df2 [a] = pd. The cut (x = df [a], bins = ps, right = False, labels = [0, 1, '1, 0'])
```

For literal class features with a small number of categories, we convert them directly to onehot encoding. For example, the gender field value is "male," "female," and "unknown" three values, we encode them into "0,0,1," "0,1,0," and "1,0,0," respectively, using the get_dummies() function of pandas:

```
one_hot = pd.get_dummies(df2[' gender '],prefix=' gender ') # one-hot
code
df2 = df2.join(one_hot)
del df2[' gender ']
```

After the one-hot encoding of the original feature is generated, the original feature column is no longer needed and should be removed. After processing all other fields, write the data to the feature.txt file, which consists of 82 columns of data, each separated by commas, in the format:

```
Customer number, feature1, feature2, feature3, feature4, ......,
feature79, feature80, whether active on mobile banking next month
   AHCEEJKEHC,1,0,0,1,......,1,0,1
   ACIJJKEFBB,0,0,0,0,......,0,1,0
   AEHEHKABHB,0,0,1,1,......,0,1,0
   ......
```

Where the customer number is the only primary key, and each column of feature and supervisory data "whether active on mobile banking next month" after the above processing, its value is either 0 or 1. The code in this section is detailed in the feature. py file. This completes the preparation of the dataset.

1.3.3 Feature Selection Code Practice

First use feature selector to select the Feature.

Set the feature column and monitor the data. Be careful to use the unique primary key as the index when reading the dataset file, so that it does not participate in the feature selection process:

```
data = pd.read_csv('features.txt', index_col=0, low_memory=False,
encoding='UTF-8') #index_col specifies the unique encoding column,
which is not used as a feature
   train_labels = data. Whether active on mobile banking next month
   train_features = data.drop(columns='whether active on mobile banking
next month')
   fs = FeatureSelector(data=train_features, labels=train_labels)
```

Show collinear features:

```
   fs.identify_collinear(correlation_threshold=0.8, one_hot=False) #
Find the features with a correlation degree greater than 80%
   collinear_features = fs.ops['collinear'] # collinear features to be
```

1.3 Case Practice

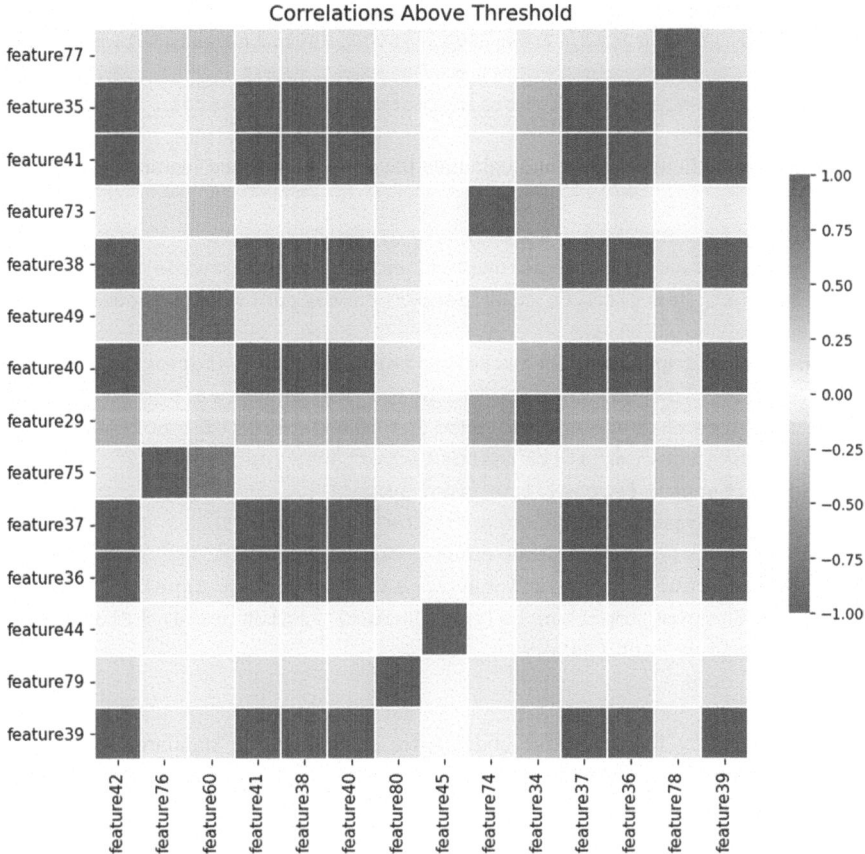

Fig. 1.26 Visual heat map of collinear features

```
removed
  print(' Features with a correlation greater than 80% =',
collinear_features)
  fs.plot_collinear()
  plt.show()
```

Figure 1.26 Collinear features visual heat map will be obtained.

Get a list of features with a correlation degree greater than 80%:

Features with a correlation degree greater than 80%=['feature34', 'feature36', 'feature37', 'feature38', 'feature39', 'feature40', 'feature41', 'feature42', 'feature45', 'feature60', 'feature74', 'feature76', 'feature78', 'feature80']

For features with large collinearity, there is a strong linear correlation between independent variables, which will lead to a decline in the predictive ability of the model and an increase in the interpretation cost of the model results. Therefore, we need to delete them.

Data. The drop (data. The columns [[33,35,36,37,38,39,40,41,44,59,73,75,77,79]], the axis = 1, inplace = True)
   fs = FeatureSelector(data=train_features, labels=train_labels)

Train the gradient elevator and calculate the zero-importance features:

```
fs.identify_zero_importance(task='classification',
eval_metric='auc', n_iterations=10, early_stopping=True)
  zero_import_feature = fs.ops['zero_importance'] # Zero importance
feature
  print(' zero importance feature =', zero_import_feature)
  fs.identify_low_importance(cumulative_importance=0.9) # Select
other features that are not included in the selection of important
features when the cumulative feature importance reaches 90%
  low_importance = fs.ops['low_importance']
  print(' low importance feature =', low_importance ')
  fs.identify_zero_importance(task='classification',
eval_metric='auc', n_iterations=10, early_stopping=True)
  fs.plot_feature_importances(threshold=0.9, plot_n=50) # Plot the
top 50 most important features
  print(fs.ops)
```

See f_select.py for the above code. After execution, the standardized feature importance score is plotted, resulting in the following ranking of important features, as shown in Fig. 1.27.

The dotted line indicates that more than 50 features are required when the cumulative importance reaches 90%. There are 80 features in feature.txt, and the average feature importance score is 1/80=0.0125. It can be seen that the importance of most features has exceeded the average score. We eliminate zero and low importance features in fs.ops, and the rest is important features. Note that the low

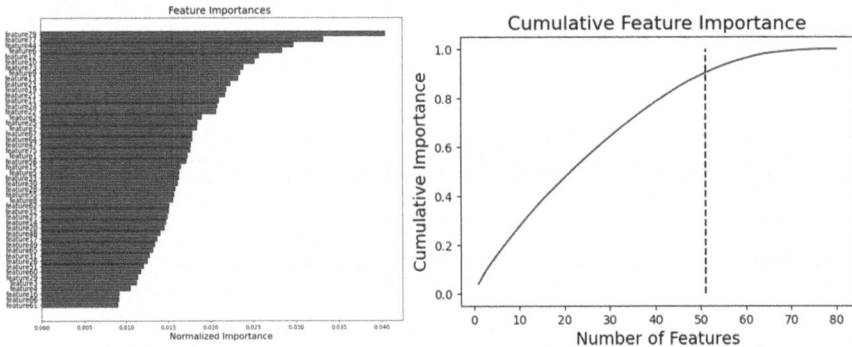

Fig. 1.27 Important feature of feature_selector

1.3 Case Practice

importance feature is related to the 90% threshold we set, which is determined by the cumulative_importance=0.9 parameter. Adjusting this parameter will result in a different selection of low importance features.

Let's compare the results of Boruta's selection again, first entry package:

```
import pandas as pd
from sklearn.ensemble import RandomForestClassifier
from boruta import BorutaPy
```

Read the dataset where X is all the feature columns and y is the supervised data column:

```
Data = pd. Read_csv (' the features. TXT, encoding = "utf-8", dtype = {'
customer number: 'STR'})
X = data.copy()
del X[' customer number ']
del X[' whether active on mobile banking next month ']
X = X.values
y = data[' whether active on mobile banking next month ']
y = y.astype('int')
y = y.values.ravel()
```

The Boruta library uses a random forest algorithm to select features. So instantiate a random forest class, turn on concurrency mode, specify balanced mode set y value will automatically adjust the weight inversely to the class frequency in the input data to n_samples/(n_classes * np.bincount(y)), the tree depth is 5:

```
rf = RandomForestClassifier(n_jobs=-1, class_weight='balanced', max_depth=5)
```

Pass the random forest instantiation object to the BorutaPy function, automatically set the number of learners based on the data set size, specify the perc parameter to set the threshold for comparison of shadow features and true features, the lower the perc, the more irrelevant features will be selected as relevant, but the fewer relevant features will be missed, and the two need to be balanced.

```
feat_selector = BorutaPy(rf, n_estimators='auto', verbose=2, random_state=1, perc=100)
```

Pass in the data set, which will train the random forest of feature selection:

```
feat_selector.fit(X, y)
```

```
BorutaPy finished running.
Iteration:        100 / 100
Confirmed:        72
Tentative:        1
Rejected:         7
feat_selector.support_n_features_ : 72
feat_selector.support_ : [ True  True  True  True  True  True  True  True  True  True  True  True
  True  True  True  True  True  True  True  True  True  True  True  True
  True  True  True  True  True  True  True  True  True  True  True  True
  True  True  True  True  True False  True  True False  True  True
  True  True  True  True False False  True  True False False False  True
  True  True  True  True  True  True  True  True  True  True False  True
  True  True  True  True  True  True  True]
feat_selector.ranking_ : [1 1 1 1 1 1 1 1 1 1 1 1 1 1 1 1 1 1 1 1 1 1 1 1 1 1 1 1 1 1 1 1 1 1 1 1
 1 1 1 1 8 1 1 4 1 1 1 1 1 1 5 7 1 1 6 3 9 1 1 1 1 1 1 1 1 1 1 1 2 1 1 1
 1 1 1 1 1 1]
n_features_ : 72
X_filtered= [[0 0 0 ... 1 0 1]
 [0 0 0 ... 1 1 0]
 [0 0 0 ... 1 0 1]
 ...
 [0 0 0 ... 0 0 1]
 [0 0 0 ... 1 0 1]
 [0 1 0 ... 1 0 1]]
```

Fig. 1.28 Boruta feature selection result

Print feature selection results and features:

```
print('feat_selector.support_n_features_ : ', feat_selector.
n_features_)
print('feat_selector.support_ : ', feat_select.support_)
print('feat_selector.ranking_ :', feat_selector.ranking_)
print('n_features_ :', feat_selector.n_features_)
X_filtered = feat_selector.transform(X)
print('X_filtered=', X_filtered)
```

See Boruta1.py for the above code. The following result is obtained after execution, as shown in Fig. 1.28.

Among the 80 features to be selected, Boruta selects 72 important features, which is basically the same as the selection result of Feature_selector. Note that the selection result of Feature_selector is determined by the importance threshold. We select some of the top non-collinear important features and divide the training set and test set according to the ratio of 75%: 25% to form the feature. train and feature. test files, respectively:

```
df = pd.read_csv('features.txt', Usecols =
[0,79,77,44,6,12,10,9,73,13,23,22,24,19,11,21,2,7,64,67,
25,75,47,15,5,56,1,55,33,28,30,8,62,27,32,14,48,20,65,17 ,49,81],
encoding='UTF-8')
df = sklearn.utils.shuffle(df) # random shuffle
df_train = df[:int(len(df)*0.75)]
df_test = df[int(len(df)*0.75):]
```

1.3 Case Practice

```
df.to_csv('features.all', index=False)
df_train.to_csv('features.train', index=False)
df_test.to_csv('features.test', index=False)
```

1.3.4 Actual Automation Modeling Code

After the feature project is completed, enter the automated modeling link. Start with version 0.9.5 of the Microsoft Flaml framework. This is the process of searching for the best solution in the pre-set model space and parameter space. First lead entry package:

```
from flaml import AutoML
Instantiate an AutoML class:
automl = AutoML()
```

Set parameters related to automatic machine learning. The time_budget parameter is the number of seconds for model training, the metric parameter is the measurement value, accuracy represents optimization based on precision, the task parameter specifies classification or regression task, classification indicates this example is a classification task, the log_file_name parameter specifies which log file to record the training process to, the training process data includes computational cost, time cost for each search, validation loss, model parameters, etc. The estimator_list parameter specifies the list of learners, hence specifying the model space. Not setting it in this example implies using all default learners for training. FLAML supports learners such as LGBM, RF (Random Forest), CatBoost, XGBoost, Extra_tree (Extra Random Tree), XGB_Limitdepth, and LRL1 (Logistic Regression with L1 regularization).

```
automl_settings = {
  "time_budget": 600,
  "metric": 'accuracy',
  "task": 'classification',
  "log_file_name": "flaml_classify.log",
  #"estimator_list": ['rf','lgbm',...],
}
```

Then get the feature matrix X_train and the supervised vector y_train from the training set file, set index_col=0 to set the unique primary key to the index, which will not participate in the model training:

```
df = pd.read_csv('features.train',index_col=0,encoding='utf-8')
X_train = df.drop(' whether active on mobile banking next month ',axis
```

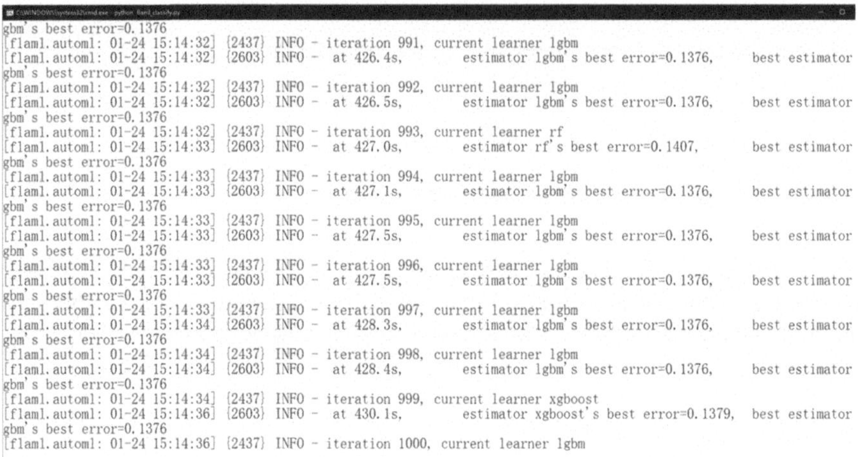

```
gbm's best error=0.1376
[flaml.automl: 01-24 15:14:32] {2437} INFO - iteration 991, current learner lgbm
[flaml.automl: 01-24 15:14:32] {2603} INFO -   at 426.4s,       estimator lgbm's best error=0.1376,    best estimator
gbm's best error=0.1376
[flaml.automl: 01-24 15:14:32] {2437} INFO - iteration 992, current learner lgbm
[flaml.automl: 01-24 15:14:32] {2603} INFO -   at 426.5s,       estimator lgbm's best error=0.1376,    best estimator
gbm's best error=0.1376
[flaml.automl: 01-24 15:14:32] {2437} INFO - iteration 993, current learner rf
[flaml.automl: 01-24 15:14:33] {2603} INFO -   at 427.0s,       estimator rf's best error=0.1407,     best estimator
gbm's best error=0.1376
[flaml.automl: 01-24 15:14:33] {2437} INFO - iteration 994, current learner lgbm
[flaml.automl: 01-24 15:14:33] {2603} INFO -   at 427.1s,       estimator lgbm's best error=0.1376,    best estimator
gbm's best error=0.1376
[flaml.automl: 01-24 15:14:33] {2437} INFO - iteration 995, current learner lgbm
[flaml.automl: 01-24 15:14:33] {2603} INFO -   at 427.5s,       estimator lgbm's best error=0.1376,    best estimator
gbm's best error=0.1376
[flaml.automl: 01-24 15:14:33] {2437} INFO - iteration 996, current learner lgbm
[flaml.automl: 01-24 15:14:33] {2603} INFO -   at 427.5s,       estimator lgbm's best error=0.1376,    best estimator
gbm's best error=0.1376
[flaml.automl: 01-24 15:14:33] {2437} INFO - iteration 997, current learner lgbm
[flaml.automl: 01-24 15:14:34] {2603} INFO -   at 428.3s,       estimator lgbm's best error=0.1376,    best estimator
gbm's best error=0.1376
[flaml.automl: 01-24 15:14:34] {2437} INFO - iteration 998, current learner lgbm
[flaml.automl: 01-24 15:14:34] {2603} INFO -   at 428.4s,       estimator lgbm's best error=0.1376,    best estimator
gbm's best error=0.1376
[flaml.automl: 01-24 15:14:34] {2437} INFO - iteration 999, current learner xgboost
[flaml.automl: 01-24 15:14:36] {2603} INFO -   at 430.1s,       estimator xgboost's best error=0.1379, best estimator
gbm's best error=0.1376
[flaml.automl: 01-24 15:14:36] {2437} INFO - iteration 1000, current learner lgbm
```

Fig. 1.29 Flaml frame search process

=1).values

y_train = df['whether active on mobile banking next month'].values

Put the feature matrix X_train, the supervised vector y_train and the model parameters into the frame to run, this will start Flaml's automatic model search mechanism, the algorithm will search for the best learner and the best model parameters in the default space of multiple learners and model parameters:

automl.fit(X_train=X_train, y_train=y_train, **automl_settings)

You will see the screen display the following, as shown in Fig. 1.29.

Finally, the predict() function is used to predict the test set, and the prediction category of each sample in the test set is obtained, and the model performance index is calculated:

df = pd.read_csv('features.test',index_col=0,encoding='utf-8')
X_test = df.drop('whether active on mobile banking next month ',axis=1).values
y_test = df['whether active on mobile banking next month'].values
y_pre = automl.predict(X_test)
from sklearn import metrics
test_auc = metrics.roc_auc_score(y_test,y_pre) # verifies the auc value on the set
print('test set accuracy=',(y_pre==y_test).sum()/len(y_test),'test set auc=',test_auc) print('beste model: ',automl.model)
print(' Best model parameters: ',automl.best_config)

1.3 Case Practice

```
[flaml.automl: 11-19 20:58:26] {2199} INFO - fit succeeded
[flaml.automl: 11-19 20:58:26] {2201} INFO - Time taken to find the best model: 417.130277633667
test accuracy= 0.8652720916755463 test auc= 0.7345315387324833
the best model:  <flaml.model.XGBoostSklearnEstimator object at 0x0000023BCF57BA88>
the best model parameter:  {'n_estimators': 364, 'max_leaves': 9, 'min_child_weight': 38.4337827026
9615, 'learning_rate': 0.06850381671614315, 'subsample': 0.9105235569084855, 'colsample_bylevel': 0
.9619311807479314, 'colsample_bytree': 1.0, 'reg_alpha': 0.0014055181562419288, 'reg_lambda': 0.046
18450504786178, 'FLAML_sample_size': 10000}
```

Fig. 1.30 Flaml frame search results

After 10 min of training, the best learner searched by Flaml was lgbm, with 86.73% accuracy on the test set and an auc indicator of 0.74, as shown in Fig. 1.30.

See flaml_classify.py for the above code. Next, let's compare to version 0.3.1 of Amazon's AutoGluon framework. Import the library package:

```
from autogluon.tabular import TabularDataset, TabularPredictor
```

To read the training set:

```
label = ' whether active on mobile banking next month '
train_data = TabularDataset('./features.train')
train_data = train_data.drop(columns=[' client number '])
Set the number of training seconds for the model:
time_limit = 600
```

We specify the auc value as the model evaluation metric, which is defined as the area under the roc curve, the larger the value, the better. After each iteration, the model will calculate the evaluation metric on the data set, and observe this metric to understand the optimization direction of the model.

```
metric = 'roc_auc' # Metric
```

Next start the automatic learning process of AutoGluon, which will use a single model such as CATBoost, ExtraTree, LightGBM, NeuralNet, RandomForestEntr, XGBoost for integrated learning, The training weight files of these models will be stored in the path specified by the path parameter, eval_metric parameter specifies the model evaluation metric, fit() function feeds the training set data to the framework, and it will use the label column of the TabularPredictor() instance as the monitoring data for supervised learning. presets='best_quality' specifies that the model will obtain the best prediction accuracy, num_bag_folds=5 indicates that a fivefold cross-validation will be performed:

```
predictor = TabularPredictor(label, path='autogluon_models_new',
eval_metric=metric).fit(train_data, time_limit=time_limit,
presets='best_quality', num_bag_folds=5)
```

Fig. 1.31 AutoGluon model performance evaluations

```
Evaluations on test data:
{
    "roc_auc": 0.903257935463595,
    "accuracy": 0.8668797786626288,
    "balanced_accuracy": 0.7399586662294096,
    "mcc": 0.5314409933737757,
    "f1": 0.6037152047369417,
    "precision": 0.691474172658202,
    "recall": 0.5357234567901235
}
```

Get data from the test set:

```
test_data = TabularDataset('./features.test')
test_data = test_data.drop(columns=[' customer number '])
test_data_nolab = test_data.drop(columns=[label])
```

Predictions are made using the predict() function, which outputs the prediction category for each sample, and the predict_proba() function, which outputs the probability for each prediction category.

```
y_pred = predictor.predict(test_data_nolab)
print("Predictions: \n", y_pred)
pred_probs = predictor.predict_proba(test_data_nolab)
print(pred_probs)
perf = predictor.evaluate_predictions(y_true=y_test, y_pred=pred_probs, auxiliary_metrics=True)
```

Use the evaluate_predictions() function to obtain model performance predictions as shown in Fig. 1.31.

We get an auc measure of 0.90 and an accuracy measure of 0.86, which is slightly better with AutoGluon compared to Flaml's results.

1.3.5 Automated Reference Code Practice

The following uses Bayesian optimization for hyperparameter tuning of gradient boosted decision tree (GBDT) classification models. First import the library package:

```
from sklearn.ensemble import GradientBoostingClassifier
from sklearn.model_selection import cross_val_score
from bayes_opt import BayesianOptimization
import pandas as pd
```

1.3 Case Practice

Where GradientBoostingClassifier is the classification class of GBDT encapsulated by the scikit-learn framework, and GBDT also has a regression class, GradientBoostingRegressor. read the training dataset file:

```
df = pd.read_csv('features.all',encoding='utf-8',index_col=0)
x = df.drop(' whether active on mobile banking next month ',axis = 1).values
y = df[' whether active on mobile banking next month '].values
```

We first use default parameters to perform fivefold cross-validation on the data set, that is, the data set is divided into five data blocks, one of which is used as the validation set each time, and the other four blocks are used as the training set. The average value of the auc index obtained during five training sessions is taken to measure the model performance:

```
result = cross_val_score(gbdt, x, y, cv=5, scoring='roc_auc').mean()
```

We get the result value of 0.865768. Here we begin Bayesian optimization by first defining the function to be optimized, which is the average of the auc metrics of a GBDT classification with uncertain parameters after a fivefold cross-validation on the dataset:

```
def gbdt_cv(n_estimators, min_samples_split, max_features, max_depth):
    res = cross_val_score(GradientBoostingClassifier(n_estimators=int(n_estimators),
    min_samples_split=int(min_samples_split),
    max_features=min(max_features, 0.999),
    max_depth=int(max_depth),
    random_state=2), x, y, scoring='roc_auc', cv=5).mean()
    return res
```

Next specify the range of values for each parameter passed into the GBDT classification, within which the Bayesian optimization algorithm will find the best parameter:

```
gbdt_op = BayesianOptimization(
    gbdt_cv,
    {'n_estimators': (10, 250),
    'min_samples_split': (2, 25),
    'max_features': (0.1, 0.999),
    'max_depth': (5, 15)}
)
```

Where the n_estimators parameter specifies the maximum number of iterations of the weak learner, the min_samples_split parameter specifies the minimum number of samples required for node partitioning, the max_features parameter specifies the maximum number of features, and the max_depth parameter specifies the maximum depth of the decision tree. Then call the maximize() function to maximize the objective function value, we start with the initial five points, and after 200 samples, finally get the optimal model parameters at the current sampling point:

```
gbdt_op.maximize(init_points=5, n_iter=200)
params = gbdt_op.max['params']
```

Use the optimal model parameters to construct a GBDT classification instance, where round_up is the integer rounding function:

```
gbdt1 = GradientBoostingClassifier(n_estimators=round_up(params['n_estimators']),
    max_depth=round_up(params['max_depth']),
    min_samples_split=round_up(params['min_samples_split']),
                max_features=params['max_features'])
We then calculate the average auc after 5-fold cross-validation for the training set:
result = cross_val_score(gbdt1, x, y, cv=5, scoring='roc_auc').mean()
```

At this point, result has been improved to 0.872804, up from 0.865768. The code can be found in the download file autog.py. So far, we have tried Flaml, AutoGluon, and Bayesian optimization three automation model techniques, respectively. We took the union of the predicted values of the three methods as the marketing target of the monthly active customers of mobile banking in the next month to carry out marketing work.

1.4 Case Summary

The author found in actual work that marketing to model-predicted samples has a success rate eight times higher than non-predicted samples. This illustrates that the model identifies non-monthly active customers with monthly active characteristics, significantly improving the marketing success rate of mobile banking, which is crucial for the bank's operations. This case demonstrates the application of automated machine learning technology in commercial banking scenarios, showcasing methods including automated data preprocessing, feature selection, modeling, algorithm selection, and parameter tuning. Readers can make slight modifications to apply these methods to other subjects. Although automated machine learning usually

1.4 Case Summary

searches for the optimal solution within a predefined algorithm and parameter space, it is highly versatile and suitable for many modeling scenarios. The entire project process requires almost no human intervention and, compared to traditional manual modeling, is efficient, accurate, and intelligent, significantly reducing manpower and time costs in modeling. It has positive implications in precise marketing, customer identification, intelligent risk control, and business decision-making. Therefore, automated machine learning has great potential. By combining it with the machine learning demonstration framework Streamlit, we can even create a real-time automated modeling system for business use, which is faster and more efficient than traditional modeling methods. This chapter serves as an introduction. Readers can delve into the specific details of the open-source frameworks in this case for higher-quality projects.

Chapter 2
Retail Potential High-value Customer Identification: Graph Neural Network Technology

It is well known that the greatest development opportunity for China's banking industry lies in retail banking. Retail banking has become a critical component of banking business transformation due to relatively weak economic cycle fluctuations, lower operational risks, lower personal loan non-performing rates, and increased contribution to operations. Industrial and Commercial Bank of China (ICBC), leveraging its extensive payroll and branch network, once proposed the strategic goal of becoming "China's top retail bank." Likewise, Construction Bank aimed for "top-tier retail banking," and China Merchants Bank positioned itself as the "king of retail." Given the relatively low capital utilization of retail banking and tightening capital regulatory requirements, commercial banks have successively introduced concepts such as retail digital transformation, large retail segments, and customer operations deepening. The competition in the retail banking sector has intensified, primarily manifesting in price competition, product competition, channel competition, as well as talent and customer competition. With the opening of financial markets and the entry of foreign and private banks, the number of domestic and international banks has shown a slight increase. According to data released by the China Banking and Insurance Regulatory Commission: as of the end of December 2018, there were 4588 banking financial institution legal entities in China; by the end of June 2019, this number had increased to 4597; and by the end of June 2021, it had risen to 4608. The more banks there are, the greater the competitive pressure in retail banking. Under circumstances of product homogenization and intensified competition in the banking industry, precision marketing becomes crucial. Banks can use artificial intelligence algorithms to identify which customers provide greater value returns, thus gaining a larger market share and enhancing competitiveness under equal conditions. The key to retail banking is customer operations, and the key to customer operations is understanding customers. However, this is not an easy task. On the one hand, the sheer number of retail customers makes one-on-one marketing via traditional methods labor-intensive, inefficient, and minimally effective. On the other hand, many clients' assets are dispersed across multiple banks, securities firms, and insurance companies. Viewing from the perspective of a single bank may not capture the full picture—a "regular client" at one bank may be a high-value client at another. Hence, we need an intelligent, comprehensive data model to identify potential high-value clients.

An ancient saying goes, "One becomes red when near vermilion, and black when near ink." Targeting individuals within social circles rich in high-value customers is a sound strategy. A sociological theory divides the population into nine classes, indicating that people usually associate with others from the same or adjacent classes. The principle that "birds of a feather flock together" implies that social circles seldom span widely across classes. Thus, the social circles of high-value customers typically consist of similar high-value individuals. From a bank's perspective, a person's social circle encompasses active transaction leads within the bank, including transaction counterparties, known social relationships, and financial behavior clues. By incorporating both the customer's attributes and their social circle's traits, our data model will achieve higher identification accuracy.

The technology that addresses these business needs is the graph neural network (GNN), an advanced data mining method that has seen rapid development in recent years. Evidence shows that by leveraging the entity characteristics and relationship data of retail bank customers to build a social knowledge graph, and using deep learning to extract the graph's semantic information, it effectively addresses scenarios such as customer value recognition, loan risk prediction, and identification of suspicious accounts.

This chapter discusses the graph convolutional network (GCN) node classification tasks based on financial characteristics and relationships of customers. It introduces GCN implementation methods to accurately identify potential high-value retail customers. This model is already in production at the author's bank. In practice, we find that the GCN model demonstrates strong generalization ability with unknown samples, handles more complex data mining tasks effectively, and yields better practical application results. We believe the graph neural network is a highly valuable data mining method for retail banking business scenarios.

2.1 Introduction to Graph Neural Network

This section mainly introduces the concept of graph neural network, the advantages of algorithm, the development history of technology and the development opportunity in the era of big data.

2.1.1 The Concept of Graph Neural Network

In essence, a graph neural network overlays a neural network onto a graph data structure. Simply put, it's a deep learning technique based on graph structures. Understanding this concept reveals that a graph neural network consists of two

2.1 Introduction to Graph Neural Network

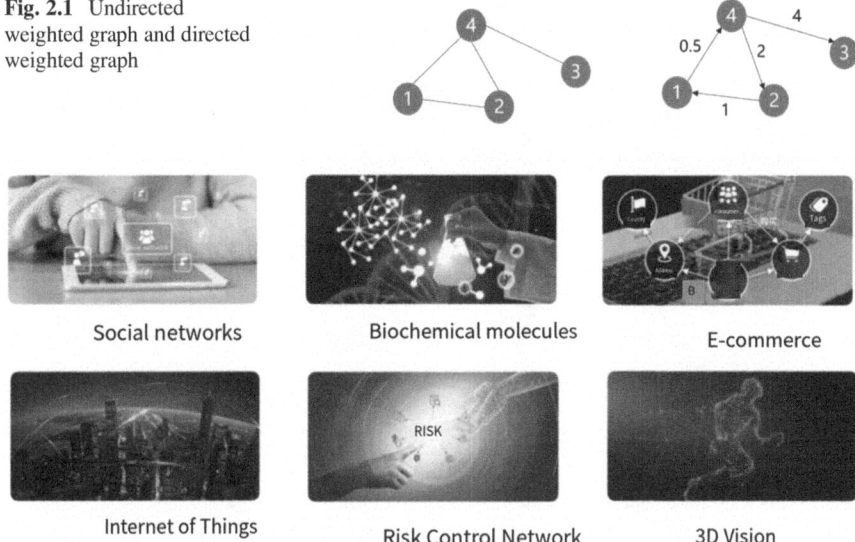

Fig. 2.1 Undirected weighted graph and directed weighted graph

Fig. 2.2 Graph data in the real world

primary components: "graph" and "neural network." The "graph" refers to the graph data structure in graph theory, typically represented as a knowledge graph. The "neural network" aspect signifies a computational network made up of multiple neurons organized hierarchically.

To elaborate, a graph is a finite, nonempty set of vertices and edges, usually denoted as $G(V, E)$. Here, G represents a graph, V represents the set of vertices in the graph G, and E represents the set of edges in the graph G. Graphs can be classified into directed and undirected based on edge directionality, and into weighted and unweighted based on the presence of edge weights, as shown in Fig. 2.1.

In real life, social networks, Internet of Things, chemical molecules, e-commerce, and other data, big to the macro world, small to the micro world, they are all "graph" structure data, as shown in Fig. 2.2. These data accounts for more than 80% of the data in the real world, covering a wide range of industries. If you can carry out deep learning on the graph data, you can dig out the value behind these massive data, which is very considerable.

Generally, neural network is a machine learning method that stacks several neurons (which can be understood as computing units) according to a certain structure level and fits the data set through forward propagation and backpropagation. Wherein, forward propagation obtains the result and loss of model inference, and back propagation uses this error to optimize the parameters of each layer of the network, and minimizes the loss through repeated iteration, so as to fit the data set. Simply put, the neural network is a distillation of feature transformation, that is, the input layer is the representation feature, and the middle layer is the semantic feature of different levels, among which the closer the part of

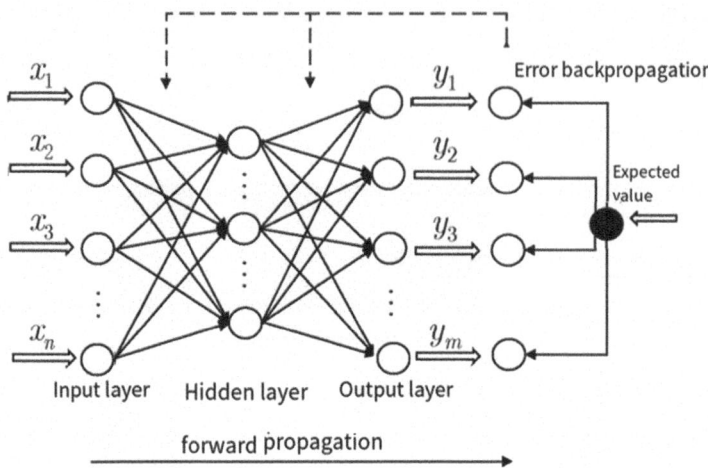

Fig. 2.3 Forward propagation and reverse propagation of neural network

the input layer, the more obvious the representation feature, the more specific the feature, the closer the part of the output layer, the more obvious the semantic feature, the more abstract the feature (that is, the closer to the predicted semantics), as shown in Fig. 2.3.

After reviewing the concepts of "graph" and "neural network," it is logical that graph neural network is to complete the "semantic feature distillation" of neural network on graph data structure. If the graph is also regarded as a network (such as social network, fund exchange network, and so on), then in the graph neural network, there are two kinds of networks, "graph network" and "neural network." Its intuitive structure is shown in Fig. 2.4, the left is the graph network and the right is the neural network.

In this case, the neural network can act on the nodes, subgraphs, and other objects of the graph structure, and the type of neural network is not limited to the fully connected neural network shown in the figure above, but also MLP (multilayer perceptron), CNN (convolutional network), RNN (recurrent network), etc., so that algorithms adapted to different tasks can be designed.

Graph embeddings are an important aspect of graph neural networks. It refers to the process of mapping a graph signal (i.e., an input signal accepted by a graph neural network, usually a high-dimensional dense matrix) to a low-dimensional dense vector. Simply put, graph embedding is the conversion of graphs into vectors or sets of vectors that represent information such as topologies, nodes, subgraphs, etc., thus providing input for subsequent machine learning tasks such as classification, regression, etc. Graph embedding is divided into two types: node embedding and subgraph embedding. The former uses vectors to represent nodes and is usually used for node-level tasks such as node classification and link prediction. The latter uses vectors to represent subgraphs, which are usually used in graph-level tasks such as graph structure classification and graph pattern recognition.

2.1 Introduction to Graph Neural Network

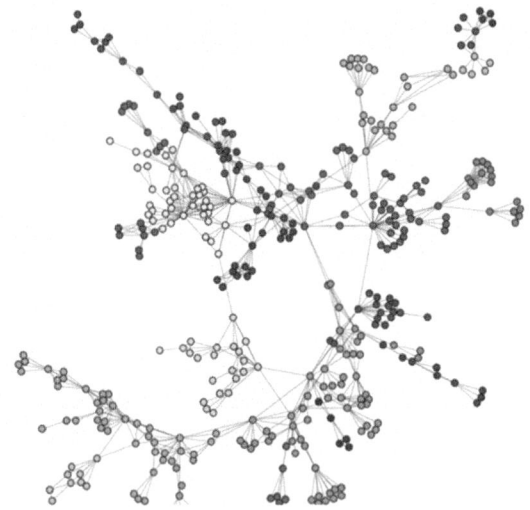

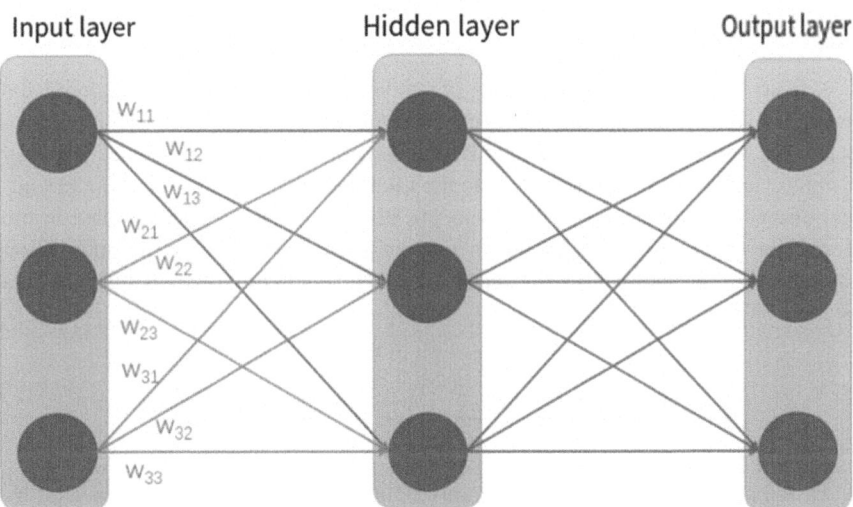

Fig. 2.4 Two kinds of network structures simultaneously existing in the neural network

2.1.2 The Advantages of Graph Neural Network

Since graph neural network is a new method in the field of data mining, let's first analyze what data types are involved in data mining. From the perspective of data arrangement, it can be divided into Euclidean data and non-Euclidean data. Euclidean data is data that can be measured by Euclidean distance, is characterized by "alignment," has "translation invariance," and can usually be represented by a matrix, as shown in Fig. 2.5.

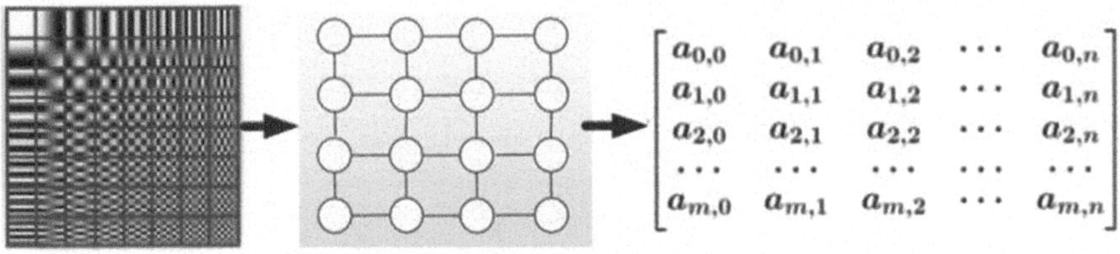

Fig. 2.5 Grid matrix representation of Euclidean data

Fig. 2.6 Non-Euclidean data: knowledge graph

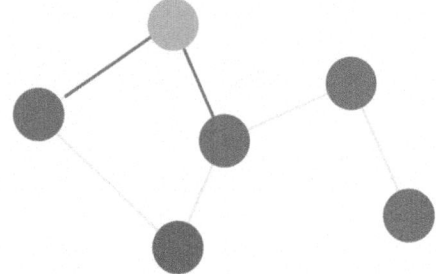

For any Euclidean data, the number of neighbor data is the same, for example, the image is two-dimensional Euclidean data and the text is one-dimensional Euclidean data, they are grid matrix data of different dimensions, and the distance between samples can be calculated according to the Euclidean distance formula. The Euclidean distance formula reflects the geometric distance of two Euclidean data in the Euclidean space, it is the Euclidean data in each dimension of the subtraction calculation of the square after the accumulation of square root obtained, the formula is

$$d(x,y) := \sqrt{(x_1 - y_1)^2 + (x_2 - y_2)^2 + \cdots + (x_n - y_n)^2} = \sqrt{\sum_{i=1}^{n}(x_i - y_i)^2}.$$

Different from Euclidean data in non-Euclidean data, it cannot be measured by Euclidean distance, it is arranged randomly, the number of neighbors of different nodes is different, and it does not have "translation invariance." The knowledge graph is one kind of non-Euclidean data, as shown in Fig. 2.6.

More than 80% of the world's data is described by the non-Euclidean structure of knowledge graph, such as money flow graph, social relationship graph, transportation route graph, investment relationship graph, risk relationship graph, and so on. In the field of retail banking, it also involves customer–product relationship graph, customer–customer relationship graph, customer–fraud relationship graph, etc. Figure 2.7 shows the bank's lost contact customer relationship graph.

Graph data reflects many aspects of the world but is difficult for traditional neural networks to learn. Traditional neural networks excel at handling sequential or grid data like speech, images, and text. Without applying neural networks to knowledge

2.1 Introduction to Graph Neural Network

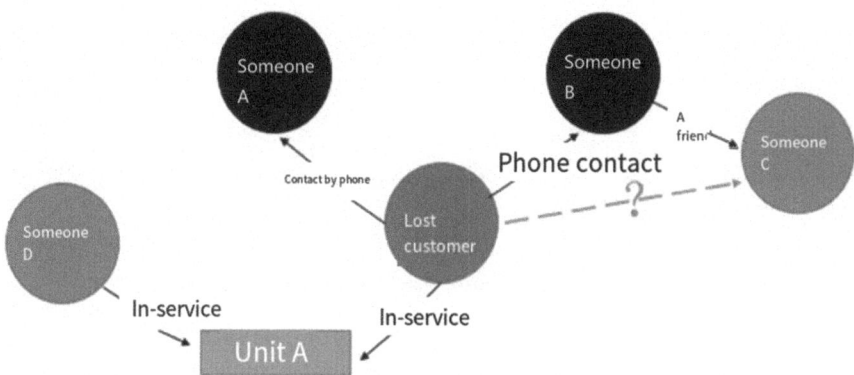

Fig. 2.7 Mapping the relationships of banks' lost customers

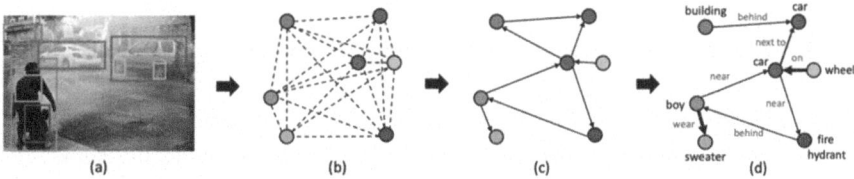

Fig. 2.8 Graph neural network generates the scene relationship according to the image

graphs, extracting semantic information from vast graph data becomes challenging, limiting its cognitive and decision-making value. Fortunately, graph neural networks (GNNs) accept graph data as input and can learn embeddings from it. In other words, GNNs enable "deep learning on knowledge graphs," transforming "data mines" into "gold," offering substantial value.

Additionally, GNNs integrate deep learning methods into graph data structures, considering both entity features and their relationships. This extra dimension of entity relationships allows more comprehensive modeling, aligning better with the real world, enhancing model explainability, and often outperforming traditional neural networks in practical applications. GNNs can perform various complex machine learning tasks, such as node classification, graph classification, link prediction, subgraph similarity, subgraph embedding, and more. They are widely used in quantitative investment, event impact assessment, biopharmaceuticals, traffic prediction, scene relationship generation, recommendation systems, risk transmission and warning, anomaly detection, behavior prediction, pattern recognition, etc. For example, in scene relationship generation, GNNs can understand relationships between entities and achieve bidirectional generation of images and scene relationships, as illustrated in Figs. 2.8 and 2.9.

In short, the benefits of graph neural networks include: turning massive graph data into "fuel" that can be learned, good interpretability, and suitable for a variety of

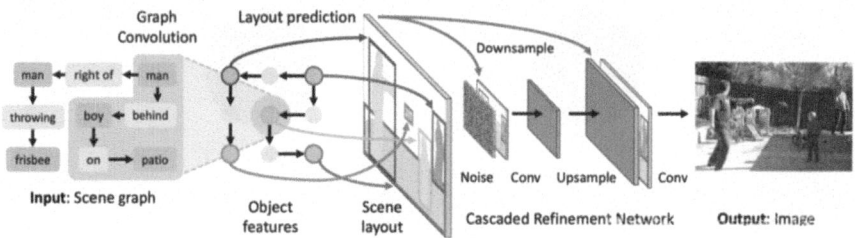

Fig. 2.9 Graph neural network generates images based on scene relationships

machine learning tasks. This makes it a very wide range of application scenarios, and the application effect is also qualitatively improved.

2.1.3 The Development of Graph Neural Networks

In the late 1990s, GNNs were introduced alongside recurrent neural networks (RNNs) and convolutional neural networks (CNNs) as applications of neural networks on different data types. Initially translated as "neural networks for graphs," GNNs did not see rapid development compared to the rapidly rising CNNs because graph data lacked widespread application scenarios at the time. After more than a decade of stagnation, the maturity of CNNs led to considerations on how to generalize the parameter-sharing nature of CNNs to non-Euclidean graph data structures. This objectively promoted the development of GNNs, which saw a boom starting in 2017, as evident from the number of academic papers published, as shown in Fig. 2.10.

Graph neural networks surged in popularity after 2018, becoming a rapidly expanding technology. The international academic summit highlights research directions and trending topics in academia. At the International Conference on Learning Representation (ICLR) in deep learning, the number of accepted papers on graph neural networks ranked 29th in 2019, rose to 7th in 2020, and secured 3rd place in 2021. The academic community favors graph neural networks because they address numerous research problems, indicating the technology's growing maturity, systematic theory, and methodology. This attracts more scholars to the field. Figure 2.11 shows the technical keyword cloud from the 2020 International Conference on Data Mining and Knowledge Discovery (KDD), a leading academic conference. In the word cloud, the larger the font size of technical keywords, the more papers and research results on that technology. Graph neural networks top the list.

The Alidama Institute forecasts the top 10 tech trends for 2019, emphasizing that "hyperscale graph neural network systems will endow machines with common sense." Graph neural networks combine end-to-end deep learning with graph reasoning, merging the benefits of pattern recognition and relational reasoning. This provides a fundamental framework for emulating complex neuronal brains and

2.1 Introduction to Graph Neural Network

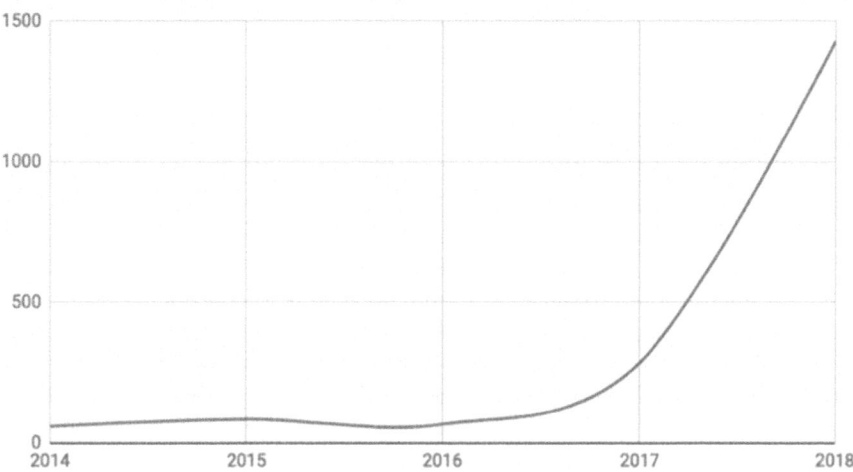

Fig. 2.10 Shows the early development of Graph neural networks

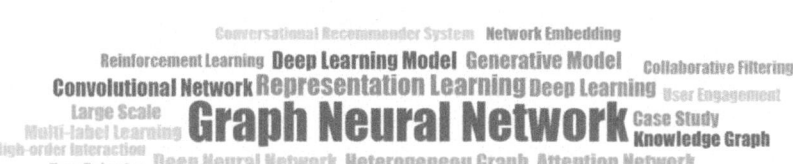

Fig. 2.11 Keyword cloud at the 2020 KDD conference

developing machine intelligence with enhanced common sense, cognition, and understanding. The technology surrounding graph neural networks is thriving, drawing top global talent to this field and spawning many branches. For instance, the GCN adapts convolution operations for graph structures, while the graph attention network (GAT) applies attention mechanisms to graphs. The gated graph neural network (GGNN) integrates gated recurrent units (GRUs) into graph structures, and GraphSAGE builds upon GCN for large-scale graphs. Recently, several frameworks have gained prominence in graph neural network development, including the DGL introduced here, the TensorFlow-based tf_geometric, the TensorFlow and PyTorch-compatible GraphGallery, AliGraph from Ali, and Facebook's PyTorch BigGraph. With Internet companies on the rise, many large-scale distributed graph neural network models have emerged, creating powerful applications within the internet ecosystem. The era of graph deep learning is now upon us.

2.1.4 Graph Neural Network as the Product of Big Data Era

Graph neural networks emerged in the big data era for several reasons. First, from a technological development perspective. The core of big data is data intelligence. Data intelligence aims to identify and evaluate relationships among numerous concepts within vast data sets, formulating mathematical representations for predictions. This requires uncovering underlying patterns and solving data representation challenges. Data intelligence evolved through expert systems, traditional machine learning, and neural networks. Input knowledge transitioned from concrete to abstract, from rules to features to patterns. Consequently, intelligent processing efficiency improved, but model explainability weakened. Traditional machine learning struggles compared to neural networks as data set sizes grow, due to limited expressiveness. Figure 2.12 compares traditional machine learning to various neural network scales, showing performance improvement with larger data and models.

Goodfellow presented his paper "Maxout Networks" (Maximum Output Networks) at ICML 2013 (International Conference on Machine Learning). In this paper, he demonstrated that Maxout Networks can be approximated infinitely to any continuous function. In other words, the neural network can fit any continuous function, and the neural network has better expressibility than the traditional machine learning.

We observe several trends: rapid growth in industry data volumes, swift advancements in GPU computational power, innovations in algorithms, prolific academic achievements, substantial capital investment, and diverse application scenarios, all of which drive the fast development of neural networks. Neural networks exhibit two developmental directions: vertical development, represented by deep neural networks (DNNs) and convolutional neural networks (CNNs), characterized by stacked

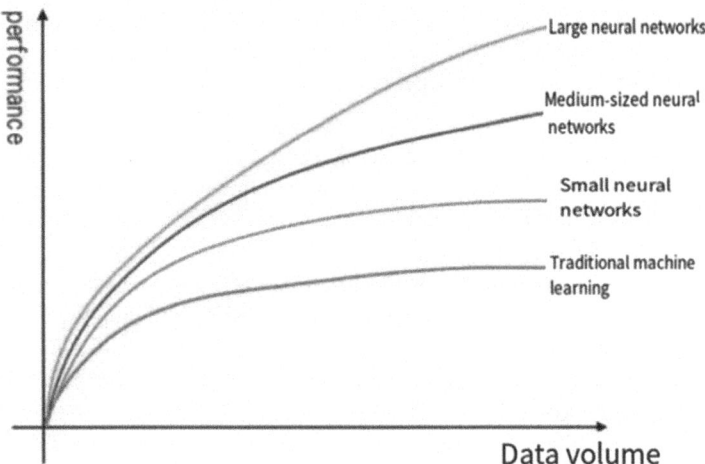

Fig. 2.12 Comparison of the performance of traditional machine learning and small-, medium-, and large-scale neural networks

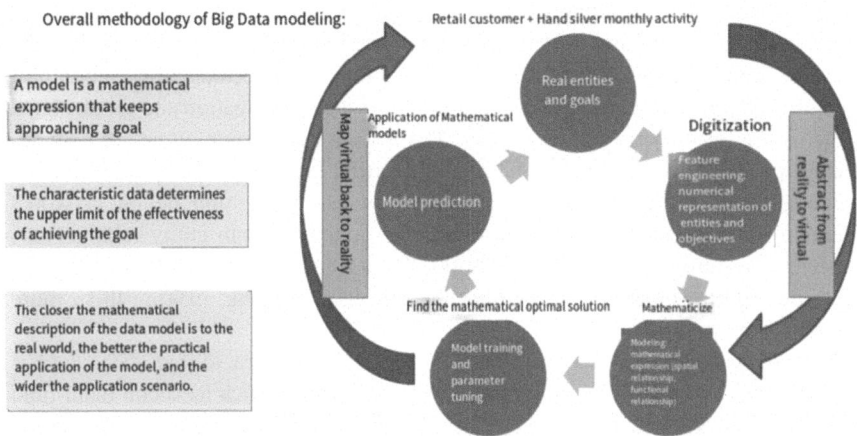

Fig. 2.13 Overall methodology of big data modeling

layers with typical applications in computer vision (CV); and horizontal development, exemplified by recurrent neural networks (RNNs), involving neuron connections with applications in natural language processing (NLP). The concurrent vertical and horizontal advancements in neural network technologies, with wide-ranging applications across multiple fields, indicate that this technology has reached maturity. Future development is likely to integrate vertical and horizontal advancements into more application domains. In fact, within graph neural networks, node message passing can be seen as horizontal iteration, while feature transformation represents vertical iteration, thus embodying their integration. Consequently, the emergence of graph neural networks is a logical progression.

From a methodological perspective in data mining, we can summarize this approach: technology abstracts real-world problems into mathematical expressions, solves them using mathematical methods at the abstract level, and maps the solutions back to real-world problems for practical guidance. Feature engineering digitizes real entities and their relationships, achieving digitization; models and algorithms establish mathematical relationships between digital representations and modeling objectives, achieving mathematization; model computation and training solve mathematical problems by finding optimal solutions in the abstract mathematical space and mapping them to real-world problems, thus creating real-world value. The overall methodology for big data modeling is illustrated in Fig. 2.13.

Data mining mirrors real-world entities and their interconnections. The more accurate this representation, the broader its applications. For instance, the Markov chain effectively models temporal objects and dependencies, making it valuable in speech recognition, machine translation, national economy analysis, and event prediction. Similarly, probability graphs depict uncertain event relationships, proving useful in anti-fraud measures and event prediction. However, both Markov chain and probability graphs have limitations, as they downplay embedded representations, losing cryptographic-meaning information. Graph neural networks address this shortfall. Graph neural networks consist of two types: the topological network, which outlines entities and their relationships, and the feature transformation neural network, which handles the feature transformation of nodes, edges, or subgraphs. The topological network facilitates horizontal message propagation and graph signal transmission, grounded in graph theory. The feature transformation neural network manages vertical feature transformation, converting original features into semantic features, based on deep learning principles. In essence, graph neural networks integrate graph theory and deep learning. They consider both entity relationships and features. Hence, graph neural networks abstract entity relationships as a "graph" while describing semantic information through deep learning, making them more powerful than traditional algorithms.

Finally, The scale of the data is very large. In the era of big data, data is characterized by "big." The graph data in the real world, such as social data, fund flow data, and so on, are very large. Under the distributed learning architecture, graph neural network can process very large data scale, which is very suitable for processing hundreds of millions of nodes of industrial data. Many Internet technology companies have spent heavily on active layout in this field and have made significant progress. For example, in Alibaba Group, graph neural network has covered Taobao recommendation search, new retail, network security, online payment, Youku, Ali Health, and other related businesses, forming a number of heterogeneous graphs with billions of edges and billions of vertices. Large-scale graph neural networks have become a winning tool for Internet companies.

To sum up, graph neural network was born in the era of big data, which is a natural thing caused by technology development, data mining methodology and data scale.

2.2 Scheme Design

The objective of this case is to establish an effective social relationship graph among retail customers according to the characteristics of each retail customer and its strong cognition relationship, and build a graph convolutional neural network on this graph to classify each customer, so as to judge the possibility that he is a potential high-value customer.

Figure 2.17 reflects the basic idea of this case: select the features that are closely related to the high-value goal to form the feature vector of each customer [x_{i1}, x_{i2}......x_{im}]. For high-value customers, the label is 1, and for other customers, the label is 0. The edges in the diagram indicate that there is a strong cognitive relationship between two customers, and the edges are directional but not heavy. We take all customers as nodes and construct a large directed authority graph. We hope to judge whether each customer node is of high value by learning the embedded representation of the whole graph in accordance with the idea of Sect. 2.1, "One becomes red when near vermilion, and black when near ink," considering the characteristics of both customers and their neighbors. Here we should distinguish

2.2 Scheme Design

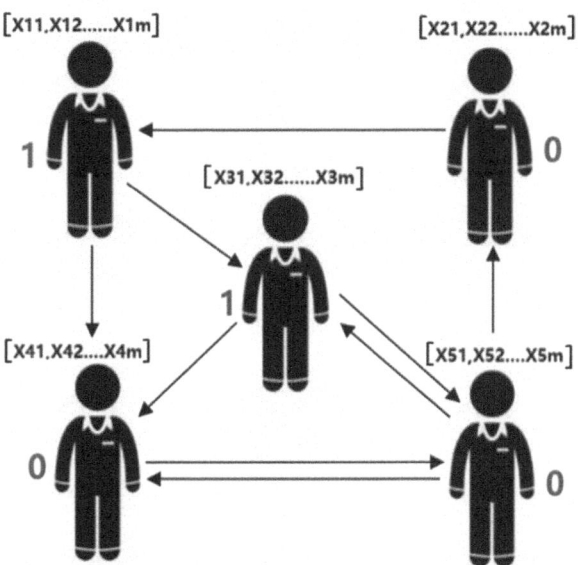

Fig. 2.14 Retail customer strong cognition relationship social graph

between weak cognition and strong cognition. Consumer transactions and electronic payment transactions with the nature of consumption, the two sides of the transaction are often weak cognition relationship, while relatives, friends, and other social relations and active transfer, electronic payment transactions, the two sides of the transaction are strong cognition relationship. It can be judged by the transaction postscript and transaction attribute data. In addition, if different customers pay to the same subscriber number (such as Shui Electric), it means that there is also a strong cognition relationship between them. We extracted the strong cognition relationship separately to form the "retail customer strong cognition relationship social graph" as shown in Fig. 2.14.

Figure 2.15 illustrates the system architecture of this case study, consisting of three layers: feature engineering, social graph, and graph convolutional neural network. The bottom layer extracts data on bank customer attributes, historical transactions, social relationships, and payment transactions. It generates raw features, then uses the FeatureTools library to compute derived features, and the Feature-Selector library identifies features strongly correlated with high-value target labels for modeling. The middle layer employs the Neo4j graph database to create a social graph based on customer node characteristics and their strong relationships. The top layer builds a deep learning environment on NVIDIA GPU hardware and its components, where graph convolutional neural network model code runs on the DGL framework.

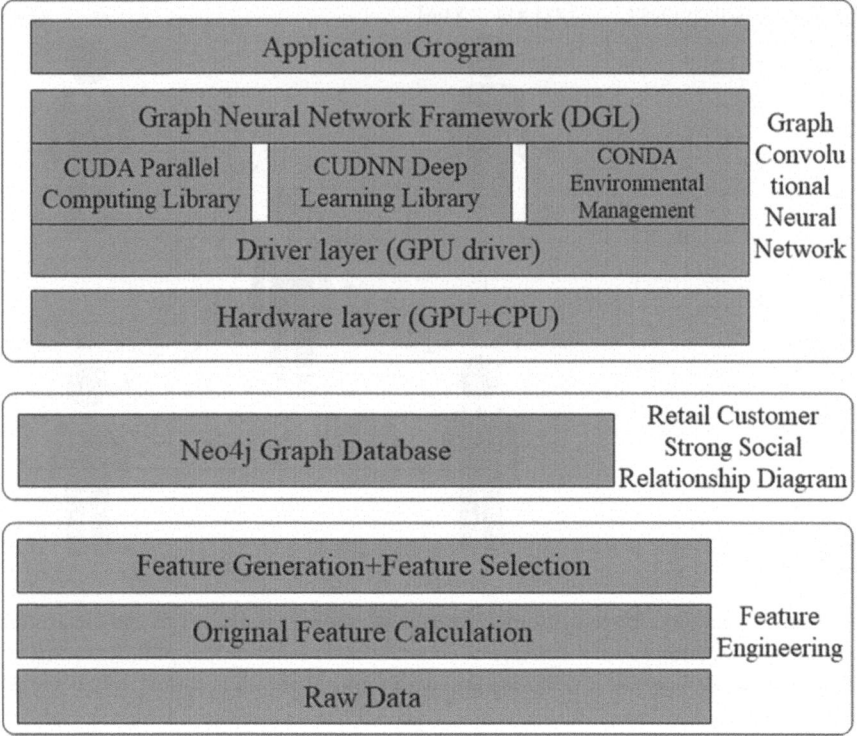

Fig. 2.15 System architecture of this case

2.3 Graph Convolutional Neural Network Algorithm

Graph convolutional neural networks (GCNNs) perform convolutional operations on graphs. In contrast, traditional convolutional neural networks (CNNs) operate on Euclidean spatial data like images, where each pixel has fixed neighboring nodes, ensuring a consistent receptive field. For graph data, the number of neighbors per node varies, making traditional CNN methods inapplicable. To address this, new methods are required. Spectral domain convolution was the initial solution. The number of edges per node is the node's degree. A matrix with degrees on its diagonal and zeros elsewhere is the degree matrix, denoted D. If an edge exists between two nodes, the corresponding element in the adjacency matrix is 1; otherwise, it is 0. The adjacency matrix is denoted A. In spectral domain convolution, the Laplacian matrix L is defined as $L = D - A$ (Fig. 2.16).

The Laplacian matrix L is a real symmetric matrix that characterizes the smoothness of the graph signal and can be understood as the derivative on the graph. It can be shown mathematically that real symmetric matrices can all be orthodiagonalized, i.e., can be expressed as the product of an orthogonal basis matrix and its transpose matrix with a frequency diagonal matrix:

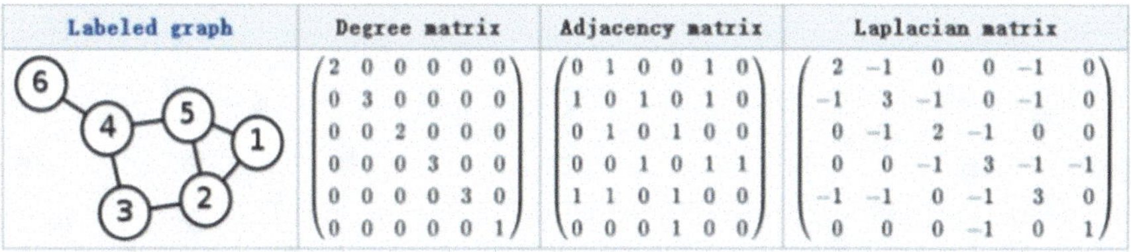

2.3 Graph Convolutional Neural Network Algorithm

Labeled graph	Degree matrix	Adjacency matrix	Laplacian matrix
(graph with nodes 6,4,5,1,3,2)	$\begin{pmatrix} 2 & 0 & 0 & 0 & 0 & 0 \\ 0 & 3 & 0 & 0 & 0 & 0 \\ 0 & 0 & 2 & 0 & 0 & 0 \\ 0 & 0 & 0 & 3 & 0 & 0 \\ 0 & 0 & 0 & 0 & 3 & 0 \\ 0 & 0 & 0 & 0 & 0 & 1 \end{pmatrix}$	$\begin{pmatrix} 0 & 1 & 0 & 0 & 1 & 0 \\ 1 & 0 & 1 & 0 & 1 & 0 \\ 0 & 1 & 0 & 1 & 0 & 0 \\ 0 & 0 & 1 & 0 & 1 & 1 \\ 1 & 1 & 0 & 1 & 0 & 0 \\ 0 & 0 & 0 & 1 & 0 & 0 \end{pmatrix}$	$\begin{pmatrix} 2 & -1 & 0 & 0 & -1 & 0 \\ -1 & 3 & -1 & 0 & -1 & 0 \\ 0 & -1 & 2 & -1 & 0 & 0 \\ 0 & 0 & -1 & 3 & -1 & -1 \\ -1 & -1 & 0 & -1 & 3 & 0 \\ 0 & 0 & 0 & -1 & 0 & 1 \end{pmatrix}$

Fig. 2.16 Laplace matrix is equal to the degree matrix minus the adjacency matrix

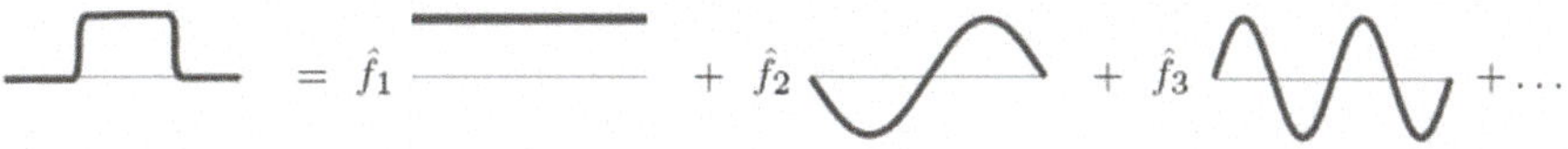

⊓⊔ = $\hat{f}_1$ ———— + $\hat{f}_2$ ∿ + $\hat{f}_3$ ∿∿ + ...

Fig. 2.17 Fourier transform

$$L = V\Lambda V^T = \begin{bmatrix} \vdots & \vdots & \cdots & \vdots \\ v_1 & v_2 & \cdots & v_N \\ \vdots & \vdots & \cdots & \vdots \end{bmatrix} \begin{bmatrix} \lambda_1 & & & \\ & \lambda_2 & & \\ & & \ddots & \\ & & & \lambda_N \end{bmatrix} \begin{bmatrix} \cdots & v_1 & \cdots \\ \cdots & v_2 & \cdots \\ & \vdots & \\ \cdots & v_N & \cdots \end{bmatrix}$$

The essence of the Fourier transform is to express any function as a linear combination of several orthogonal functions (composed of sin and cos), as shown in Fig. 2.17.

In spectral convolution, both the graph signal and the convolution kernel are projected into the frequency component space. Mathematically, this is the original matrix multiplied by the v vector to obtain the projection matrix after Fourier transforms. This process projects the non-translation invariant spatial graph signal matrix into the translation-invariant spectral domain signal matrix of Fourier space. Consequently, it converts non-Euclidean spatial data into Euclidean spatial data. Thus, the convolution operation occurs in the spectral domain. Using the convolution theorem, an inverse Fourier transformation returns from the spectral domain to the spatial domain to indirectly compute the spatial domain convolution result. The convolution theorem states that the Fourier transform of the convolution of two functions equals the convolution of their Fourier transforms. That is $f_1(t) \leftrightarrow F_1(\omega)$, $f_2(t) \leftrightarrow F_2(\omega)$, if F represents the Fourier transform, then $F[f_1(t) * f_2(t)] = F_1(\omega) \cdot F_2(\omega)$. During the transformation of the original graph signal and the convolution graph signal in the base space, the graph signal is transitioned from the spatial domain to the spectral domain corresponding to the Fourier base. The inverse transformation then returns it to the spatial domain, resolving the issue. However, this introduces new challenges: the large node size of graphs and the dense nature of the Fourier base entail high computational complexity for graph signal projection. Additionally, the orthogonal diagonalization decomposition of the Laplace matrix is computationally intensive, making it impractical to train the spectral domain convolution model on standard computers. Consequently, a simplified approach, often referred to as the "spatial convolution" method, is commonly used. In spatial convolution, each node updates its features by aggregating the features of its K-order neighbors after

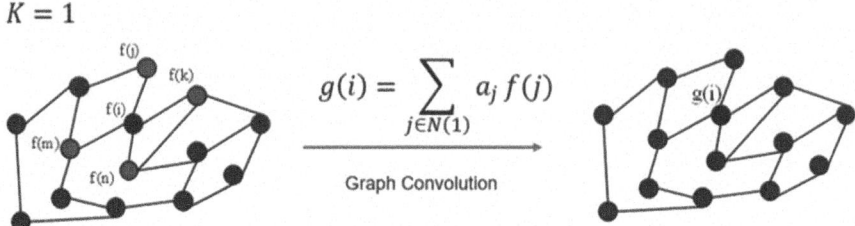

Fig. 2.18 Spatial convolution based on first-order neighbors

transforming its own features. For instance, the feature smoothing process based on first-order neighbors is illustrated in Fig. 2.18.

2.4 Development Framework

This section introduces the underlying technology and development framework for building graph neural networks.

2.4.1 Neo4j Graph Database

Neo4j is an open-source NoSQL graph database developed with Java and Scala languages, which is the most popular graph database at present. Neo4j is developed by Neo Technology, an American graphic database service provider. Neo4j's official website address is https://neo4j.com/ and the code repository is https://github.com/neo4j/neo4j.

The graph database constructs data using a graph structure, employing nodes, edges, and attributes to represent and store data. Neo4j stands as a robust, network-oriented database, featuring a high-performance graph engine. It surpasses relational databases in efficiency for relational scenarios. For instance, to find all movies funded by a particular person (as shown in Fig. 2.19), accessing neighbor nodes in a graph database is straightforward. Conversely, a relational database would require creating separate tables for movies and investors, and performing complex queries across multiple tables. Generally, graph databases exhibit execution efficiency an order of magnitude higher than relational databases in managing complex entity relationships. Thus, in the "relational" scenario, the graph database truly acts as a "relational" database.

Neo4j has enterprise Server version and community Server version. The differences are as follows.

2.4 Development Framework

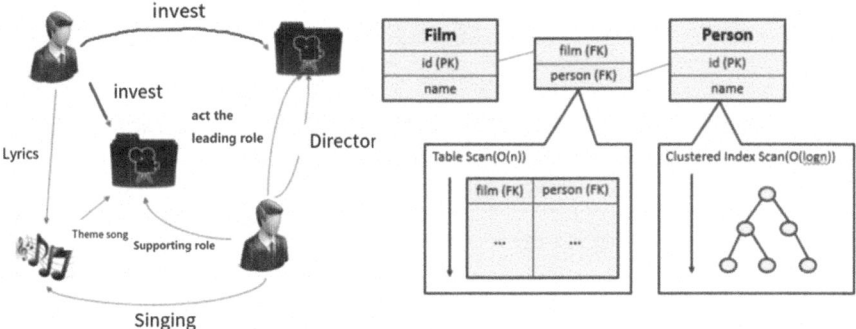

Fig. 2.19 Compares a graph database to a relational database

1. **Capacity**: The community version supports up to 32 billion nodes, 32 billion relationships and 64 billion attributes, while the enterprise version does not have this limitation.
2. **Concurrency**: Community Edition can only be deployed as a single instance, not as a cluster. The enterprise edition can be deployed as a high availability cluster or causal cluster, which can solve the problem of high concurrency.
3. **Disaster recovery**: Since the enterprise Edition supports clusters, the failure of some instances will not affect the normal operation of the entire system.
4. **Hot backup**: The community edition only supports cold backup, that is, backup can be performed only after the service is stopped, while the enterprise edition supports hot backup, with full backup for the first time and incremental backup for the next.
5. **Performance**: The community edition uses a maximum of 4 cores, while the enterprise can use all the cores, and the performance is carefully optimized.
6. **Support**: Enterprise version customers can get 10 hours of phone support on working days, including phone, email, WeChat, and so on.
7. **Plugins**: The enterprise version can use Bloom, ETL, and other tools, the community version does not support.

In this case, Neo4j3.5.17 Community Edition is used to build a social graph of customer strong awareness relationship, which will serve as the input data source for the graph neural network.

2.4.2 DGL Graphical Neural Network Framework

Deep graph library (DGL) is a graph neural network development framework jointly open-sourced by New York University, NYU Shanghai, AWS Shanghai Research Institute, and AWS MXNet Science Team. It is a highly popular development framework in the field of graph neural networks. Its official website is https://www.dgl.ai/, the code repository is https://github.com/dmlc/dgl, following the

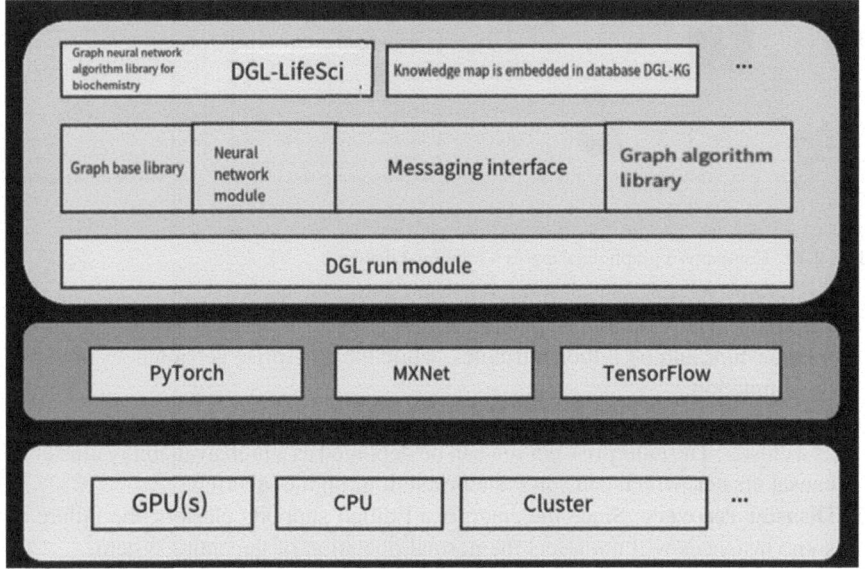

Fig. 2.20 Business LOGO of DGL and its frame structure

Apache-2.0 open-source license agreement. The commercial logo and framework structure of DGL are shown in Fig. 2.20.

The framework is designed based on the following three principles:

1. It works seamlessly with the current mainstream deep learning frameworks (Pytorch, MXNet, TensorFlow, etc.). Thus, the transformation from traditional tensor arithmetic to graph arithmetic can be realized freely.
2. Provide the minimum API to reduce the learning threshold of users.
3. Efficiently and transparently parallel graph computations, which can be easily scaled to large graphs.

DGL has three main features. First, DGL functions as a "frame on a frame," as depicted in Fig. 2.21. To avoid unnecessary duplication, DGL adopts a Keras-like approach by building on existing deep learning frameworks. DGL supports both MXNet and PyTorch frameworks. Unlike Keras though, DGL doesn't constrain users to its own syntax. Instead, DGL allows for flexible syntax usage between the two frameworks. For instance, users can implement common convolution layers and attention layers using their framework of choice, switching to DGL for graph-related tasks. When a user calls a function through DGL, the optimized system still uses the underlying framework's computation and automatic differentiation capabilities, allowing for efficient development and debugging.

2.5 Case Practice

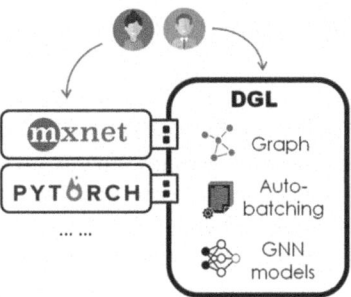

Fig. 2.21 DGL is a graph neural network framework based on two deep learning frameworks

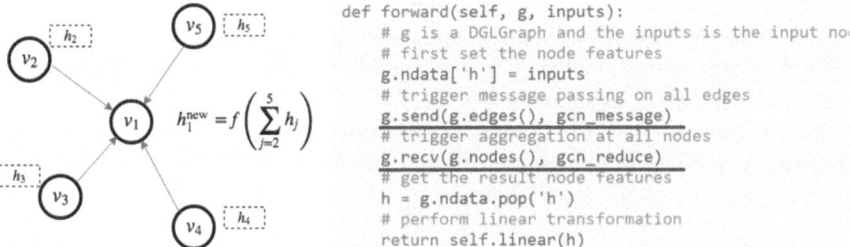

Fig. 2.22 DGL's easy-to-use "messaging" programming approach

Second, DGL provides an easy-to-use "messaging" programming approach, as shown in Fig. 2.22. Developers can customize Message functions and Reduce functions of nodes to construct new models. The developer uses the Pytorch syntax throughout the application, while DGL only provides supplementary messaging APIs such as mailbox, send, recv, etc. to help the user complete the graph calculation.

Third, DGL supports hyperscale graph neural networks, providing distributed architecture to train networks of tens of billions of nodes. This case uses DGL to build the graph convolutional neural network, including model definition, model training, model prediction, and other links.

2.5 Case Practice

This section introduces the case development and operation environment construction, data preparation, social graph construction, feature engineering, model code writing, operation evaluation, etc.

2.5.1 Environment Preparation

Server Hardware: Intel i9-10900K, 128GB RAM, 2TB SSD.
GPU: GTX 1080 Ti 11GB, single card setup.
Operating system: Windows10 64-bit.
Operating System: Windows 10 64-bit.
Graph Database: Neo4j 3.5.17 Community Edition.
Machine Learning Framework: Anaconda.
Development Language: Python 3.6.7.

First, set up the operating environment for the graph database Neo4j. Download JDK 8 from the Oracle official website. The author used jdk-8u301-windows-x64.exe, available at https://www.oracle.com/java/technologies/downloads/#java8-windows. After installation, double-click to run and execute java -version to confirm the successful installation, as shown in Fig. 2.23.

After finding Neo4j 3.5.17 at https://neo4j.com/download-center/, download the Community Services Edition and unzip it to E:\. You will be able to see the main directory structure of Neo4j as follows.

```
bin directory: stores the executable programs of Neo4j;
conf directory: the configuration file that controls Neo4j startup;
data directory: stores the core database files;
plugins directory: stores Neo4j plug-ins;
```

Set the NEO4J-HOME environment variable, whose value is the installation path, as shown in Fig. 2.24.

Add e:\neo4j-3.5.17\bin\neo4j.bat to the PATH environment variable. Put the apoc-3.5.0.7-all.jar from the attachment to this book into the neo4j-3.5.17\plugins directory and add the following to the e:\neo4j-3.5.17\conf\ neo4j.conf configuration file:

Fig. 2.23 Installing the JDK8

```
C:\>java -version
java version "1.8.0_301"
Java(TM) SE Runtime Environment (build 1.8.0_301-b09)
Java HotSpot(TM) 64-Bit Server VM (build 25.301-b09, mixed mode)
```

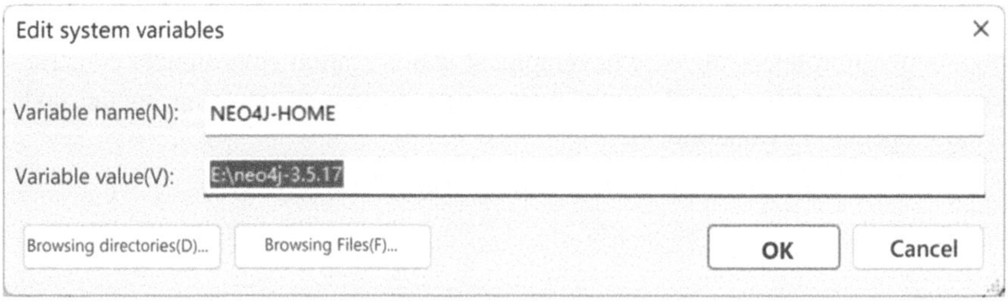

Fig. 2.24 Setting the Neo4j environment variable

2.5 Case Practice

Fig. 2.25 Starting and stopping the Neo4j service

```
(dgl) E:\>neo4j uninstall-service
Neo4j service uninstalled

(dgl) E:\>neo4j install-service
Neo4j service installed

(dgl) E:\>neo4j start
Neo4j service started

(dgl) E:\>neo4j status
Neo4j is running

(dgl) E:\>neo4j stop
Neo4j service stopped
```

```
dbms.security.procedures.unrestricted=apoc.*
apoc.export.file.enabled=true
```

Complete the apoc control installation and allow data export. Set dbms.active_database=XX in the e:\neo4j-3.5.17\conf\neo4j.conf file where XX represents the name of the newly created database. After Neo4j is started, this database will be created automatically if it does not exist, and if it does, this database is used. Execute neo4j install-service to register neo4j as a system service; Run neo4j uninstall-service to uninstall neo4j service; Run neo4j start to start neo4j; Run neo4j stop to stop neo4j; Do neo4j console to start the neo4j console; Run neo4j restart to restart neo4j; Run neo4j status to display the neo4j status, as shown in Fig. 2.25.

Next, we need to set up the graph database. First build the data set. We create csv files for node data and edge data respectively for storage. The format of the node csv file is:

```
Name of node number: ID, attribute 1, attribute 2, ... attribute N
```

Take the customer node as an example, the csv file is named customer.csv and its first action is:

Customer Number: ID, Customer Name, Account Opening Institution, Customer Name, ID Number, Phone Number, Mobile Number, Mailing Address, Residential Address, Business Address.

The subsequent rows are the values of each field in turn. The customer number represents the unique number of the node, and the letter ID after the colon indicates that Neo4j will create the node according to this field.

The format of the relational csv file is:

```
Start node number: START_ID, end node number: END_ID, edge TYPE: TYPE,
edge attribute 1, edge attribute 2, ... , edge attribute M.
```

Take the fund transaction relationship between customers as an example, the csv file is named relationship1.csv, and its first behavior is as follows:

```
Customer number: START_ID, peer customer number: END_ID,
relationship TYPE: TYPE, transaction amount, and number of
transactions.
```

The subsequent rows are the values of each field in turn. Where the customer number represents the start node of the edge, the customer number of the other side represents the end node of the edge, the relationship type field fills in various transactions and investment relationships, and the transaction amount and number of transactions are the transaction statistics from the start node to the end node. In this way, Neo4j will create a directed edge. We continue to add the edge of payment knowledge and social relationship, and the csv file is named relationship2.csv, in the format of:

```
Customer ID: START_ID, customer ID: END_ID, relationship TYPE: TYPE
```

Transaction types are filled with "payment awareness" and social relationships like colleagues, friends, and so on. Prepare the node CSV and relationship CSV. We stop the Neo4j service by executing 'neo4j stop' and then run the data import command in the command line.

```
neo4j-admin import --mode=csv --database=cq2 --id-type STRING --
nodes: retail customers ="customer.csv" --relationships="
relationship1.csv" --relationships=" relationship2.csv" --ignore-
extra-columns=true --ignore-duplicate-nodes=true --ignore-missing-
nodes=true
```

In the command, cq2 indicates the name of the graph database to be created, id-type indicates the data type of the node number, nodes specifies the node csv file, relationships specifies the node relationship csv file, ignore-extra-columns indicates that redundant columns are ignored. ignore-duplicate-nodes indicates that duplicate nodes are ignored (that is, multiple records with the same node number); ignore-missing-nodes indicates that non-existent nodes are ignored (that is, node numbers exist in relationship1.csv and relationship2.csv. But does not exist in customer.csv). At this point, Neo4j will create a graph database called cq2, showing the following prompt, as shown in Fig. 2.26.

As can be seen from Fig. 2.26, 8623974 nodes are imported into the graph database, 30675355 edges and 121038875 attributes are created, and the whole database construction process only takes 1 min and 7 s, which is very efficient. At this time, cq2 directory has been created under E:\neo4j-3.5.17\data\databases, which is the storage directory of graph database. We specify the name of the database in the E:\neo4j-3.5.17\conf\ neo4j.conf configuration file:

```
dbms.active_database=cq2
```

2.5 Case Practice

```
..........  ..........  ..........  ..........  ..........   75%  ?0ms
..........  ..........  ..........  ..........  ..........   80%  ?200ms
..........  ..........  ..........  ..........  ..........   85%  ?0ms
..........  ..........  ..........  ..........  ..........   90%  ?200ms
..........  ..........  ..........  ..........  ..........   95%  ?150ms
..........  ..........  ..........  ..........  ..........  100%  ?0ms

IMPORT DONE in 1m 7s 149ms.
Imported:
  8623974 nodes
  30675355 relationships
  121038875 properties
Peak memory usage: 1.11 GB
There were bad entries which were skipped and logged into G:\neo4j\import.report
```

Fig. 2.26 Neo4j creates the graph database

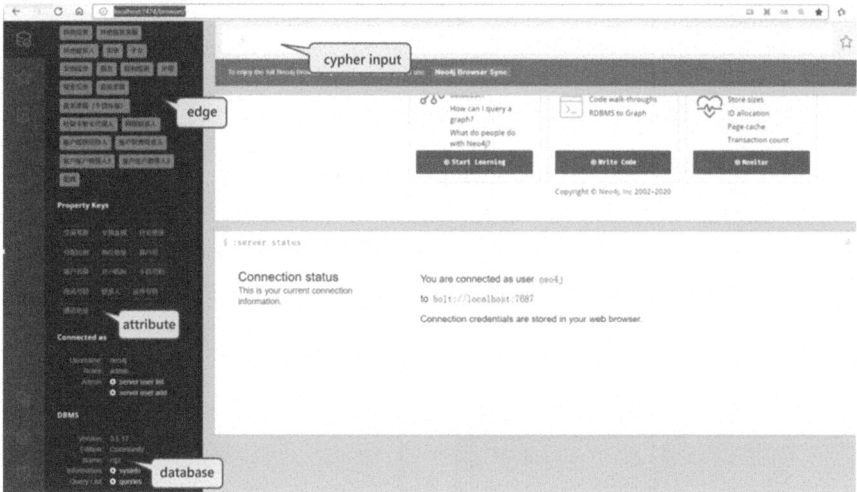

Fig. 2.27 Neo4j database page

Perform "neo4j start" starts neo4j database service, the browser type http://localhost:7474/browser/, the default user name and password are neo4j. Click the database icon, and you can view the edge relationship, properties, and database name of the current database, as shown in Fig. 2.27.

Neo4j uses cypher statements to perform data queries. Enter the cypher statement: match(n: 'retail customer')-[r]-(m: 'retail customer') return n,r,m limit 100, which returns the relationship between 100 retail customers, as shown in Fig. 2.28.

Select a certain edge and display the type data (transactions), id number and other attribute data (number of transactions and transaction amount) for this edge in the lower left corner. At this point, the graph database creation is complete.

Next, build the deep learning environment of graph neural network. You need to prepare an NVIDIA graphics card and install the graphics card driver. Since the GPU is required to train the neural network, two kits, CUDA and CUDNN, must also be

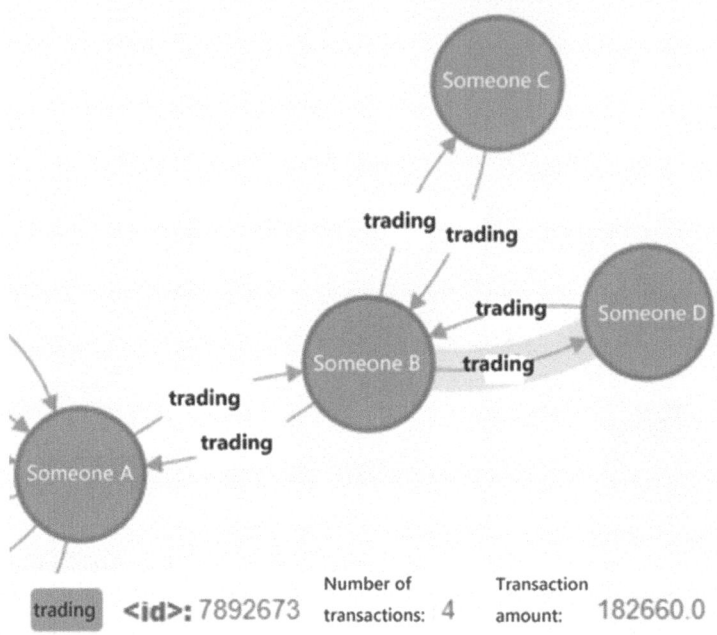

Fig. 2.28 Neo4j data query

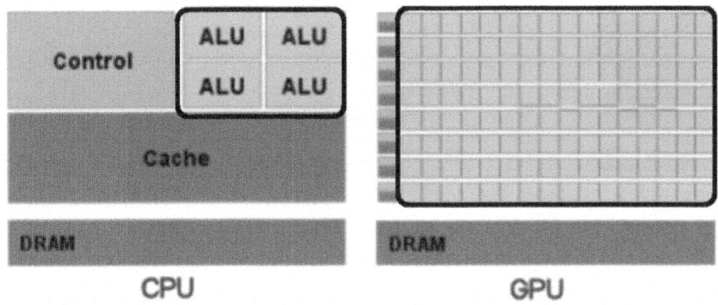

Fig. 2.29 Comparison of the CPU and GPU chip structures

installed. The versions chosen by the authors are CUDA 10.0 and CUDNN 7.6.4. CUDA (Compute Unified Device Architecture) is a GPU parallel computing framework from Nvidia Corporation.

Figure 2.29 compares the chip architecture of the CPU and GPU. The sss in the black box represents the computing core of the CPU and GPU, respectively. As you can see, the control unit and storage of the CPU occupy most of the chip area, and the computing core area is less; While the control unit and storage of GPU occupy a small proportion, the computing core occupies most of the chip area.

CUDA Toolkit 10.0 Archive

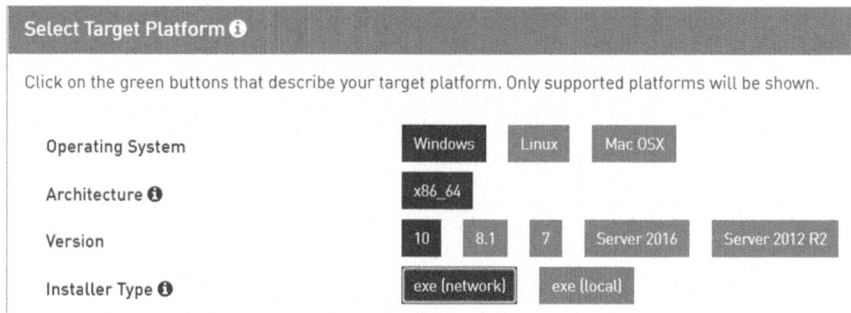

Fig. 2.30 CUDA parallel computing kit download

```
sudo chmod +x ~/cuda_10.0.130_410.48_linux.run
sudo ~/cuda_10.0.130_410.48_linux.run
```

Fig. 2.31 Installing CUDA on the Linux CLI

Because the number of CUDA computing cores in a GPU far exceeds the number of computing cores in a CPU, there is often an order of magnitude difference between a GPU and a CPU in terms of floating-point computing speed. The role of the CUDA framework is to distribute complex computing tasks to multiple CUDA computing cores, thus achieving high-performance parallel computing. CUDA website to download address: https://developer.nvidia.com/cuda-toolkit-archive. Select the CUDA version and operating system to download. It is worth noting that RTX30 series graphics cards currently only support CUDA11 and above. The author uses the RTX20 series graphics card and has CUDA10.0 installed. Figure 2.30 shows the download interface.

In Linux, run the following command to install the CUDA, as shown in Fig. 2.31.

The CUDA graphics card Driver may have compatibility problems. Therefore, do not select the driver option during the installation, as shown in Fig. 2.32.

When done, in the current user's $HOME directory, add the following environment variables to the .bashrc file:

```
export PATH=/usr/local/cuda-10.0/bin:$PATH
export LD_LIBRARY_PATH=/usr/local/cuda-10.0/lib64:$LD_LIBRARY_PATH
export CUDA_HOME=/usr/local/cuda
```

Then log in to the system again and run the nvcc-v command. If the version number is displayed, the installation is successful, as shown in Fig. 2.33.

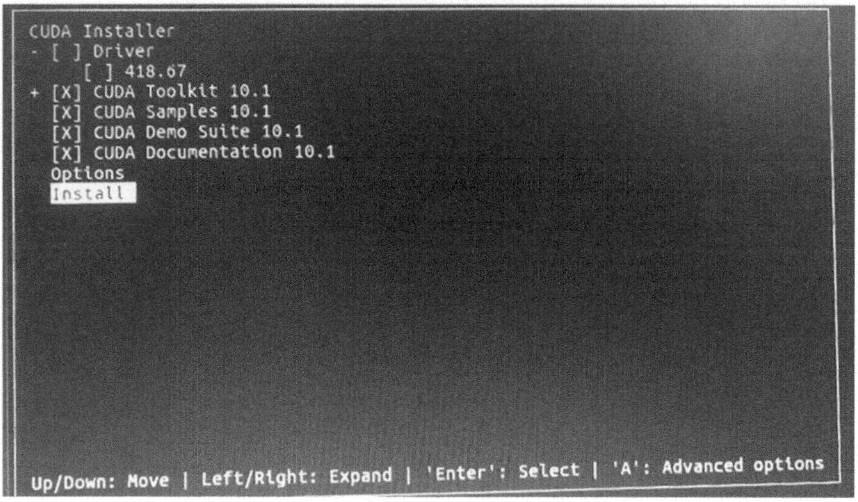
CUDA Installer
- [ ] Driver
[ ] 418.67
+ [X] CUDA Toolkit 10.1
[X] CUDA Samples 10.1
[X] CUDA Demo Suite 10.1
[X] CUDA Documentation 10.1
Options
Install
Up/Down: Move | Left/Right: Expand | 'Enter': Select | 'A': Advanced options

Fig. 2.32 CUDA installation page on Linux

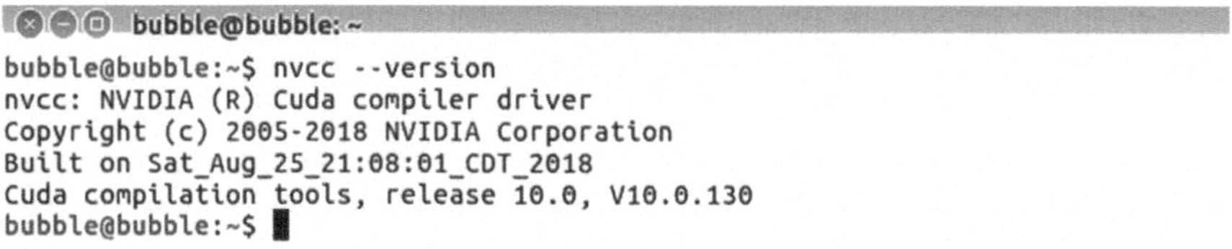
bubble@bubble: ~
bubble@bubble:~$ nvcc --version
nvcc: NVIDIA (R) Cuda compiler driver
Copyright (c) 2005-2018 NVIDIA Corporation
Built on Sat_Aug_25_21:08:01_CDT_2018
Cuda compilation tools, release 10.0, V10.0.130
bubble@bubble:~$

Fig. 2.33 CUDA installation in Linux is complete

If it is Windows, select CUDA only when executing the installation package and do not select other options, which may cause the installation to fail due to some compatibility issues, as shown in Fig. 2.34.

After the installation is complete, check whether the CUDA_PATH and CUDA_PATH_V10_0 environment variables take effect, as shown in Fig. 2.35.

Check again whether the following is set in the PATH environment variable, as shown in Fig. 2.36.

After the preceding operations are complete, run nvcc-v on the CLI. If CUDA 10.0 is displayed, CUDA is successfully installed, as shown in Fig. 2.37.

CUDNN is a GPU-accelerated library for deep neural networks from Nvidia Corporation. It encapsulates many of the underlying functions of deep learning, such as convolution, pooling, and other functions, and optimizes the computing speed. The relationship between CUDA and CUDNN can be seen as the relationship between platform and tool, the platform solves the concurrency support, the tool solves the neural network support. CUDNN website to download the address is: https://developer.nvidia.com/cudnn-download-survey. Log in and select the CUDNN suite and operating system that matches the CUDA version. The author has CUDNN7.6.4 installed, as shown in Fig. 2.38.

2.5 Case Practice

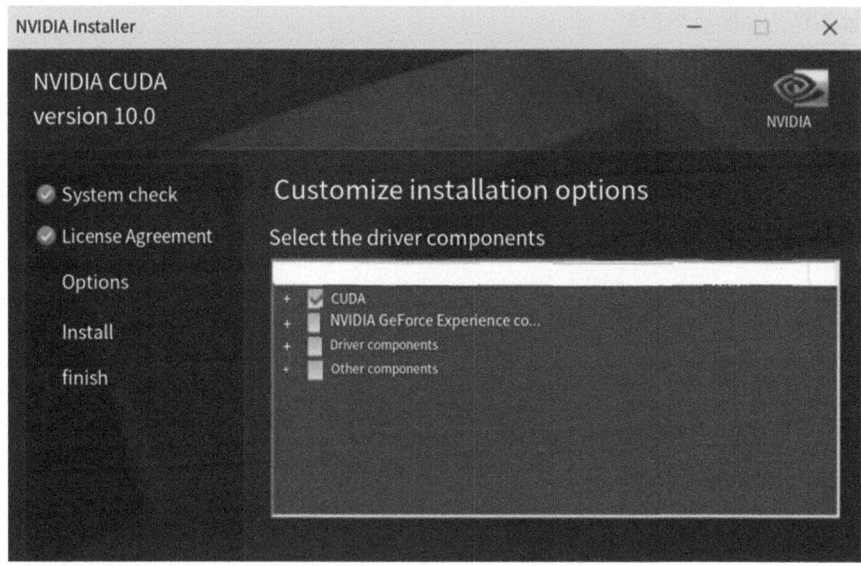

Fig. 2.34 CUDA installation screen in Windows

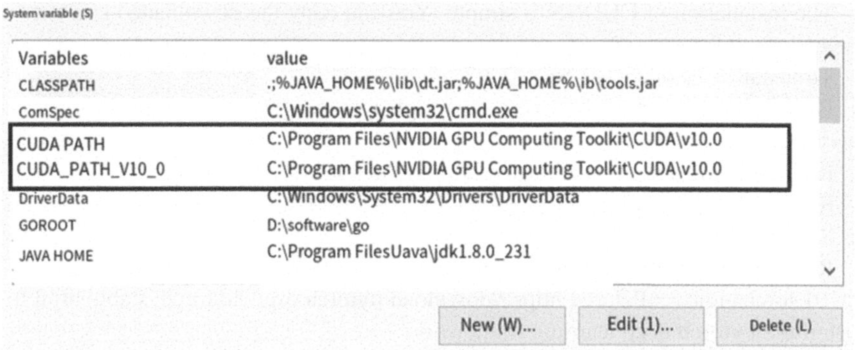

Fig. 2.35 CUDA environment variables in windows

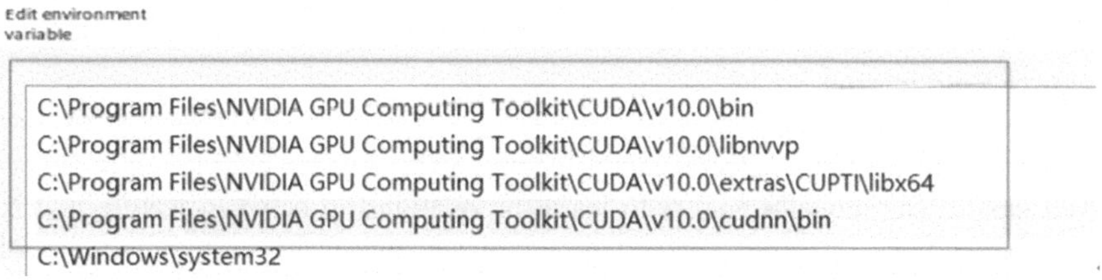

Fig. 2.36 PATH environment variable in windows

Fig. 2.37 Verifying CUDA installation

```
C:\Users>nvcc -V
nvcc: NVIDIA (R) Cuda compiler driver
Copyright (c) 2005-2018 NVIDIA Corporation
Built on Sat_Aug_25_21:08:04_Central_Daylight_Time_2018
Cuda compilation tools, release 10.0, V10.0.130
```

cuDNN Archive

NVIDIA cuDNN is a GPU-accelerated library of primitives for deep neural networks.

Download cuDNN v7.6.4 (September 27, 2019), for CUDA 10.1

Download cuDNN v7.6.4 (September 27, 2019), for CUDA 10.0

Download cuDNN v7.6.4 (September 27, 2019), for CUDA 9.2

Download cuDNN v7.6.4 (September 27, 2019), for CUDA 9.0

Fig. 2.38 Downloading the CUDNN suite

The installation of CUDNN is simple. You can copy the downloaded content to the cuda installation path.

To install Anaconda, see Sect. 2.1.3.

Run the command of conda create -n dgl python=3.7.9 to create a dgl virtual environment.

Run the command of conda activate dgl to activate the environment.

Run the command of pip install openpyxl matplotlib dgl-cu101==0.4.2 to install DGL framework for GPU version.

Run the command of pip install torch==1.7.1+cu101 torchvision==0.8.2+-cu101 torchaudio==0.7.2 -f https://download.pytorch.org/whl/torch_stable.html to install the Pytorch deep learning framework.

2.5.2 Code Practice

Feature Engineering

For supervised learning, select original features with a strong correlation to supervised labels. The model's goal is to learn the relationship between feature matrix X and label y. In this case study, labels indicate high-value customers. Choose features highly correlated to these labels. Consult domain experts to select features based on business knowledge and verify statistical data. Due to constraints, the feature set here lacks diversity, and readers should enhance it further. In practice, the selected features are:

2.5 Case Practice

High-Value Mark of Other Banks, Credit Card Level, Credit Card Retained Limit, Maximum Transaction Amount in the Past Year, Maximum Nine Asset Values in the Past Year, Nature of Workplace, Industry Code, Occupation Code, Position Code, Personal Salary Income, Family Salary Income, Flag for Housing Loan Over 3 Million, Flag for Historical Large Transfers, Flag for Business Owner Merchant.

Here is an explanation of the original features listed above. If a customer's historical transactions include interbank transfers with the same name, credit card-related repayments, or personal loan repayment card-related transfers, then we determine the card bin number of the other party's bank card. If the card bin reflects a customer level higher than our bank's level, we set the "high value from other bank" field to 1, otherwise, we set it to 0. The credit card retention limit equals the total credit card limit minus the customer's historical maximum overdraft amount. The larger this indicator, the more ample the funds the customer has, and the greater the potential for exploration. If the customer has a large home loan amounting to more than 3 million, it indicates a high probability of high value. Other fields, such as credit card levels, transactions, income, and other data, reflect the customer's value from different perspectives.

The range of values for each feature varies. For example, transaction amounts and the nine asset value ranges are continuous real numbers, while credit card levels and work unit types are discrete data with a limited number of values. This does not mean that the feature with the larger numerical value contributes more to the prediction of the target field. Thus, binning operations need to be performed on the original features to categorize them, and then the category data is used as features input into the model, ensuring the comparability of features within different value ranges. We use the decision tree binning algorithm. For a detailed introduction to this algorithm, refer to Sect. 2.1.3 of this book. We use the scikit-learn machine learning library to train the decision tree. The key code is as follows:

```
# Obtaining a list of boundary values for the optimal bins using a
decision tree
def optimal_binning_boundary(x, y):
    boundary = []  # List of bounding values of the bins to be returned
    x = x.values  # Filling in missing values
    y = y.values
    clf = DecisionTreeClassifier(criterion='entropy',  # Entropy
increase minimization criterion division
        max_leaf_nodes=8,    # Maximum number of leaf nodes
        min_samples_leaf=0.05) # Minimum percentage of leaf node sample size
    clf.fit(x.reshape(-1, 1), y)  # Training decision trees
    n_nodes = clf.tree_.node_count
    children_left = clf.tree_.children_left
    children_right = clf.tree_.children_right
    threshold = clf.tree_.threshold
    for i in range(n_nodes):
```

```
# Obtain the division boundary values on the decision tree nodes
    if children_left[i] != children_right[i]:
     boundary.append(threshold[i])
    boundary.sort()
    min_x = x.min()
    max_x = x.max() + 0.1  # +0.1 is to consider samples that contain the
maximum value of the feature for subsequent groupby operations
    boundary = [min_x] + boundary + [max_x]
    return boundary
# Replace the original feature column with an array of categories
ps = optimal_binning_boundary(x=df[a], y=df['flag'])
if len(ps) == 3:  # The original features are split into 2 categories
    df2[a] = pd.cut(x=df[a], bins=ps, right=False, labels=
['0,1','1,0'])
    if len(ps) == 4:  # The original features were split into 3 categories
     df2[a] = pd.cut(x=df[a], bins=ps, right=False, labels=
['0,0,1','0,1,0','1,0,0'])
```

After the decision tree is partitioned, the values of the original features are replaced with N categories. The values are '0,1' or '1,0' if replaced with two categories, and '0,0,1', '0,1,0', or '0,0,1' if replaced with three categories. The decision tree bin splitting algorithm determines the most appropriate number of categories and boundary points for each category. In this way, after each feature is split into multiple features (categories), its value can only be 0 or 1, which unifies the range of values, forming a total of 55 columns of category values. We take the category data formed after the decision tree is divided into boxes as the features of each customer node, namely the vector $[x_{i1}, x_{i2}......x_{i55}]$ in Fig. 2.17, and input it into the neural network model. Save the data to the data\features.txt file in the model directory, its format is:

```
Customer number xi1 xi2 ...... xi55 Whether high value label (0 or 1)
```

Next, we export the customer's strong knowledge relationship from the graph database. The apoc control is called in Neo4j to complete the installation of apoc is very simple, just need to download APOc-3.5.0.7-all.jar and put it in the plugins directory of the neo4j installation directory. Execute Cypher statement: CALL apoc. export.csv.query("MATCH (a: retail customer)-[r]-(m: retail customer) RETURN a. Customer number, type(r), m. customer number ", "Relations.txt", {}), will generate the relations.txt file under E:\neo4j-3.5.17\import, copying it to the data directory under the model directory. The effect of this operation is to export all the edges in the graph database. Relations.txt file reflects the strong recognition relationship between customers in the following data format:

```
customer number 1 \t customer number 2
customer number 2 \t customer number 1
......
```

2.5 Case Practice

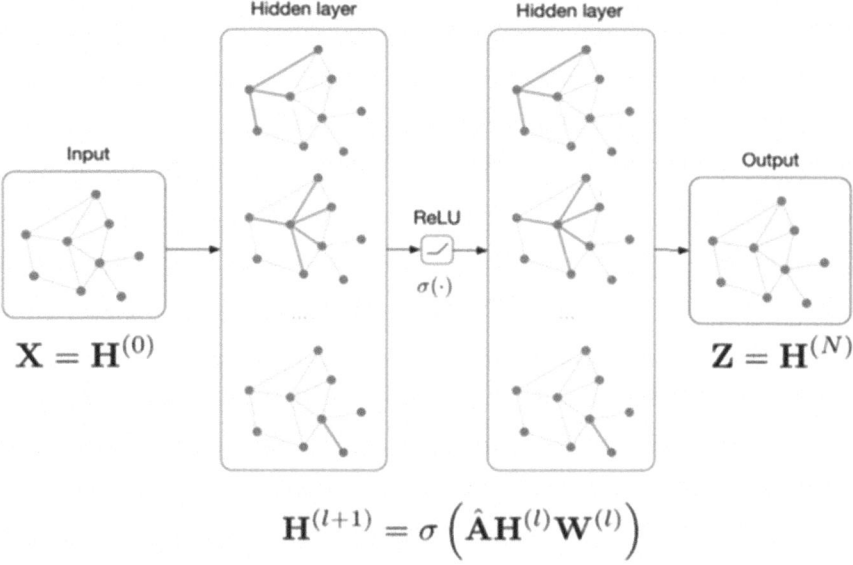

Fig. 2.39 GCN network structure

The first field represents the start node number, while the second field signifies the end node number, forming a directed graph. The first row shows that customer 1 knows customer 2; the second row indicates that customer 2 knows customer 1. Notably, despite the directed nature of the relationship, we aim to derive an undirected relationship. This is because whether A knows B or B knows A, they likely belong to the same class, thus requiring consideration of both instances. Since in subsequent modeling tasks, we will create directed graphs to perform convolution operations, in this step, "The representation that A knows B" will be derived from two records, namely "A \t B" and "B \t A."

Constructing GCN Graph Convolutional Neural Networks

Since spatial convolution aggregates neighbor features at each graph layer, each aggregation diminishes the nodes' intrinsic features. In a deep network, this leads to convergence toward the overall average, causing the "fixed point" issue. To prevent this, graph neural networks using the aggregate spatial method typically have only two or three layers. In this case, we constructed a two-layer GCN model. The model's overall structure appears as shown in Fig. 2.39.

The key codes for building the GCN graph structure are as follows:

```
from dgl import DGLGraph # Import the framework
def InitData(): # Create graph and associated data
    data = LoadData() # Load the output vector data of the feature project,
the feature vector length of each node is 55
```

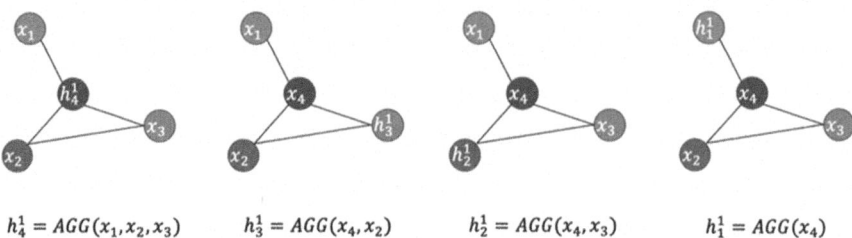

$h_4^1 = AGG(x_1,x_2,x_3)$ $h_3^1 = AGG(x_4,x_2)$ $h_2^1 = AGG(x_4,x_3)$ $h_1^1 = AGG(x_4)$

Fig. 2.40 GCN aggregation process

```
features = th.FloatTensor(data.features) # Set the input feature
matrix of the network
    labels = th.LongTensor(data.labels) # Set the label matrix
    g = DGLGraph(data.graph) # Create a directed graph by constructing
nodes and edges according to the retail customer's cognitive
relationship
    return g, features, labels
```

The node feature update process for each layer is as follows: First, the input vector of each node of the neural network of this layer is propagated forward to obtain the transformed Embedding vector (that is, the high-dimensional vector is mapped to the low-dimensional vector, the same below), and then use an AGG function to aggregate the first-order neighbor node embedding vector of each node vi, and obtain the vector h_i^k coding after the k round aggregation of vi. Then, vi of the adjacent point N(vi) encoding average, and the coding results of the previous round of a weighted combination, to get a new coding results. Its mathematical expression is: $h_v^k = \sigma\left(W_k \sum_{u \in N(v)} \frac{h_u^{k-1}}{|N(v)|} + B_k h_v^{k-1}\right)$. In this example, we are building an unweighted graph where there is an edge between customers if they know each other, and the weight of the edge is always 1. The aggregation function chooses the simplest averaging method, that is, the idea that birds of a feather flock together, where the features of each node are the average of the features of its first-order neighbors: $X_i^* = \sum_{j \in \text{neighbor}(i)} A_{ij}X_j = \sum_{j=1}^{N} A_{ij}X_j$, $A_{ij} = 0$ or $A_{ij} = 1$. For all nodes, the update process expressed by the matrix is: $X^* = AX$, $AGG = \sum_{u \in N(v)} \frac{h_u^{k-1}}{|N(v)|}$, as shown in Fig. 2.40.

The key codes for neighbor aggregation are as follows:

```
import dgl.function as fn # Import frame
    gcn_msg = fn.copy_src(src='h', out='m') # Define the message
propagator function: node features propagate by edge, from h to m
    gcn_reduce = fn.mean(msg='m', out='h') # Define the message
```

2.5 Case Practice

aggregation function: the node takes the average value of its neighbors as its own features

```
def forward(self, g, feature): # Spatial convolution over graph g:
message propagation
   with g.local_scope():
      g.ndata['h'] = feature # Message propagation h= feature vector for
each node
      G.uppdate_all (gcn_msg, gcn_reduce) # Completes the message
propagation of all nodes in the layer network in the airspace
      h = g.ndata['h']
      return self.linear(h) #h is the node feature after message
propagation in this layer is complete
```

Next, we define a fully connected feedforward neural network with feature transformation for each node, which is responsible for converting the feature vector (i.e. model input) processed by feature engineering into binary output of high-value customers or not. Its structure is shown in Fig. 2.41.

In the GCN model, the node feature transformation is carried out layer by layer. In this example, the output of the feature project is a vector of length 55, and GCN has 55 neurons in the input layer, 32 neurons in the hidden layer, and 2 neurons in the output layer. The key codes for the feedforward network are as follows:

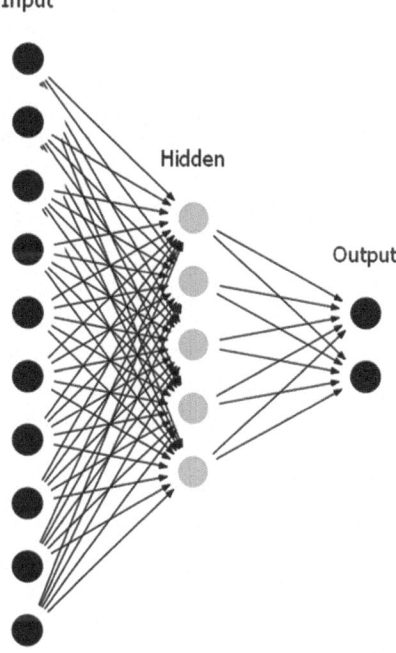

Fig. 2.41 Node feature transformation feedforward neural network structure of GCN

```
# Define a graph convolution layer class for propagating the feature
transformation of a node in the current layer to the next layer
class GCNLayer(nn.Module):
    def __init__(self, in_feats, out_feats):
        super(GCNLayer, self).__init__()  # Call the init function of the
parent class
        self.linear = nn.Linear(in_feats, out_feats) # Constructing fully
connected networks, defining the number of input and output neurons
# Define a GCN network with a 2-layer structure
class Net(nn.Module):
    def __init__(self):
        super(Net, self).__init__()
        self.layer1 = GCNLayer(55, 32)  # 55 neurons in the input layer and
32 neurons in the hidden layer
        self.layer2 = GCNLayer(32, 2)  # 2 neurons in the output layer
    # feed-forward network communication
    def forward(self, g, features):
        x = F.relu(self.layer1(g, features))  # The activation function
uses relu
        x = self.layer2(g, x)
        return x
```

Training GCN Graph Convolutional Neural Networks

We split the dataset into a training set and a validation set with an 8:2 ratio. We randomly sample from the training set and cycle these to the network. GCN Layer 1 conducts feature transformation and neighbor aggregation sequentially, converting 55 input features into 32 hidden features, which are then activated by the Relu function before being passed to Layer 2. Layer 2 performs further feature transformation and neighbor aggregation, converting 32 hidden features into 2 output features. We compute the loss using the cross-entropy function to compare the predicted and true values, then use backpropagation to derive the gradients. The Adam optimizer adjusts the weights of each feature transformation network layer. We resample the dataset and repeat the iterations for multiple epochs until the model converges to an acceptable range.

The key code for model training is as follows:

```
optimizer = th.optim.Adam(net.parameters(), lr=float(args.learn))  #
Using the Adam Optimizer
    for epoch in range(int(args.epochs)):  # Train for a specified number of
rounds
        logits = net(g, features)  # Input graph structure and all node feature
vectors to the graph network to get the output of logits second layer
network
```

2.5 Case Practice

```
    logp = F.log_softmax(logits, 1) # The output of the softmax
activation function is then solved for the log value
    loss = F.nll_loss(logp[train_mask], labels[train_mask]) # On the
training set, the loss of softmax prediction to the true value is
obtained
    optimizer.zero_grad() # The optimizer sets the gradient to zero, that
is, it makes the derivative of the loss with respect to the weights zero
        loss.backward() # Backpropagation update weights
        optimizer.step() # Updated models
```

During the training process, we also monitor the performance of the graph neural network on the validation set. We do this by calculating the accuracy and loss of the network's prediction on the validation set in each epoch. The key code is as follows:

```
# Evaluating Prediction Accuracy on a Validation Set
  def evaluate(model, g, features, labels, mask):
    model.eval() # Converting models to test mode
    with th.no_grad(): # Test network, the following code without
gradient descent and without backpropagation
        logits = model(g, features) # Get the category classification of all
nodes
        logits = logits[mask] # Only the classification predictions from the
validation set are taken, each with out_feats predictions, and
out_feats is the predicted number of outputs from the layer 2 network
        labels = labels[mask] # Verification set of true classification
labels
        _, indices = th.max(logits, dim=1) # Take the largest of the
predicted outcome categories
        correct = th.sum(indices == labels)
        return correct.item() * 1.0 / len(labels)
```

In the model training iteration process, once the accuracy of the verification set is higher than the historical maximum, it indicates that the current model weight is optimal, and we save the current model as a model.pth file:

```
  if val_acc > val_acc_max:
    print('validation set precision from {:.4f} to {:.4f}, save model'.
format(val_acc_max,val_acc))
    val_acc_max = val_acc
    th.save(net.state_dict(), args.model)
```

For the actual training results of this example, the model quickly fits the training set, and when training to 300 epochs, the verification set accuracy reaches 91.4% and the loss is 0.24. Command line execution:

```
conda activate dgl
```
After activating the environment, perform the model training:
```
python main.py --train -epochs=300
```

During model training, use GPU-Z software to observe GPU computing core and video memory usage. You can see that the compute core usage is 94% and the video memory usage is 2G, as shown in Fig. 2.42.

Finally, we get the training results shown in Fig. 2.43.

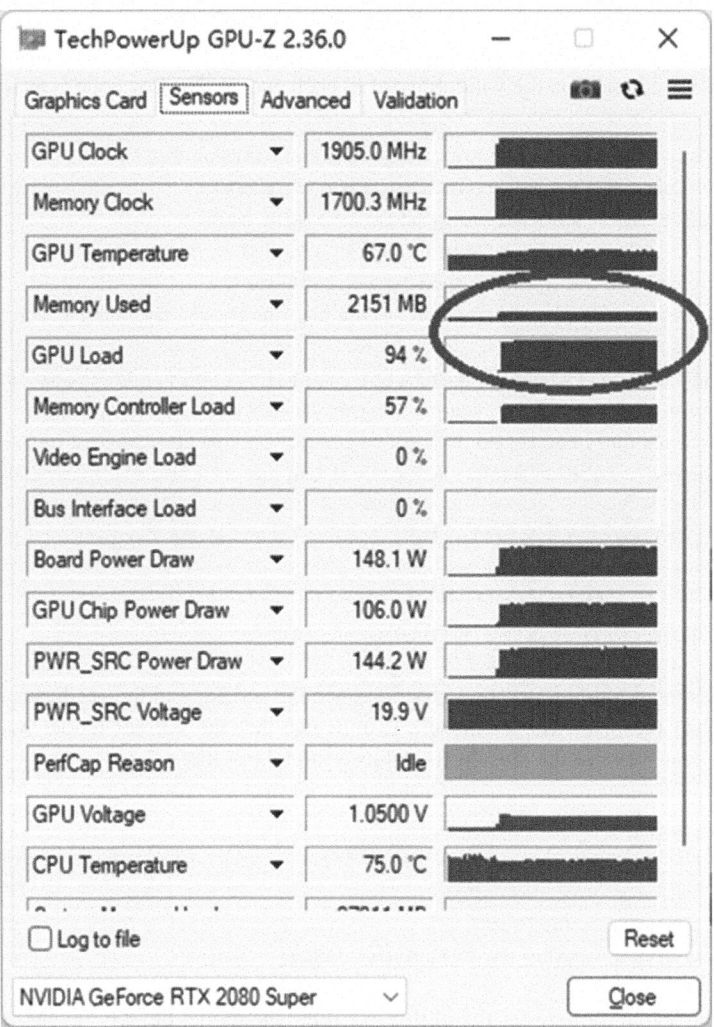

Fig. 2.42 GPU usage during model training

2.5 Case Practice

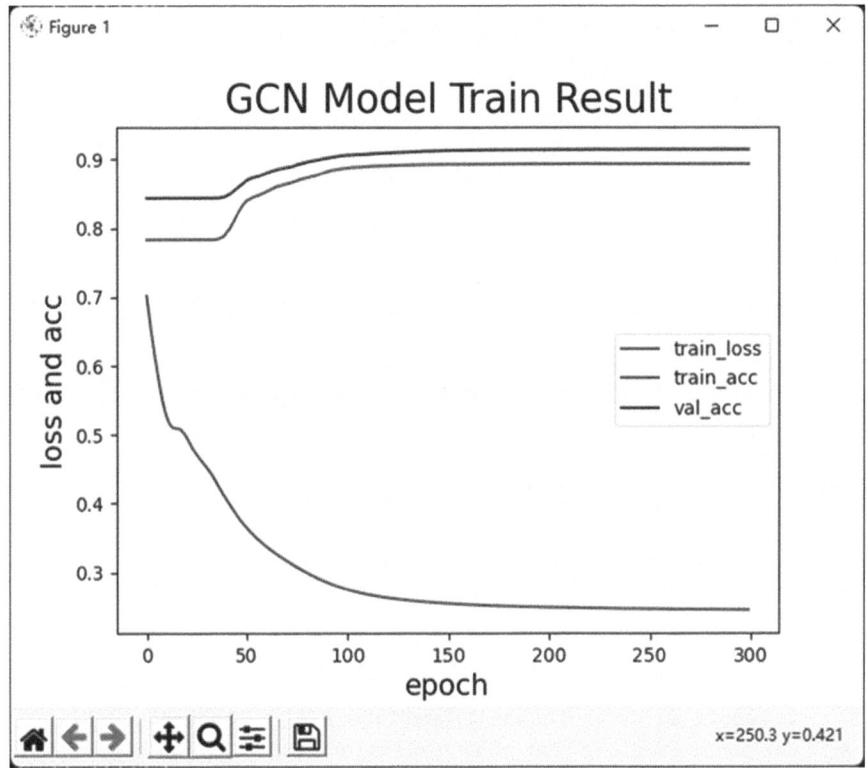

Fig. 2.43 GCN training results

Now proceed to the model inference phase. First load the model:

```
net = Net()
```
Load the network weights from the previously saved model file model.pth:
```
net.load_state_dict(th.load(args.model))
```

To place the model in video memory:

```
net = net.to(th.device('cuda:0'))
```

We do model inference, not training, so we need to set the model state to inference:

```
net.eval()
```

Get customer strong knowledge relation network, each node eigen matrix, each node high-value label vector, training set mask, and verification set mask:

```
g, features, labels, train_mask, val_mask = InitData()
```

Put relevant data into video memory in order to take advantage of GPU parallel computing to speed up inference:

```
features = features.to(th.device('cuda:0'))
labels = labels.to(th.device('cuda:0'))
g = g.to(th.device('cuda:0'))
Make predictions for all nodes of the graph:
logits, result = predict(net, g, features)
```

If the customer is labeled as non-high value and the model predicts high value (i.e., confidence >0.5), the program outputs the customer number and its confidence for high value:

```
sum_ok = 0
f = np.genfromtxt("data/features.txt", dtype=np.dtype(str))
custids = sp.csr_matrix(f[:, 0], dtype=np.int64).toarray()
for i in range(len(result)):
    if result[i] == labels[i]:
        sum_ok += 1
    else:
        if result[i] == th.tensor(1) and labels[i] == th.tensor(0):
# If the model predicts that it is high value, and in fact it is not, output
# The second element is the probability that the model predicts a high-value customer, or confidence
            v = logits[i].tolist()[1]
            print(custids[0][i], v)
print(' precision =', sum_ok/len(result))
```

Execute python main.py–test to load the model you just trained and make high-value predictions for each client on the full graph data. The program outputs the customer number and the probability that the model predicts that the customer is of high value. A higher probability means a higher probability that the customer is a high-value customer. The output result is shown in Fig. 2.44.

Fig. 2.44 Forecast results of GCN high-value customers

```
9048945  0.7236385941505432
9048953  0.6810771822929382
9048954  0.7198455333709717
9048969  0.5498154759407043
9048990  0.6810771822929382
9048991  0.7794248461723328
9048994  0.6810771822929382
9048996  0.5536487102508545
accuracy= 0.8537436483480603
```

We simply market in order of confidence from highest to lowest. See the download file main.py for the code.

2.6 Case Summary

In practical applications, this case identified 18,136 potential high-value customers (i.e., deemed high-value by the model but not in reality). We targeted the top 4000 customers with high confidence, assisting daily retail assets to increase by 67 million yuan. This case is merely an attempt at applying graph neural networks (GNNs) in a banking marketing scenario. There remains significant room for improvement. Due to various constraints, the original features were inadequate, and derived features were not utilized, impacting model performance. Readers are encouraged to enhance these aspects further. This case serves as an initial attempt to explore problem-solving approaches and experimental exploration. Data project outputs closely relate to data and feature engineering, and readers can refine data features based on this chapter to better address related issues.

"Graphs" serve as mathematical models that abstract real-world entity relationships into "nodes" and "edges." GNNs apply deep learning techniques to these non-Euclidean data structures, integrating graph theory with neural networks. In recent years, GNNs have rapidly advanced in data mining. Their strength lies in learning graph embeddings to extract semantic features of graph structures, considering node features, neighbor nodes, and topology–capabilities that traditional neural networks lack. Consequently, GNNs exhibit superior semantic representation, enhancing their effectiveness in solving real-world problems. Recent industry applications demonstrate GNNs' impressive achievements in molecular drug discovery, traffic planning, online social networks, e-commerce, natural language processing, and image processing.

Graph neural networks excel in tasks like node classification, edge classification, link prediction, and subgraph recognition. They account for entity features and relationships, facilitating the learning of semantic information within graphs. These networks are ideal for data-intensive industries, such as banking. In practice, they deliver promising results. Valuable applications in banking include bidirectional recommendation tasks on the "customer-financial product" graph, recommending products to customers and vice versa. For risk transfer evaluation based on business chains, graph neural networks can assess risk transmission in a customer relationship graph, identifying implications within subgraphs formed by interactions in areas like joint guarantees and supply chains. Recognizing subgraph patterns and assessing risk transfer in dynamic transactional relations are crucial for early bank credit risk warnings. Additionally, in scenarios like transaction anomaly detection and anti-money laundering, these networks significantly enhance prediction accuracy through the message-passing mechanism. Gated graph neural networks (GGNNs) also have established uses in natural language sequence processing, improving semantic understanding and intent recognition. They offer vast possibilities in automated customer service and employee training. Unlike traditional image recognition that may overlook entity relationships, graph neural networks can learn relationship semantics within knowledge graphs, enhancing semantic understanding in image processing. For instance, in a bank lobby monitoring system identifying both an LV bag and a retail customer entity, a graph neural network can instantaneously derive semantic representations of luxury items and high-value customers from the knowledge graph, prompting immediate targeted marketing actions by branch staff, as shown in Fig. 2.45.

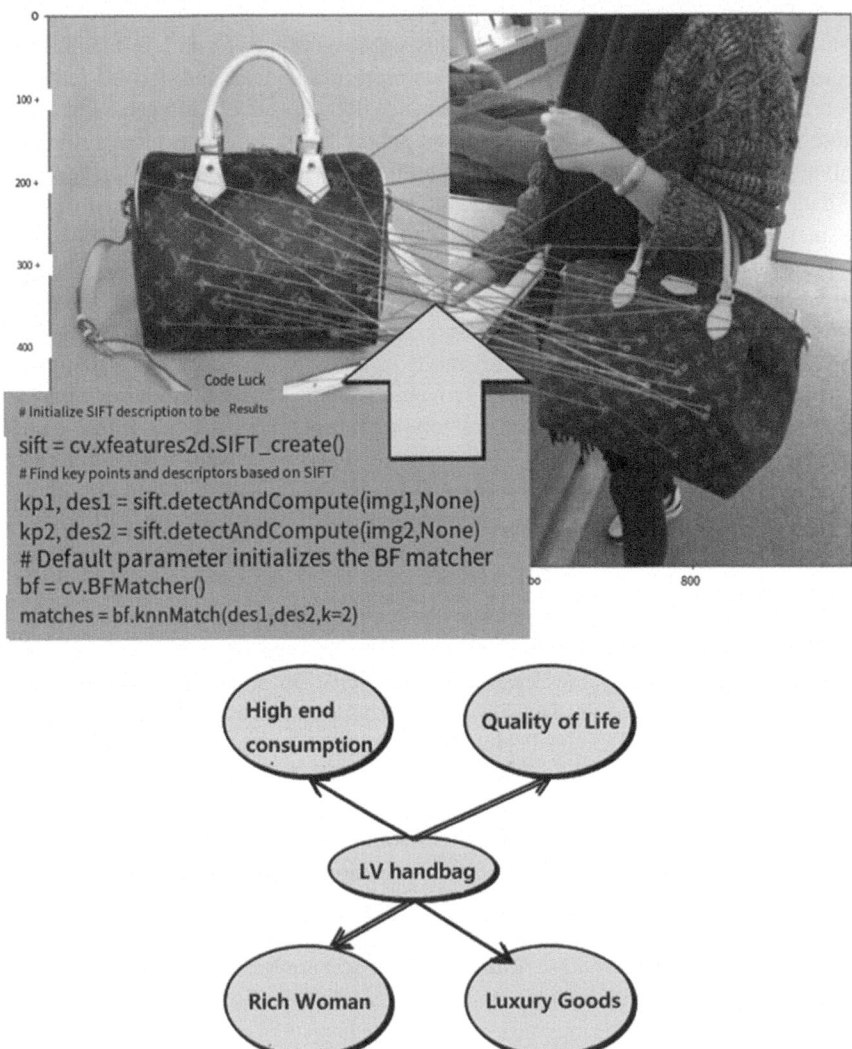

Fig. 2.45 OpenCV feature matching + Graph neural network knowledge association

2.6 Case Summary

Recent studies have shown that graph neural networks have a strong ability in complex graph structure modeling, which pushes machine learning to a new height. It is believed that in the near future, graph neural network will play an important role in more scenarios of banking operations.

Chapter 3
Accurate Recommendation for Banking: Recommender System

In the modern era, banks either have adopted the Internet or are transitioning toward it. The recommendation system stands as the pivotal component for Internet application. Once operational online, businesses integrate Internet infrastructure, scenarios, financial products, and customer services, creating a vast data landscape. This data includes elements like scene context, customer behavior, interests, supply value, and product characteristics. By extracting insights from this data, new marketing opportunities can emerge. The recommendation system scrutinizes the relationships between customers and products/services. It identifies the most appropriate target customers for businesses and vice versa. For banks, this leads to increased sales and performance. For customers, it enhances their experience by providing tailored products and services. Furthermore, recommendation systems support the development of profit models like advertising and value-added services. For instance, by predicting individual advertising preferences, banks can display personalized ads in mobile banking. This customization boosts ad exposure and click-through rates, funneling valuable data to advertisers. This data becomes essential for competitiveness in the Internet sector. An example includes setting both Chinese and Western medicine service providers in a value-added service segment in mobile banking. The recommendation system predicts preferences, offering Chinese medicine to one customer and Western medicine to another, based on their needs. This approach enhances customer retention, engagement, and loyalty in value-added services. App usage frequency remains a challenge for banks. A robust recommendation system combined with strong content operations can significantly enhance app engagement and user base. Increased app usage can lay the groundwork for open banking. Open banking integrates banking services into non-banking online contexts, becoming a part of daily life. This future banking model emphasizes open formats, platforms, and ecosystems. Thus, the recommendation system is critical for customer engagement and bank development. Short-term benefits include improved advertising outcomes, boosted sales of banking products, increased value-added service provider prosperity, and expanded customer scale and activity. The recommendation system is invaluable for a bank's operations.

© The Author(s), under exclusive license to Springer Nature Singapore Pte Ltd. 2025
L. Shao et al., *AI in Banking*, https://doi.org/10.1007/978-981-96-3837-6_3

3.1 Introduction to the Recommendation System

In 1995, Stanford University introduced the personalized recommendation system LIRA at the Conference on Artificial Intelligence in the USA. Subsequently, academic institutions and corporate giants like Yahoo, Google, NEC Research Institute, and IBM joined the research and launched their own recommendation systems. This milestone marked the beginning of the rapid development phase for recommendation systems as an independent technology. Broadly speaking, recommendation systems include the following technologies: content-based recommendation, collaborative filtering-based recommendation, association rule-based recommendation, graph algorithm-based recommendation, and recommendations based on deep learning and graph neural networks.

Content-based recommendation leverages metadata of products or content to discover their relevance and subsequently recommends similar products to users based on their previous preferences. Initially, content-based recommendations found application primarily in information retrieval systems. This technology boasts two primary advantages: it does not require data from other users, thus avoiding cold start and sparsity issues, and its models are highly interpretable. However, the downside is that both the products and user interests must be expressed in structured features, making it impossible to utilize information from other users, as shown in Fig. 3.1.

Collaborative Filtering Recommendation is one of the earliest and most successful technologies applied in recommendation systems, including user-based collaborative filtering recommendation and project-based collaborative filtering recommendation. User-based collaborative filtering recommendation is to first find

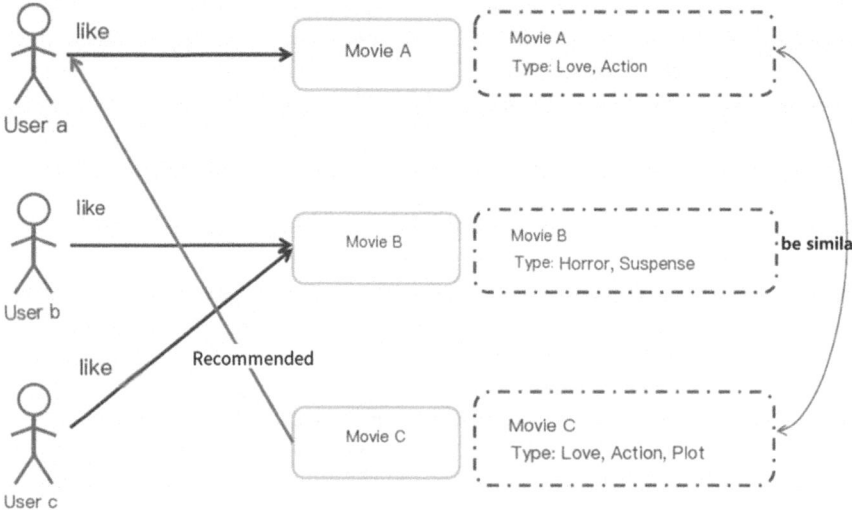

Fig. 3.1 Content-based recommendation

3.1 Introduction to the Recommendation System

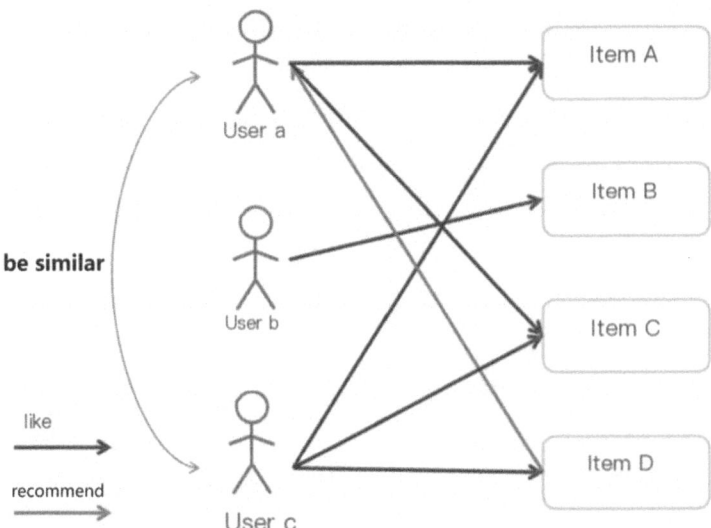

Fig. 3.2 User-based collaborative filtering recommendation

users who have the same preferences as target users, and then recommend products to target users according to their preferences. In essence, it makes use of the correlation of user preferences to recommend products to each other, as shown in Fig. 3.2.

Project-based collaborative filtering recommendation is to first find the similarities between goods and goods, and then recommend related goods to users according to user preferences, as shown in Fig. 3.3.

Recommendation based on association rules utilizes statistical or probabilistic methods to identify association clues in recommendation rules. For example, Walmart found that "customers who buy diapers are likely to buy beer," which is a classic association rule.

Recommendation based on graph algorithms represents the relationships between users and products as a graph, where vertices represent users and products and edges represent customer preferences. This data structure efficiently and intuitively reflects the analysis target and facilitates recommendations for target customers by leveraging information from other users and products, such as the PersonalRank algorithm.

Recommendation based on deep learning has developed rapidly in recent years due to its significant application effects. This method uses the powerful feature extraction capabilities of neural networks to establish embeddings of users and products for recommendation prediction. Typical models include Google's Wide&Deep model based on Embedding+MLP, the DeepFM model based on feature combinations, the dual-tower model based on independent user and product features, and the Transformer model based on behavior sequences.

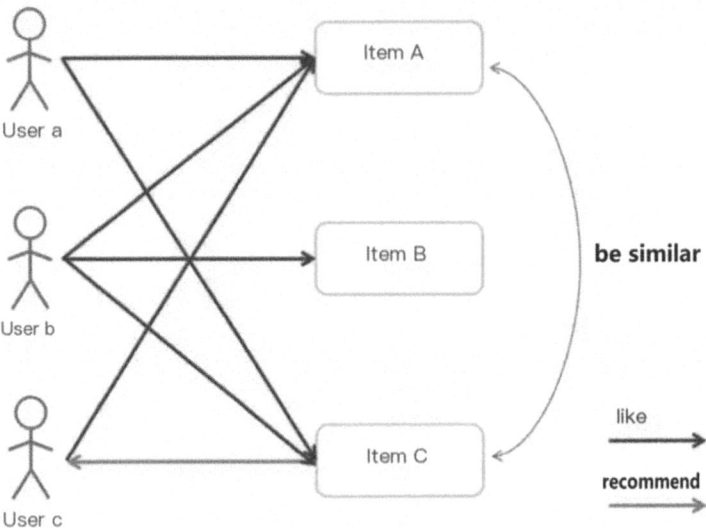

Fig. 3.3 Project-based collaborative filtering recommendation

Recommendation based on graph neural networks integrates deep learning capabilities with knowledge graphs, considering both feature transformations and topological structures to build recommendation models on larger information bases, such as algorithms like LR-GCCF, LightGCN, GNN_MF, and so on.

3.2 Recommendation Algorithm

We analyze retail customers' banking usage data to develop a recommendation system. Banking services can include banks' financial products, marketing activities, or third-party services authorized by banks. For illustration, we use financial product recommendations to explain the modeling process. Generally, one recommendation technology suffices. But, banks prefer creating a diverse recommendation baseline to maximize product outreach. Thus, we employ three methods: collaborative filtering, PersonalRank, and a two-tower neural network. These methods represent classical algorithms, graph algorithms, and deep learning models, respectively. We then combine their results for the final recommendation output.

3.2.1 Collaborative Filtering Algorithm

Collaborative filtering is a kind of cryptic model (also known as latent factor model), the core idea is based on matrix factorization technology to generate a certain number of hidden feature factors to fit the observed correlation data between

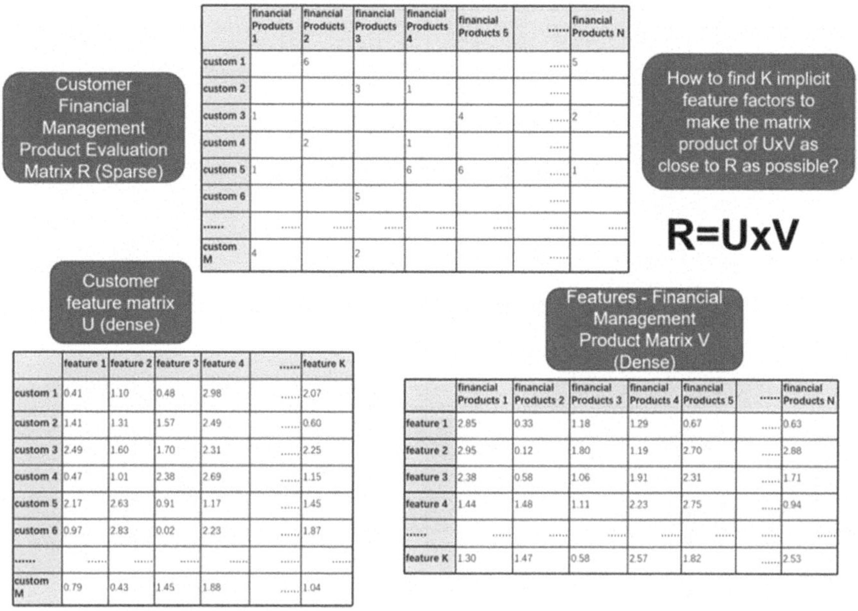

	financial Products 1	financial Products 2	financial Products 3	financial Products 4	financial Products 5	......	financial Products N
custom 1		6				......	5
custom 2			3	1		......	
custom 3	1				4	......	2
custom 4		2		1		......	
custom 5	1			6	6	......	1
custom 6			5			......	
......	......	......	......	......	......	......	......
custom M	4		2			......	

	feature 1	feature 2	feature 3	feature 4	......	feature K
custom 1	0.41	1.10	0.48	2.98	......	2.07
custom 2	1.41	1.31	1.57	2.49	......	0.60
custom 3	2.49	1.60	1.70	2.31	......	2.25
custom 4	0.47	1.01	2.38	2.69	......	1.15
custom 5	2.17	2.63	0.91	1.17	......	1.45
custom 6	0.97	2.83	0.02	2.23	......	1.87
......	......	......	......	......	......	......
custom M	0.79	0.43	1.45	1.88	......	1.04

	financial Products 1	financial Products 2	financial Products 3	financial Products 4	financial Products 5	......	financial Products N
feature 1	2.85	0.33	1.18	1.29	0.67	......	0.63
feature 2	2.95	0.12	1.80	1.19	2.70	......	2.88
feature 3	2.38	0.58	1.06	1.91	2.31	......	1.71
feature 4	1.44	1.48	1.11	2.23	2.75	......	0.94
......	......	......	......	......	......	......	......
feature K	1.30	1.47	0.58	2.57	1.82	......	2.53

Fig. 3.4 The decomposition of customer-financial management matrix

customers and products. First of all, we take the number of customers' active purchase of financial products as the score of the correlation strength between customers and products, and construct a "customer-financial rating matrix *R*" with the full customer data and the full financial product data. The row title of the matrix is the wealth management product number, the column title is the customer number, and the value of the row i column j is the customer I's score on the wealth management product j. The matrix decomposition technique decomposes the matrix R into a "customer-feature matrix U" and a "feature-finance matrix V," and makes R equal to the matrix product of U and V as much as possible, that is, $R = U*V$, as shown in Fig. 3.4.

It's important to note that this feature lacks an explainable meaning and serves merely as a mathematical factorization component, often termed as the implicit feature factor. The number of features acts as a hyperparameter of the model, influencing performance to some extent. Post matrix decomposition, the original sparse matrix R (with many empty elements) is divided into two dense matrices U and V (with mostly non-empty elements). An algorithm determines the values in U and V such that the matrix product UV closely matches the non-empty elements of the R matrix. This allows the elements in the UV product matrix to predict customer scores for products. Essentially, during matrix decomposition, U and V values are defined by all elements in R, making U and V latent factor representations of R. Hence, customer scores for financial products decompose into multiple data points, each containing score information for all customers and products. The matrix decomposition is illustrated in Fig. 3.5.

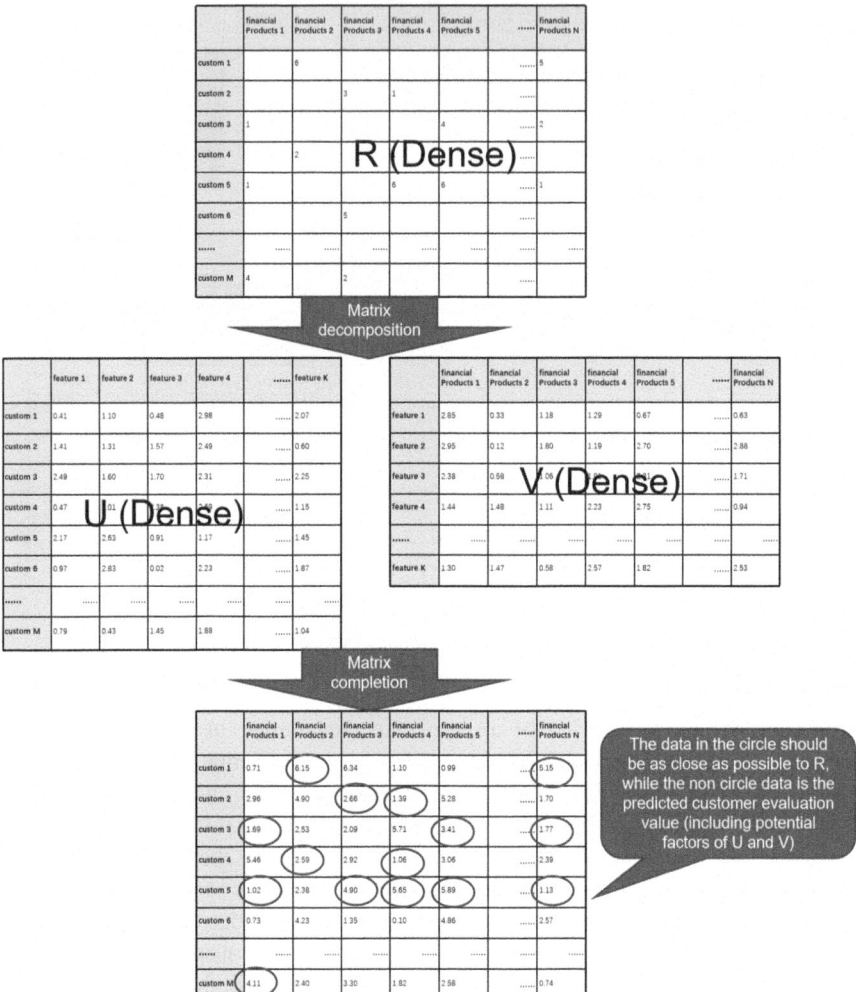

Fig. 3.5 Customer-finance matrix decomposition diagram

For the matrix calculation of U and V, the algorithm used is alternating least squares (ALS). Before introducing this algorithm, let's first introduce the least squares method with linear regression as an example. As shown in Fig. 3.6, the problem with linear regression is how to fit a line from several observation points such that each point on the line has the smallest sum of squares of distance from the observation point.

For the linear regression problem, its loss function $= \min \sum$ (observed value − fitted value)^2, its mathematical expression is: suppose the fitted line is $y = bx + a$, solve b and a, so that the value of the $D = \sum_{i=1}^{n} d_i^2 = \sum_{i=1}^{n} (y_i - a - bx_i)^2$ minimum.

3.2 Recommendation Algorithm

Fig. 3.6 Linear regression schematic

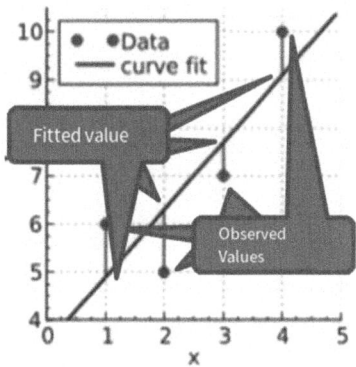

This is a function, mathematical knowledge tells us that the maximum value of a continuous function can be solved by the extreme value, and the method of solving the extreme value is to make the partial derivative equal to zero, so take the derivative of a and b respectively, and solve the equation to get the values of a and b.

$$\frac{\partial D}{\partial a} = -2\left(\sum_{i=1}^{n} y_i - \sum_{i=1}^{n} a - b \sum_{i=1}^{n} x_i\right) = 0 => a = \bar{y} - b\bar{x}$$

$$\frac{\partial D}{\partial b} = -2\left(\sum_{i=1}^{n} x_i y_i - n a \bar{x} - b \sum_{i=1}^{n} x_i^2\right) = 0 => b = \frac{\sum_{i=1}^{n}(x_i - \bar{x})(y_i - \bar{y})}{\sum_{i=1}^{n}(x_i - \bar{x})^2}$$

From a spatial point of view, the problem of minimizing the loss function is translated into the problem of solving the shortest distance from Y to the plane. The geometric meaning of the least square method is to solve the observed vector $Y = (y1, y2, y3... yn)$ to $X = (x1, x2, x3... xn)$ and $1 = (1,1,1... 1)$ the square of the module of the normal vector of the subspace (hyperplane) formed by two vectors, as shown in Fig. 3.7.

Now back to the matrix decomposition problem, we use the root-mean-square error to calculate the loss of two matrices of UV to R, which is mathematically expressed as: $RMSE = \sqrt{\frac{1}{n} \sum_{u,v} |(p_{u,v} - r_{u,v})^2|}$, where $p_{u,v}$ is the predicted value of U's score against V and $r_{u,v}$ is the observed value of U's score against V. Notably, $p_{i,j} = \langle u_i, v_j \rangle$ using the implicit feature factor in the training calculation, $r_{i,j} = \langle u_i, v_j \rangle$ use the actual value of the customer-finance matrix, u_i, v_j means each element in the U and V matrices is represented. The matrix decomposition problem is then transformed into solving for K implicit feature factors so that the $(u_i, v_j) = \min_{u,v} \sum_{(u,i) \in K} (p_{u,v} - r_{u,v})^2$ values are minimized. To avoid overfitting

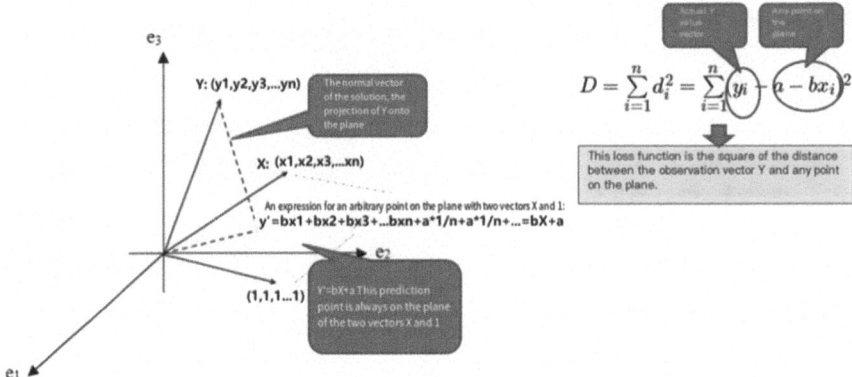

Fig. 3.7 Geometric meaning of least square method

problems, it is usually necessary to add regularization parameters λ, so the model can be expressed as: $(u_i, v_j) = \min\limits_{u,v} \sum\limits_{(i,j) \in K} (p_{i,j} - r_{i,j})^2 + \lambda \left(\|u_i\|^2 + \|v_j\|^2 \right)$.

With this in mind, it's time to introduce alternating least squares. The specific iterative steps are as follows:

Step 1: Initialize "customer-feature matrix U" and "feature-finance matrix V";
Step 2: Fix V and solve U using least square method;
Step 3: Fix the U output of step 2 so that the least square method solves V;
Step 4: Repeat step 2, 3 until U and V converge.

3.2.2 PersonalRank Diagram Recommendation

Utilizing a graph as a data structure for making recommendations proves advantageous, particularly due to the typically sparse nature of user interaction data with products. Financial institutions, for instance, may offer a multitude of product types, yet individual users engage with only a limited subset of them. The major drawback of the collaborative filtering algorithm lies in its dependence on a highly sparse matrix for predictions, relying on minimal observational data to infer a significant amount of unknown information. This limitation hinders the generation of precise recommendations, especially for users with sparse interaction history. To address this challenge, the PersonalRank graph recommendation algorithm was developed, drawing inspiration from the well-known PageRank algorithm. Originally designed for web value assessment, the principles of PersonalRank and PageRank are applicable to product value assessment without any fundamental discrepancies, thus facilitating their adaptation to the recommendation domain.

3.2 Recommendation Algorithm

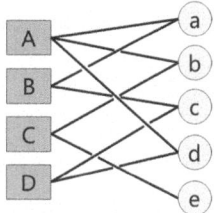

Fig. 3.8 Customer-product two-part diagram

Consider an example with four users labeled A, B, C, and D, and five financial products labeled a, b, c, d, and e. An undirected edge denotes the user's interaction with a product. This setup can be visualized as a bipartite graph where vertices represent two distinct sets: users and products. Each edge connects vertices from these two sets, forming a structure as depicted in Fig. 3.8. Given this graph, the recommendation challenge is to determine whether product c or product e is more suitable for user A.

First, consider the number of paths between the user and the product. A-c is connected through two paths, A-a-B-c and A-d-D-c, respectively, and A-e is connected through only one path, A-b-C-e. Therefore, from the point of view of the number of paths, A-c is more correlated than A-e, and we should recommend c. Then examine the length of the path between the user and the product. The path of A-c is A-a-B-c and A-d-D-c, both of which pass through 3 nodes and have length 3. The path of A-e is A-b-C-e, also have length 3. Therefore, in terms of path length, the correlation between e and c is the same. Then consider the sum of the exitance of the vertices on the connected path, the exitance refers to the number of connected edges of the vertices. On the path A-A-B-C, the output of A is 3, the output of a is 2, the output of B is 2, the output of c is 2, and the output sum is 9. On the path A-D-D-C, A comes out with 3, d comes out with 2, D comes out with 2, c comes out with 2, and the sum is 9. On the path A-B-C-E, A comes out with 3, b comes out with 2, C comes out with 2, e comes out with 1, and the sum comes out with 8. Therefore, in terms of exit, A-E has A higher correlation than A-C, and we should recommend e. In the PersonalRank algorithm, the priority is decreased according to the order of "number of paths, path length, exit sum," so in summary, c should be recommended.

Through the above example, we summarize the three principles of PersonalRank algorithm:

1. To recommend a product to a target customer, actually from the product set and customer set associated with the customer to find the most relevant product;
2. The vertices with more paths are more important, the vertices with shorter total path length are more important, and the smaller the sum of all vertices on the path are more important;
3. In terms of priority, number of paths > path length > exit sum.

In PersonalRank's view, to recommend products for target users is to calculate the relevance of each product to the target users. How does that work? Take Fig. 3.8 as an example and assume that the target user is node A. In the initial state, the

Fig. 3.9 PersonalRank vertex importance calculation formula

$$PR(j) = \begin{cases} \alpha * \sum_{i \in in(j)} \dfrac{PR(i)}{|out(i)|} & if(j \neq u) \\ 1-\alpha + \alpha * \sum_{i \in in(j)} \dfrac{PR(i)}{|out(i)|} & if(j \neq u) \end{cases}$$

correlation degree of node A is set to 1, and the correlation degree of other nodes is set to 0, that is, A is associated with itself but not with other nodes. We use symbol PR to represent the correlation degree, then the initial state can be expressed as: PR $(A) = 1$, PR(B) = PR(C) = PR(D) = PR(a) = PR(b) = PR(c) = PR(d) = PR(e) = 0. Then start from node A and walk upstream of the graph, each time starting from the node where PR is not 0 and walking one step forward. The probability of continuing to walk is α, and the probability of stopping at the current node is $1-\alpha$. For the first walk, since node A has three sides, it goes from node A to a, b, and d with 33% probability each, so that a, b, and c share the partial importance of A, PR(a) = PR(b) = PR(d) = α*PR(A)*0.33. Finally, PR(A) becomes $1-\alpha$. Nodes whose PR is not 0 after the end of the first walk are A, a, b, and d, which are all the starting nodes of the next walk. The second walk, starting at nodes A, a, b, and d, respectively, takes one step forward. In this way, node a gets the importance of A $1/3 * \alpha$, node b gets the importance of A $1/3 * \alpha$, node d gets the importance of A $1/3 * \alpha$, node A gets the importance of a $1/2 * \alpha$, node B gets the importance of a $1/2 * \alpha$, node A gets the importance of b $1/2 * \alpha$, node A gets the importance of B $1/2 * \alpha$, Node C is assigned the importance of b $1/2 * \alpha$, node A is assigned the importance of d $1/2 * \alpha$, and node D is assigned the importance of d $1/2 * \alpha$. Finally PR(A) is assigned $1-\alpha$. The process is then repeated over and over again, and the importance of each node can be expressed as the formula shown in Fig. 3.9.

In this context, u signifies the target user node within the graph. PR(i) denotes the significance or ranking of item i, while out(i) represents the outward links of item node i. The parameter alpha characterizes the probability of continued navigation or access. Through rigorous mathematical proof, it can be demonstrated that, following iterative applications of the prescribed formula, the importance value for each node will reach a stable state or converge to a constant. Consider the implications of the formula: vertices with a higher number of inbound paths receive greater value allocations from other vertices, thereby enhancing their significance. Each vertex along a path contributes to the cumulative importance of preceding vertices, albeit with a diminishing effect over extended paths. Specifically, the longer the path, the more substantial the attenuation of information, resulting in a reduced proportion of importance allocation. Consequently, vertices with numerous emanations and connections experience greater informational decay, diminishing the assigned importance proportion. This formula effectively encapsulates the three foundational principles of PersonalRank. These principles include the propagation of importance through extensive node connections, the impact of path length on information decay, and the influence of outbound link quantity on importance attenuation. Ultimately, by calculating the convergence importance for each node, one can generate a prioritized list of recommended items or products based on their relevance and ranking within the graph.

3.2 Recommendation Algorithm

3.2.3 Text Convolutional Neural Network

Introducing text convolutional neural networks stems from the observation that names often mirror their content in banking products or marketing campaigns. Similar names typically indicate similar content. For example, marketing campaigns like "Get a 30-yuan coupon for activating your credit card" and "Special gift for first credit card purchase" show that similarity in names mirrors similarity in content. By using text convolutional neural networks to analyze product names, we can effectively extract embedded vectors from text, enabling semantic comparisons between texts. This method allows us to pinpoint other products with similar names and likely similar content, enhancing recommendation effectiveness and customer acceptance. For instance, to extract an embedded vector for a financial product named "New Customer Monthly Stable Income Gain," we can structure a text convolutional neural network as shown in Fig. 3.10.

In this network, after segmenting the product name, each word's vector is represented by a row vector of shape 1*5. These seven word vectors stack vertically to form a 7 × 5 matrix, fed into the input layer. In the convolutional layer, three kernel sizes are used: 2 kernels of size 4*5, 2 kernels of size 3*5, and 2 kernels of size 2*5. In text convolutional neural networks, the kernel width matches the word vector dimension. During convolution, the kernel shifts only along the height dimension. Each kernel generates feature vector outputs: 2 vectors of size 1*4, 2 of size 1*5, and 2 of size 1*6, using VALID convolution without padding. The output of the convolutional layer undergoes max-pooling in the pooling layer, with each output vector's maximum value representing that vector. This allows varying input text lengths and column vector dimensions to be transformed into consistent vector outputs. The pooling results are stacked vertically to form a vector representing the product name's embedding, such as "New Customer Monthly Stable Income Gain," capturing the Chinese semantic. This vector serves as input for downstream tasks. Text convolutional networks share structural differences with image convolutional networks but utilize similar operations, including convolution, pooling, forward propagation, and backpropagation.

3.2.4 Two-Tower Model

In recent years, deep learning-based recommendation systems have made great progress. In 2013, Microsoft published the paper "Learning Deep Structured Semantic Models for Web Search using Clickthrough Dara," and the two-tower model came out. Later, many first-tier companies such as Google, Facebook, and Baidu joined in this technology research, launched their own models and applied them in their respective ecosystems. In 2019, Google published a paper "Samply-bias-corrected Neural Modeling for Large Corpus Item Recommendations" at the RecSys

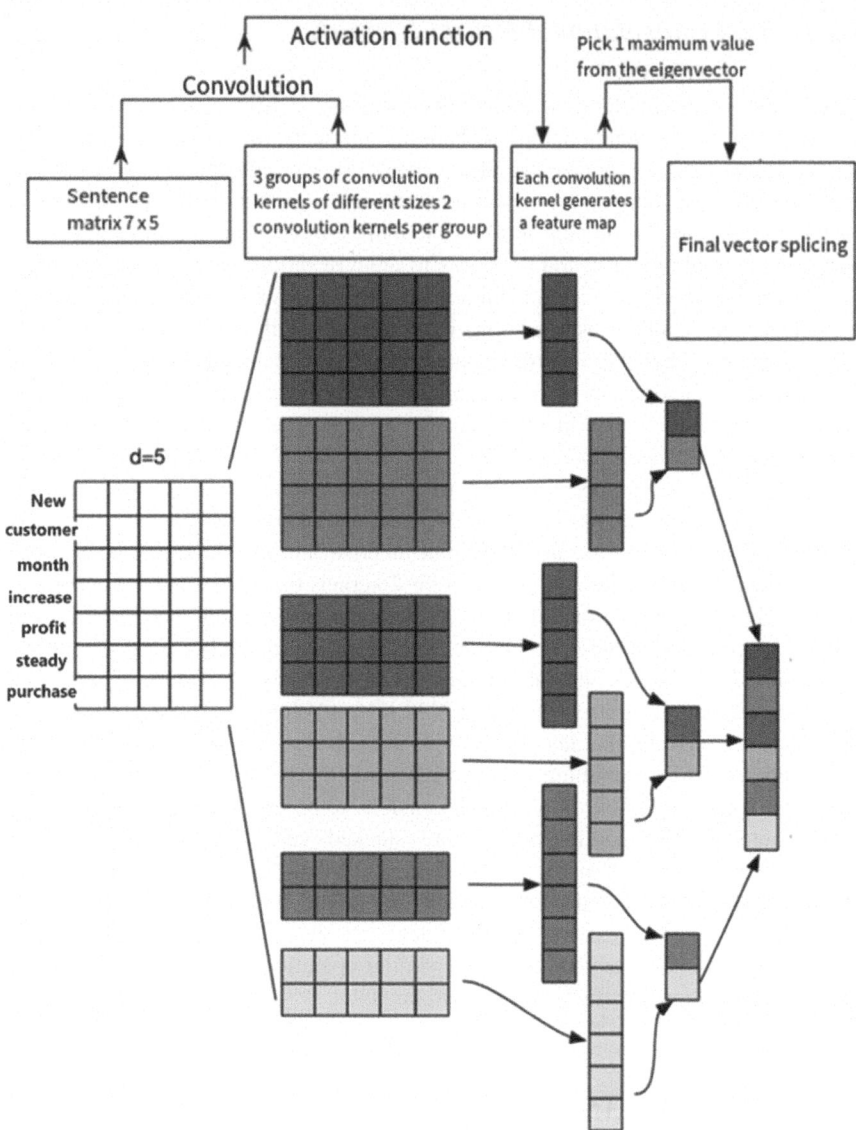

Fig. 3.10 Structure of text convolutional neural network

Conference on Recommendation Systems sponsored by ACM. It introduced the results of using the two-tower model to do the recall in the large-scale recommendation system, which attracted high attention in the industry. After the continuous progress of technology in recent years, the two-tower model has finally become a classic deep learning recall model.

3.2 Recommendation Algorithm

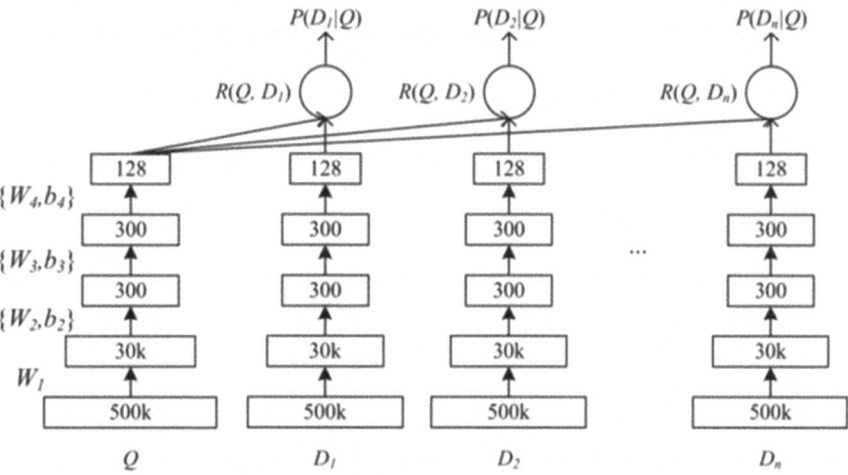

Fig. 3.11 Two-tower model structure proposed by Microsoft

The Twin Tower model, also called deep structured semantic matching models (DSSMs), was initially implemented in natural language processing to determine the semantic similarity of texts. Here's the concept: during the training phase, leverage extensive exposure and click log data from user search queries in search engines. Use a neural network to develop an embedding vector for queries and one for records. Then, compute the cosine similarity between these two semantic vectors to indicate semantic distance during online inference. Figure 3.11 illustrates the model structure.

In this Microsoft paper, the input DSSM consists of high-dimensional vectors. After multiple neural network layers, the output becomes low-dimensional vectors. These represent the user's query embedding vector and the record embedding vector. Cosine similarity calculates the similarity between these sides. Softmax activation function sorts different records to find the best match, known as the original DSSM. Semantic matching is a sorting problem, aligned with recommendation scenarios, thus DSSM fits into recommendation fields. By replacing records with items, DSSM becomes a recommendation model. The model uses two separate networks to build user and item embedding vectors, hence it's called the two-tower model.

A typical two-tower recommendation model includes independent "user tower" and "product tower" networks, with input, presentation, and matching layers, as illustrated in Fig. 3.12.

This is referred to as a "typical" structure because this technique has evolved into multiple variants. We selected a commonly used structure for illustration. The "user tower" and "product tower" respectively receive raw features of users and recommended items. After forward propagation through both networks, they generate user embedding vectors and item embedding vectors. At the input layer, each feature is independently encoded into its embedding vector (using neural networks, one-hot encoding, etc.), such as user attributes, user history with recommended items, item names, and item attributes. These embedding vectors are concatenated

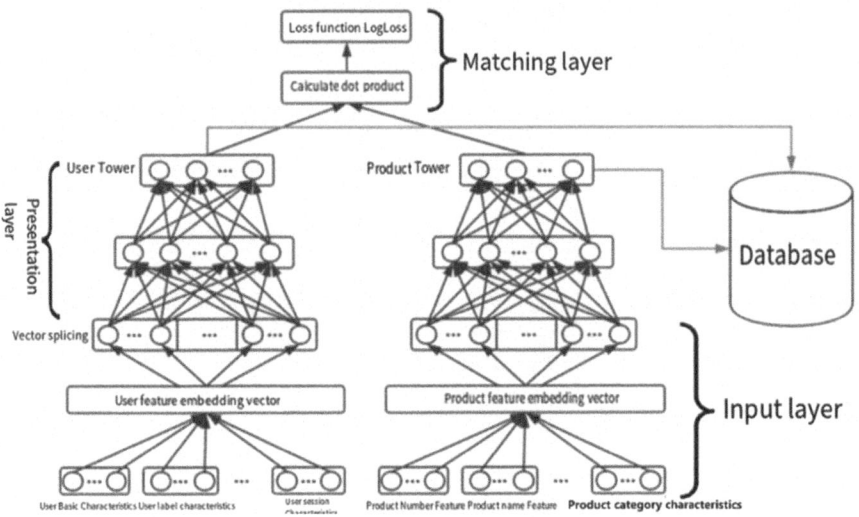

Fig. 3.12 Dual-tower recommendation model structure

into a new vector and sent to the representation layer. In the representation layer, after passing through two fully connected layers, both user and item features are converted into fixed-length vectors. This results in user embedding and item embedding vectors of the same dimension. Although the network layers and dimensions inside the user tower and product tower can differ, the dimensions of the output embedding vectors must match to perform operations at the matching layer. The representation layer's structure shown here is a simple implementation, using only two fully connected layers as feature extractors. In practice, there are many variants such as CNN, LSTM, and Transformer. The output embedding vectors of users and items are stored in Redis, an in-memory database. The dot product of the two embedding vectors is then calculated, and loss is computed with true labels using the Logloss function. During the training phase, forward and backward propagation are iterated to fit the network weights of the user tower and product tower to the dataset based on the actual user-item association labels. When recommending products to a target user, you only need to compute the dot products between the target user's embedding vector and all item embedding vectors to generate scores. Select the top-ranked items based on these scores. Training is offline, while recommendations are done in real-time in memory, thus balancing performance and speed.

The dual-tower model structure in this case largely matches the above illustration. The primary difference is the use of the mean squared error (MSE) loss function, as this case aims for the network's final output to closely match users' ratings of financial products. This regresses the dot product of user and item embedding vectors to known user ratings for financial products. Different financial products receive different ratings from the same user, and aligning calculated values with ratings is a good approach. Additionally, for extracting embedding vectors of financial product names at the input layer, a text convolutional network with dropout layers is used to prevent overfitting. The model structure is as shown in Fig. 3.13.

3.3 Development Framework

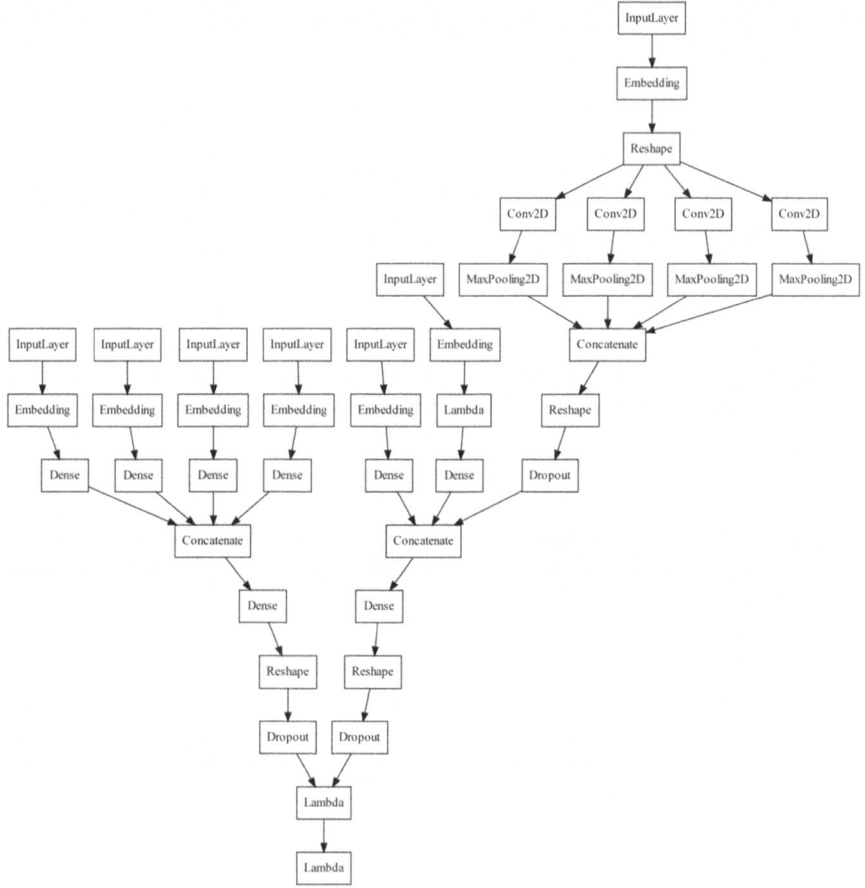

Fig. 3.13 Network structure of the two-tower model in this case

3.3 Development Framework

This section describes the open source frameworks used in this case, including PySpark, Pkuseg, Tensorflow, and Keras.

3.3.1 Computing Framework: PySpark

Apache Spark is a distributed cluster computing engine for large-scale and high-speed data processing and is the mainstream computing framework in the field of big data. It is one of the top projects, the Apache foundation has top-level domain https://spark.apache.org/ and a very influential open source community. Compared to

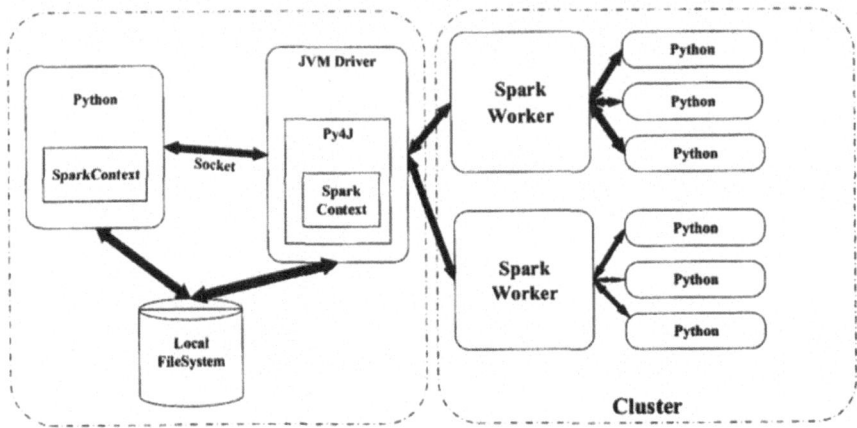

Fig. 3.14 Working architecture of PySpark

Hadoop, Spark has higher performance and operation speed than MapReduce, and is more fault tolerant and versatile, and provides a machine learning development library that users can easily call.

PySpark is the Python Shell of Spark. Similar to the relationship between shell and Unix kernel, PySpark provides apis encapsulated by Python and interacts with integrated Py4J to write Spark applications in Python. Figure 3.14 shows the overall architecture of PySpark.

This case uses the machine learning capabilities provided by PySpark to implement collaborative filtering recommendations.

3.3.2 Word Segmentation Framework: Pkuseg

Pkuseg is a Chinese word segmentation framework developed by the Language Computing and Machine Learning Research Group of Peking University. Compared with the established Jieba segmentation, Pkuseg has advantages such as higher segmentation accuracy, support for more vocabulary in professional fields and support for user-defined model training. It is worth mentioning that the use of Pkuseg is very simple, and it only takes three lines to complete the task of word segmentation almost perfectly:

```
import pkuseg
seg = pkuseg.pkuseg()            # Load model with default configuration
text = seg.cut('Everyday
Wealth Management for New Clients')   # perform participle processing
```

This will get the output: [' 'Everyday ',' Wealth Management ',' for ', New Clients '].

In the data preprocessing stage of this case, Pkuseg was used to segment Chinese names of financial products to provide conditions for the subsequent calculation of word embedding.

3.3.3 Deep Learning Framework: TensorFlow and Keras

TensorFlow is a deep learning framework launched by Google, which has been dominant in the industry due to Google's influence in the industry. TensorFlow is available in two large versions, 1.x and 2.x. Compared to PyTorch, the deep learning framework that dominates the academic world, TensorFlow 1.x has flaws in terms of not supporting dynamic graphs, code readability, ease of use, etc. Currently, the TensorFlow 2.x version addresses these issues.

Keras is an open source neural network computing library mainly developed by Python language, originally written by Francois Chollet, it uses more packaging techniques, can quickly and efficiently complete the work of neural networks. Keras can use TensorFlow, Theano, CNTK, and other frameworks to complete the gradient calculation and other work at the back-end, while the front-end uses the API provided by Keras to complete the network layer construction and other work. In TensorFlow 2.x version, Keras serves as the only interface for TensorFlow's high-level API, replacing the tf.layers interface that came with TensorFlow 1 version. In other words, only the Keras interface can now be used to build neural network layers.

In this case, TensorFlow 2.3.1 and Keras 2.4.3 are used to build the neural network of the two-tower model.

3.4 Case Practice

This section introduces the whole process of the implementation of this case, including data preparation, development and operation environment preparation, modeling code writing, and operation evaluation.

3.4.1 Data Preparation

This section does not use real product data but differentiates items by serial numbers, which does not affect the project demonstration. The collaborative filtering and PersonalRank algorithms in this case require simple data fields: customer ID, product name, and customer-product relevance score. The data extraction criteria for the customer-product relevance score should align with business needs. In practice, different events are assigned different weight coefficients, such as repeated

Table 3.1 Age codes

Age coding	Business meaning
1	Under 18
18	18–24
25	25–34
35	35–44
45	45–49
50	50–55
56	56 or older

purchases, first-time purchases, consultations via electronic channels, and browsing clicks. The total score, which reflects the customer's interest in a financial product, is obtained by multiplying event frequencies by their respective weights and summing the results. The dataset is located in the data/data.csv file in the following example format:

```
userId,title,rating
70614, financial product 1 , 5
265054, financial product 2 , 5
265054, financial product 3 , 5
265054, financial product 4 , 4
264349, financial product 5 , 4
```

The data source of the two-tower model in this case has three files, which are customer-feature file data/users_data.txt, financial product feature file data/products_data.txt, and customer-financial score file data/ratings_data.txt. Among them, the users_data file has 4 columns of data, which are customer number, gender, age code and occupation number (the code is customized according to the bank data). Table 3.1 lists the service meanings of the age code field.

The users_data file sample data is:

```
1,F,1,10
2,M,56,16
3,M,25,15
4,M,45,7
```

The products_data file has three columns of data, which are the financial product number, financial product name, and property of financial product. Among them, the property of financial product is a variable length field, that is, a financial product can have multiple attributes of an uncertain number, separated by vertical lines. The example data is as follows:

```
1,Financial Products 1,Floating income|Steadiness|Adequate quota
2,Financial Products 2,Rolling investment|Steadiness|Redemption at any time
3,Financial Products 3,Adequate quota|Fixed income
......
```

3.4 Case Practice

ratings_data file has three columns of data, which are customer number, financial product number, and customer-financial score, respectively. The example data is as follows:

```
1835,1676,4
3669,1783,3
1146,2367,3
3612,1641,2
```

3.4.2 Environment Preparation

In the Windows environment, refer to Sect. 1.4.1 for installing Anaconda. Execute "conda create -n tf21 python=3.6.7" to create a virtual environment. Activate the virtual environment by running "conda activate tf21." Install the dependencies by running "pip install tensorflow-gpu pandas scikit-learn tqdm pkuseg matplotlib py4j pyspark==3.0.1."

Extract the "spark-3.0.1-bin-hadoop2.7.tgz" file to "e:\spark" directory, set the "HADOOP_HOME" environment variable to "E:\spark\spark-3.0.1-bin-hadoop2.7, " add "E:\spark\spark-3.0.1-bin-hadoop2.7\bin" to the "PATH" environment variable. Install JDK 1.8.0_301 version, configure the "JAVA_HOME" and "PATH" environment variables. Copy the "pyspark" folder from the directory where Spark is located (mine is "E:\spark\spark-3.0.1-bin-hadoop2.7\python") to the "python" folder under the "conda" environment (mine is "C:\Anaconda\envs\tf21\Lib\site-packages"). With this, the environment setup is complete.

3.4.3 Code Practice

Collaborative Filtering Code Practice

Start by creating an app instance using SparkSession:

```
from pyspark.sql import SparkSession
spark = SparkSession.builder.appName('Collaborative_Filtering').getOrCreate()
```

Read the data source and get a DataFrame (that is, table type data, consisting of several rows and columns, the same below) :

```
df = spark.read.csv('./data/data.csv',inferSchema=True, header=True)
```

Because the name of the financial product in the data is Chinese, it cannot be calculated directly, so we need to convert the Chinese into numbers. Here, the index number is directly used to represent the Chinese name. The StringIndexer() method defines the input and output fields of the Chinese to numeric conversion; The fit() method learns the pattern of these Chinese data and returns the StringIndexerModel object; The transform() method uses the returned StringIndexerModel object to generate the converted DataFrame:

```
stringIndexer = StringIndexer(inputCol="title",
outputCol="title_new")
 model = stringIndexer.fit(df)
 indexed = model.transform(df)
```

Next, we divide the data into the training set and the test set in a ratio of 75%: 25%

```
train,test=indexed.randomSplit([0.75,0.25])
```

Since there are two hyperparameters in the collaborative filtering model, the number of hidden factors and regularization, their values will determine different model performance. Therefore, we need to find the most suitable hyperparameters under the conditions of the given data set. The idea is to iterate through different hyperparameters, calculate the rmse error under different hyperparameter conditions, and take the hyperparameter with the least error. PySpark provides a machine learning module named ml, short for machine learning, in which there is a recommendation algorithm module named Recommendation, which encapsulates the alternating least squares ALS algorithm and can be used directly. The implementation code is as follows:

```
from pyspark.ml.recommendation import ALS
 params1 = [10,20,30,40] # Define a range of values for the number of implied factors
 params2 = [0.01, 0.05, 0.08, 0.11, 0.15, 0.2, 0.3] # Define the range of values for the regularization parameter
 for p1 in params1:
   for p2 in params2:
  rec=ALS(maxIter=10,rank=p1,regParam=p2,userCol='userId',
itemCol='title_new',ratingCol='rating',nonnegative=True,coldStartStrategy="drop")
     # Execute the ALS model on the training set
     rec_model=rec.fit(train)
     # Get a predicted score
     predicted_ratings=rec_model.transform(test)
     predicted_ratings.orderBy(rand()).show(10)
```

3.4 Case Practice

```
# In order to evaluate model performance, RMSE root-mean-square error evaluation model is established
evaluator=RegressionEvaluator(metricName='rmse', predictionCol='prediction',labelCol='rating')
    # Perform RMSE error calculations on predicted ratings
    rmse=evaluator.evaluate(predicted_ratings)
```

Generate a visual graph of RMSE error under different hyperparameter conditions. The horizontal axis represents the regularization parameter value, the vertical axis represents the RMSE error, and the four curves correspond to different implied factor quantities, respectively. The implementation code is as follows:

```
plt.plot(x, y1, label=str(params1[0]), color='r')
plt.plot(x, y2, label=str(params1[1]), color='g')
plt.plot(x, y3, label=str(params1[2]), color='b')
plt.plot(x, y4, label=str(params1[3]), color='y')
plt.xlabel(' regularization parameter ', fontproperties="LiSu") # LiSu
plt.ylabel('RMSE error ', fontproperties="LiSu") # LiSu
plt.title('ALS parameter selection graph ', fontproperties="SimHei") # SimHei
plt.legend()
plt.show()
```

After execution, we get the visualized result shown in Fig. 3.15.

As can be seen, the greater the number of hidden factors, the greater the minimum RMSE error of its curve, indicating that the greater the amount of data computed using the sparse matrix, the greater its generalization error. As shown in Fig. 3.15, when the number of hidden factors is 10 and the regularization parameter is 0.11, the error value is the smallest, so we use these two hyperparameters to model.

```
rec=ALS(maxIter=10,rank=10,regParam=0.11,userCol='userId', itemCol='title_new',ratingCol='rating',nonnegative=True, coldStartStrategy="drop")
 rec_model=rec.fit(train)
 # Get a predicted score
 predicted_ratings=rec_model.transform(test)
 predicted_ratings.orderBy(rand()).show(10)
 # In order to evaluate model performance, RMSE root-mean-square error evaluation model is established
 evaluator=RegressionEvaluator(metricName='rmse', predictionCol='prediction',labelCol='rating')
  # Perform RMSE error calculations on predicted scores
  rmse=evaluator.evaluate(predicted_ratings)
```

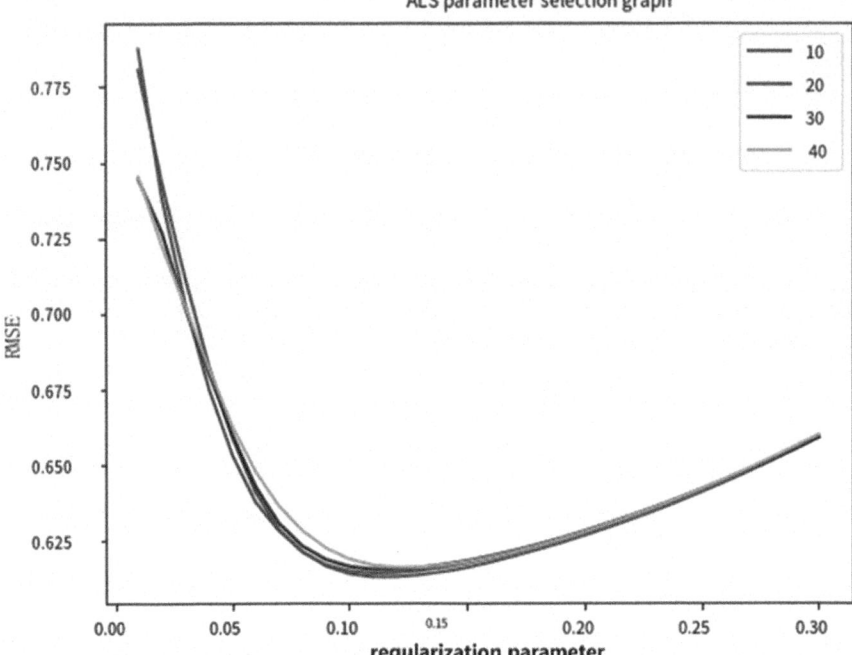

Fig. 3.15 Collaborative filtering hyperparameter search

We get an RMSE error of about 0.6, which completes the modeling. Next, we will use this model to recommend a wealth management product to a customer number specified by the command line argument sys.argv[1]. First get a list of all the products:

```
unique_product=indexed.select('title_new').distinct()
product = unique_product.alias('product')
Get the target customer number to be recommended:
user_id = int(sys.argv[1])
```

Get a list of wealth management products that the target customer has rated:

```
buy_product = indexed.filter(indexed['userId'] == user_id).select
('title_new').distinct()
b = buy_product.alias('b')
```

Get a list of wealth management products that can be recommended to target customers, i.e., a list of products that do not have rating behavior. The idea is to

3.4 Case Practice

concatenate all the product lists horizontally with the list of rated products to get the wealth management products whose scores are listed as empty.

```
total_product = product.join(b, product.title_new == b.title_new,
how='left')
 remaining_product=total_product.where(col("b.title_new").isNull
()).select(product.title_new).distinct()
```

Add a column to the recommended DataFrame data to store the target customer number:

```
remaining_product = remaining_product.withColumn("userId",lit(int
(user_id)))
```

pyspark calculates a predicted score for each recommended wealth management product, ranking this score in descending order to get a list of recommendations:

```
recommendations=rec_model.transform(remaining_product).orderBy
('prediction',ascending=False)
```

Finally, translate the recommended product number into the corresponding Chinese name and store the data in the new column title. Take the first ten recommended products as the final recommendation list:

```
product_title=IndexToString(inputCol="title_new",
outputCol="title",labels=model.labels)
 final_recommendations = product_title.transform(recommendations)
 df = final_recommendations.limit(10)
```

See Collaborative_Filtering.py for the code for this section. Run python Collaborative_Filtering.py 265054 to recommend to customer 265054. Figure 3.16 shows the result.

Code Practice of PersonalRank

First, we construct a bipartit graph from the ratings_data file, which is expressed in dictionary form with the format {UserA: {itemb:1,itemc:1},itemb:{UserA:1}}. The dictionary is divided into two parts, the first part describes the edges between the user node and the product node, and the second part describes the edges between the product node and the user node. The implementation code is as follows:

```
def build_graph(input_file):
  if not os.path.exists(input_file):
```

Fig. 3.16 Collaborative filtering recommendation result

```
rmse = 0.608914716528310
+---------+------+---------+
|title_new|userId|prediction|
+---------+------+---------+
| 1567.0  |265054|6.4009333|
| 2020.0  |265054|6.2704544|
| 1827.0  |265054|5.7094197|
| 1190.0  |265054|5.5828605|
| 595.0   |265054|5.5140657|
+---------+------+---------+
only showing top 5 rows
Recommended Results:
   userId    prediction    title
0  265054    6.400933    Financial Products 2121
1  265054    6.270454    Financial Products 2073
2  265054    5.709420    Financial Products 1756
3  265054    5.582860    Financial Products1505
4  265054    5.514066    Financial Products 284
5  265054    5.486214    Financial Products 865
6  265054    5.335969    Financial Products 831
```

```
    return {}
graph={}
linenum =0
# A rating higher than score_thr is considered a valid relationship
score_thr=2.0
fp = open(input_file, encoding='utf-8')
for line in fp:
  if linenum ==0:
    linenum +=1
    continue
  item = line.strip().split(",")
  if len(item)<3:
    continue
  userid,itemid,rating =item[0],"item_"+item[1],item[2]
  if float(rating)<score_thr:
    continue
  if userid not in graph:
    graph[userid] ={}
  graph[userid][itemid]=1
  if itemid not in graph:
    graph[itemid]={}
  graph[itemid][userid] = 1
fp.close()
return graph
```

3.4 Case Practice

Next, assess the significance of each vertex using the calculation formula in Fig. 3.9. During each iteration, utilize two dictionaries to document the importance of each financial product vertex before and after the iteration, formatted as {Product Number: Vertex Importance}. If the two dictionaries match entirely after iteration, the algorithm has converged. The iteration then concludes, and the products are ranked in descending order of vertex importance for recommendations. We will develop a PersonalRank algorithm, inputting the user product graph and the users for recommendation, performing random walks with an alpha probability, specifying iteration rounds and the number of recommended products, and returning a recommended product result dictionary in the format: {key: product number, value: importance}. The implementation code is as follows:

```
def personal_rank(graph,root,alpha,iter_num,recom_num=10):
    rank = {}
    rank = {point:0 for point in graph}# Initialize all vertices except the root vertex to 0, killing two birds with one stone and automatically de-duplicating them.
    rank[root] = 1# root vertex initialized to 1
    recom_result = {}
    for iter_index in range(iter_num):
        tmp_rank = {}
        tmp_rank = {point:0 for point in graph}# Store the importance of other vertices with respect to the root vertex
        for out_point,out_dict in graph.items():
            for inner_point,value in graph[out_point].items():
                tmp_rank[inner_point] +=round(alpha*rank[out_point]/len(out_dict),4)
                if inner_point == root:
                    tmp_rank[inner_point] +=round(1-alpha,4)
        if tmp_rank ==rank:
            print('iterate'+str(iter_index)+'convergence after times ')# See if the iteration ends early
            break
        rank = tmp_rank
    right_num = 0# Define a counter
    # Sort the structure rank according to the score of the importance value and filter out the items that User vertices and root vertices have already acted on
    for zuhe in sorted(rank.items(),key=operator.itemgetter(1), reverse=True):
        point,pr_score =zuhe[0],zuhe[1]
        if 'item_' not in point:# If it's not an item vertex, filter it out.
            continue
        if point in graph[root]:
```

```
        continue
    recom_result[point] = pr_score # Load the results into the dataset
    right_num += 1
    if right_num >= recom_num:
      break
  return recom_result
```

Execute the following master code to complete the two-part graph construction and get the recommendation results for a particular target customer:

```
user = " 265054" # Target customer number to be referred
alpha = 0.8
graph = build_graph("./data/data.csv")
iter_num = 100
recom_result=personal_rank(graph,user,alpha,iter_num,
recom_num=5)
# Print out the items of interest to the user to analyze the results
print(' Target users hold:')
for itemid in graph[user]:
  print(itemid, graph[user][itemid])
print(' Recommendations to targeted users:')
for itemid in recom_result:
  print(itemid, recom_result[itemid])
```

See personalRank.py for the code in this section. Execute python personalRank.py, we get the recommendation result to customer No. 265054, showing the recommended wealth management product name and importance, as shown in Fig. 3.17.

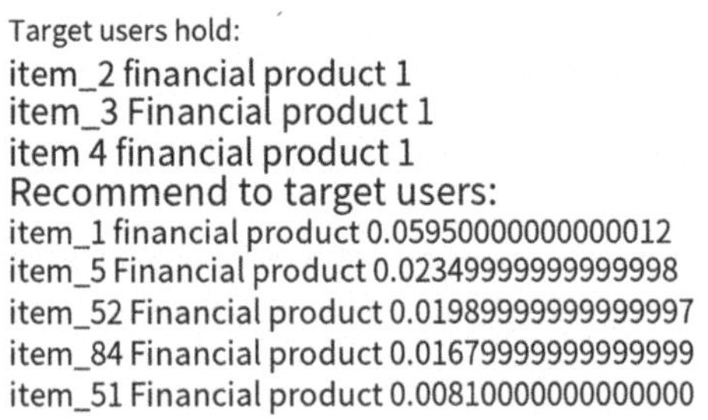

Fig. 3.17 PersonalRank recommendation result

Target users hold:
item_2 financial product 1
item_3 Financial product 1
item 4 financial product 1
Recommend to target users:
item_1 financial product 0.05950000000000012
item_5 Financial product 0.02349999999999998
item_52 Financial product 0.01989999999999997
item_84 Financial product 0.01679999999999999
item_51 Financial product 0.00810000000000000]

3.4 Case Practice

Twin Tower Recommendation Model Code Practice

First, import the associated package (omitted) and read three data files, respectively:

```
# Read user data
users_name = ['UserID', 'Gender', 'Age', 'JobID']
users = pd.read_csv('./data/users_data.txt', header=None, names=users_name)
users.head()
# Read product data
products_name = ['ProductID', 'ProductName', 'Attrib']
products = pd.read_csv('./data/products_data.txt', header=None, names=products_name)
products.head()
# Read the rating data
ratings_name = ['UserID','ProductID', 'Rating']
ratings = pd.read_csv('./data/ratings_data.txt', header=None, names=ratings_name)
ratings.head()
```

During the data preprocessing stage, the process of handling the raw data includes: converting the gender field from 'F' and 'M' to 0 and 1; converting the age group field into seven numbers based on categories, ranging from 0 to 6; for the financial product name field and product attribute field, tokenization is performed first, followed by creating a text-to-number dictionary, then converting them into a list of numbers, and padding based on the maximum field length to ensure each data record has a uniformly sized vector, facilitating subsequent network computations. The empty spaces are filled with the corresponding number for "<PAD>." Finally, the three tables are concatenated into a wide feature table, the feature matrix is divided, and the score values are taken as the target vector, serving as the supervised data for the neural network. The key code snippet is as follows:

```
def data_Pretreatment():
......
   age_map = {val:ii for ii,val in enumerate(set(users['Age']))}
   users['Age'] = users['Age'].map(age_map)
......
   # Product Type to Numeric Dictionary
   attrib_set = set()
   for val in products['Attrib'].str.split('|'):
      attrib_set.update(val)
   attrib_set.add('<PAD>')
   attrib2int = {val:ii for ii, val in enumerate(attrib_set)}
......
```

```
# ProductName to Numeric Dictionary
name_set = set()
seg = pkuseg.pkuseg() # Use PKU pkuseg to achieve Chinese participle,
the effect is better than jieba
max_words_len = 0
......
name_set.add('<PAD>')
name2int = {val:ii for ii, val in enumerate(name_set)}
......
# Consolidation of three tables
data = pd.merge(pd.merge(ratings, users), products)
```

Below, we define the network structure as described in Fig. 3.13. First, construct the input layer structure, which includes two parts: user characteristics and product characteristics. User characteristics include ID number, gender, age, occupation number, product characteristics include id number, product category attribute, product name, each field is input as an independent vector. Use the tf.keras.layers.Input() method to define the input layer by specifying the shape, type, and network layer name of the input data. The code is implemented as follows:

```
def input_layers():
    uid = tf.keras.layers.Input(shape=(1,), dtype='int32',
name='uid')
    user_gender = tf.keras.layers.Input(shape=(1,), dtype='int32',
name='user_gender')
    user_age = tf.keras.layers.Input(shape=(1,), dtype='int32',
name='user_age')
    user_job = tf.keras.layers.Input(shape=(1,), dtype='int32',
name='user_job')
    product_id = tf.keras.layers.Input(shape=(1,), dtype='int32',
name='product_id')
    product_categories = tf.keras.layers.Input(shape=(18,),
dtype='int32', name='product_categories')
    product_names = tf.keras.layers.Input(shape=(name_count,),
dtype='int32', name='product_names')
    return uid, user_gender, user_age, user_job, product_id,
product_categories, product_names
```

On top of the input layer, define an embedding layer. In TensorFlow 2.x, the "tf.keras.layers.Embedding()" method is provided to re-encode features into a dense vector for input to subsequent network layers. It is important to note that the 'Embedding()' method can only be used to construct the first layer of the network; simply specify the name of the input layer it connects to. The implementation code is as follows:

3.4 Case Practice

```
# Define the Embedding layer of the User tower
 def user_Embedding_layers(uid, user_gender, user_age, user_job):
   uid_embed_layer = tf.keras.layers.Embedding(uid_max, embed_dim,
input_length=1, name='uid_embed_layer')(uid)
   gender_embed_layer = tf.keras.layers.Embedding(gender_max,
embed_dim // 2, input_length=1, name='gender_embed_layer')
(user_gender)
   age_embed_layer = tf.keras.layers.Embedding(age_max, embed_dim //
2, input_length=1, name='age_embed_layer')(user_age)
   job_embed_layer = tf.keras.layers.Embedding(job_max, embed_dim //
2, input_length=1, name='job_embed_layer')(user_job)
   return uid_embed_layer, gender_embed_layer, age_embed_layer,
job_embed_layer
 # Define the Embedding layer for product ids
 def product_id_embed_layers(product_id):
   product_id_embed_layer = tf.keras.layers.Embedding
(product_id_max, embed_dim, input_length=1,
name='product_id_embed_layer')(product_id)
   return product_id_embed_layer
 # Embedding layer that defines product attributes
 def product_categories_layers(product_categories):
   product_categories_embed_layer = tf.keras.layers.Embedding
(product_categories_max, embed_dim, input_length=18,
name='product_categories_embed_layer')(product_categories)
   product_categories_embed_layer = tf.keras.layers.Lambda(lambda
layer: tf.reduce_sum(layer, axis=1, keepdims=True))
(product_categories_embed_layer)
   return product_categories_embed_layer
```

Below define the text convolutional neural network for the product name. The network receives the input after word segmentation and number conversion, converts it into an embedded vector, uses four different convolution kernels to extract text features, and splices the results into a long vector after maximum pooling. A percentage of neurons are randomly deleted during training to avoid overfitting (i.e., adding a dropout layer), resulting in an output of 32 length. The implementation code is as follows:

```
 def product_cnn_layer(product_names):
   # Get the embedding vector of each word corresponding to the product
name from the embedding matrix
   product_name_embed_layer = tf.keras.layers.Embedding
(product_name_max, embed_dim, input_length=name_count,
name='product_name_embed_layer')(product_names)
   sp=product_name_embed_layer.shape
```

```
    product_name_embed_layer_expand = tf.keras.layers.Reshape([sp
[1], sp[2], 1])(product_name_embed_layer)
    # Use different convolution kernels for the text embedding layer to
output results in maximized pooling
    pool_layer_lst = []
    for window_size in window_sizes:
        conv_layer = tf.keras.layers.Conv2D(filter_num, (window_size,
embed_dim), 1, activation='relu')(product_name_embed_layer_expand)
        maxpool_layer = tf.keras.layers.MaxPooling2D(pool_size=
(sentences_size - window_size + 1 ,1), strides=1)(conv_layer)
        pool_layer_lst.append(maxpool_layer)
    pool_layer = tf.keras.layers.concatenate(pool_layer_lst, 3, name
="pool_layer")
    max_num = len(window_sizes) * filter_num
    pool_layer_flat = tf.keras.layers.Reshape([1, max_num], name =
"pool_layer_flat")(pool_layer)
    # Add a Dropout layer
    dropout_layer = tf.keras.layers.Dropout(dropout_rate, name =
"dropout_layer")(pool_layer_flat)
    return dropout_layer
```

The full connection layer for the user tower and the product tower is defined below. Build a full connection layer by calling the tf.keras.layers.Dense() method, specifying the embedding dimension, network layer name, activation function, and the name of the embedded layer to be interacted with. The implementation code is as follows:

```
# Define the full connection layer of the User tower
def user_feature_layers(uid_embed_layer, gender_embed_layer,
age_embed_layer, job_embed_layer):
    # Layer 1 Full connection
    uid_fc_layer = tf.keras.layers.Dense(embed_dim,
name="uid_fc_layer", activation='relu')(uid_embed_layer)
    gender_fc_layer = tf.keras.layers.Dense(embed_dim,
name="gender_fc_layer", activation='relu')(gender_embed_layer)
    age_fc_layer = tf.keras.layers.Dense(embed_dim,
name="age_fc_layer", activation='relu')(age_embed_layer)
    job_fc_layer = tf.keras.layers.Dense(embed_dim,
name="job_fc_layer", activation='relu')(job_embed_layer)
    # Layer 2 full connection
    user_combine_layer = tf.keras.layers.concatenate([uid_fc_layer,
gender_fc_layer, age_fc_layer, job_fc_layer], 2)  #(? , 1, 128)
    user_combine_layer = tf.keras.layers.Dense
(200, activation='tanh')(user_combine_layer)  #(? , 1, 200)
```

3.4 Case Practice

```
    user_combine_layer_flat = tf.keras.layers.Reshape([200],
name="user_combine_layer_flat")(user_combine_layer)
    user_combine_layer_flat_drop = tf.keras.layers.Dropout
(dropout_rate, name = "user_combine_layer_flat_drop")
(user_combine_layer_flat)
    return user_combine_layer_flat_drop
  # Define the full connection layer of the product network
  def product_feature_layers(product_id_embed_layer,
product_categories_embed_layer, dropout_layer):
    # Layer 1 Full Connection
    product_id_fc_layer = tf.keras.layers.Dense(embed_dim,
name="product_id_fc_layer", activation='relu')
(product_id_embed_layer)
    product_categories_fc_layer = tf.keras.layers.Dense(embed_dim,
name="product_categories_fc_layer", activation='relu')
(product_categories_embed_layer)
    # Layer 2 Full Connection
    product_combine_layer = tf.keras.layers.concatenate
([product_id_fc_layer, product_categories_fc_layer, dropout_layer],
2)
    product_combine_layer = tf.keras.layers.Dense
(200, activation='tanh')(product_combine_layer)
    product_combine_layer_flat = tf.keras.layers.Reshape([200],
name="product_combine_layer_flat")(product_combine_layer)
    product_combine_layer_flat_drop = tf.keras.layers.Dropout
(dropout_rate, name = "product_combine_layer_flat_drop")
(product_combine_layer_flat)
    return product_combine_layer_flat_drop
```

Once you've finished defining the network structure, you can connect them together to form a complete two-tower network. We define a dssm_network class, in the initialization function __init__() of this class, define the hierarchical relationship between the previous network layers and the corresponding relationship of data transfer, and use the plot_model() interface to draw the network structure diagram, define the loss function and storage parameters. The implementation code is as follows:

```
  def __init__(self, batch_size=256):
    self.batch_size = batch_size
    self.best_loss = 9999
    self.losses = {'train': [], 'test': []}
    # Get input placeholder
    uid, user_gender, user_age, user_job, product_id,
product_categories, product_names = input_layers()
```

```
# Get the 4 embedding vectors of User
uid_embed_layer, gender_embed_layer, age_embed_layer,
job_embed_layer = user_Embedding_layers(uid, user_gender, user_age,
user_job)
# Get user characteristics
user_combine_layer_flat = user_feature_layers(uid_embed_layer,
gender_embed_layer, age_embed_layer, job_embed_layer)
# Get the embedding vector of the product ID
product_id_embed_layer = product_id_embed_layers(product_id)
# Get the embedding vector for the product type
product_categories_embed_layer = product_categories_layers
(product_categories)
# Get the feature vector of the product name
dropout_layer = product_cnn_layer(product_names)
# Get product features
product_combine_layer_flat = product_feature_layers
(product_id_embed_layer, product_categories_embed_layer,
dropout_layer)
# Matrix multiply user characteristics and product characteristics
to get a predictive score
inference = tf.keras.layers.Lambda(lambda layer: tf.reduce_sum
(layer[0] * layer[1], axis=1), name="inference")
((user_combine_layer_flat, product_combine_layer_flat))
inference = tf.keras.layers.Lambda(lambda layer: tf.expand_dims
(layer, axis=1))(inference)
self.model = tf.keras.Model(inputs=[uid, user_gender, user_age,
user_job, product_id, product_categories, product_names], outputs=
[inference])
self.model.summary()
self.optimizer = tf.keras.optimizers.Adam(learning_rate)
# mean square error loss function
self.ComputeLoss = tf.keras.losses.MeanSquaredError()
# average absolute error
self.ComputeMetrics = tf.keras.metrics.MeanAbsoluteError()
```

In the dssm_network class, define network forward propagation and gradient computation:

```
def train_step(self, x, y):
    with tf.GradientTape() as tape:
        logits = self.model([x[0], x[1], x[2], x[3], x[4], x[5], x[6]],
training=True)
        loss = self.ComputeLoss(y, logits)
        self.ComputeMetrics(y, logits)
```

3.4 Case Practice

```
    grads = tape.gradient(loss, self.model.trainable_variables)
    self.optimizer.apply_gradients(zip(grads, self.model.
trainable_variables))
    return loss, logits
```

In the network training stage, a batch of training sample data is sent to the user tower and the product tower, respectively. After forward propagation, the respective embedding vector is calculated and the evaluation index of model training is defined. The key codes are as follows:

```
avg_loss = tf.keras.metrics.Mean('loss', dtype=tf.float32)
acc_meter = tf.keras.metrics.Accuracy('accuracy', dtype=tf.float32)
for batch_i in range(batch_num):
    x, y = next(train_batches)
    ...
    loss, logits = self.train_step(......)
    avg_loss(loss)
    acc_meter.update_state(np.reshape(y, [self.batch_size, 1]).
astype(np.float32),
        np.reshape(logits_around, [self.batch_size, 1]).astype(np.
float32))
```

Then instantiate the dssm_network class, and pass the splicted features and supervised vectors into the network, specify epochs parameter (if the number of iterations of training is equal to the number of data bars in the data set, called 1 epoch, in this example, the parameter is 100, it means that the number of iterations of training is 100 times the number of data bars in the data set). Then start the training:

```
dssm_net = dssm_network()
dssm_net.training(features, targets_values, epochs=100)
```

You will see the training screen as shown below, as shown in Fig. 3.18.

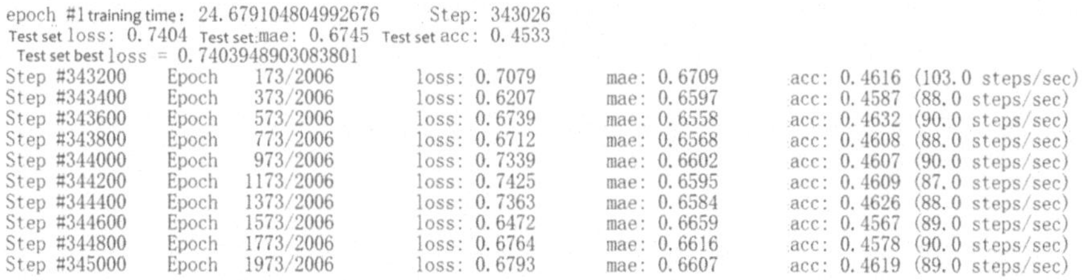

```
epoch #1 training time: 24.679104804992676    Step: 343026
Test set loss: 0.7404  Test set mae: 0.6745  Test set acc: 0.4533
  Test set best loss = 0.7403948903083801
Step #343200      Epoch    173/2006      loss: 0.7079      mae: 0.6709      acc: 0.4616  (103.0 steps/sec)
Step #343400      Epoch    373/2006      loss: 0.6207      mae: 0.6597      acc: 0.4587  (88.0 steps/sec)
Step #343600      Epoch    573/2006      loss: 0.6739      mae: 0.6558      acc: 0.4632  (90.0 steps/sec)
Step #343800      Epoch    773/2006      loss: 0.6712      mae: 0.6568      acc: 0.4608  (88.0 steps/sec)
Step #344000      Epoch    973/2006      loss: 0.7339      mae: 0.6602      acc: 0.4607  (90.0 steps/sec)
Step #344200      Epoch   1173/2006      loss: 0.7425      mae: 0.6595      acc: 0.4609  (87.0 steps/sec)
Step #344400      Epoch   1373/2006      loss: 0.7363      mae: 0.6584      acc: 0.4626  (88.0 steps/sec)
Step #344600      Epoch   1573/2006      loss: 0.6472      mae: 0.6659      acc: 0.4567  (89.0 steps/sec)
Step #344800      Epoch   1773/2006      loss: 0.6764      mae: 0.6616      acc: 0.4578  (90.0 steps/sec)
Step #345000      Epoch   1973/2006      loss: 0.6793      mae: 0.6607      acc: 0.4619  (89.0 steps/sec)
```

Fig. 3.18 Training process of twin tower model

After the model is trained, the trained network can be used to make recommendations. We can predict a user's rating on a financial product by inputting the pre-processed user characteristics and product characteristics into the network and conducting a forward propagation. The implementation code is as follows:

```
# Input a user feature and a product feature into the network, and get a score through forward propagation
def rating_product(dssm_net, user_id_val, product_id_val):
    categories = np.zeros([1, 18])
    categories[0] = products.values[productid2idx[product_id_val]][2]
    names = np.zeros([1, sentences_size])
    names[0] = products.values[productid2idx[product_id_val]][1]
    inference_val = dssm_net.model(......)
    return (inference_val.numpy())
```

To facilitate real-time recommendations, we can save the trained user embed vector and the product embed vector to binary files separately. In the future, when we do the recommendation task, we only need to load the embedded vector from the file into the memory to get the embedded vector, and calculate the product of the two embedded vectors directly to get the predicted score, without training the network. The implementation code is as follows:

```
# Generate product feature matrix and save to local
product_layer_model = keras.models.Model(inputs=[dssm_net.model.input[4], dssm_net.model.input[5], dssm_net.model.input[6]],
    outputs=dssm_net.model.get_layer("product_combine_layer_flat").output)
product_matrices = []
for item in products.values:
    categories = np.zeros([1, 18])
    categories[0] = item.take(2)
    names = np.zeros([1, sentences_size])
    names[0] = item.take(1)
    product_combine_layer_flat_val = product_layer_model([np.reshape(item.take(0), [1, 1]), categories, names])
    product_matrices.append(product_combine_layer_flat_val)
pickle.dump((np.array(product_matrices).reshape(-1, 200)), open('product_matrices.p', 'wb'))
# Generate user feature matrix, save to local
user_layer_model = keras.models.Model(inputs=[dssm_net.model.input[0], dssm_net.model.input[1], dssm_net.model.input[2], dssm_net.model.input[3]],
```

3.4 Case Practice

```
outputs=dssm_net.model.get_layer("user_combine_layer_flat").
output)
users_matrics = []
```

for item in users.values:

```
user_combine_layer_flat_val = user_layer_model(......)
users_matrics.append(user_combine_layer_flat_val)
pickle.dump((np.array(users_matrics).reshape(-1, 200)), open
('users_matrics.p', 'wb'))
```

Having product embedded vectors, we can conveniently find other products similar to a specified product at the semantic level by calculating the product feature vector's dot product with the entire product feature matrix, sorting the resulting product of the multiplication from high to low, and selecting the top_k values to obtain the most similar top_k products. In practical applications, introducing some randomness ensures variations in each recommendation. Here is the implementation code:

```
def recommend_similar_product(product_id_val, top_k = 5):
    norm_product_matrics = tf.sqrt(tf.reduce_sum(tf.square
(product_matrics), 1, keepdims=True))
    normalized_product_matrics = product_matrics /
norm_product_matrics
    # Recommend the same type of product
    probs_Embeddings = (product_matrics[productid2idx
[product_id_val]]).reshape([1, 200])
    probs_similarity = tf.matmul(probs_Embeddings, tf.transpose
(normalized_product_matrics))
    sim = (probs_similarity.numpy())
    print("The product you looked at is: {}".format(products_orig
[productid2idx[product_id_val]]))
    print("Similar products are: ")
    p = np.squeeze(sim)
    p[np.argsort(p)[:-top_k]] = 0
    p = p / np.sum(p)
    results = set()
    while len(results) != top_k:
       c = np.random.choice(2179, 1, p=p)[0]
       results.add(c)
    for val in (results):
       print(products_orig[val])
    return results
```

For recommending financial products to a specific user, it is quite straightforward. Simply calculate the matrix product of the embedded vector of that user and the embedded vectors of all products, then sort the product of the products from high to low, and select the top_k values to obtain the top_k most recommended products. The computational cost involves only one matrix multiplication, additionally incorporating a certain random factor. The implementation code is as follows:

```
def recommend_user_product(user_id_val, top_k = 5):
    probs_Embeddings = (users_matrics[user_id_val-1]).reshape([1, 200])
    probs_similarity = tf.matmul(probs_Embeddings, tf.transpose(product_matrics))
    sim = (probs_similarity.numpy())
    print(" Here's a recommendation for you: ")
    p = np.squeeze(sim)
    p[np.argsort(p)[:-top_k]] = 0
    p = p / np.sum(p)
    results = set()
    while len(results) ! = top_k:
      c = np.random.choice(2179, 1, p=p)[0]
      results.add(c)
    for val in (results):
      print(products_orig[val])
    return results
```

To predict which users who like this financial product will also like other products, one can first multiply the embedding vector of this product by the matrix composed of all user embedding vectors. By sorting the product of the results in descending order, one can obtain the top-k individuals who like this product the most. Then, calculate the ratings of these top-k individuals for all products and recommend the product with the highest rating for each individual, incorporating a random selection factor. The implementation code is as follows:

```
def recommend_other_user_product(product_id_val, top_k = 5):
    probs_product_Embeddings = (product_matrics[productid2idx[product_id_val]]).reshape([1, 200])
    probs_user_favorite_similarity = tf.matmul(probs_product_Embeddings, tf.transpose(users_matrics))
    favorite_user_id = np.argsort(probs_user_favorite_similarity.numpy())[0][-top_k:]
    print(" Target product is: {}".format(products_orig[productid2idx[product_id_val]]))
    print(" People who like the target product are: {}".format(users_orig[favorite_user_id-1]))
```

3.4 Case Practice

```
    probs_users_Embeddings = (users_matrics[favorite_user_id-1]).
reshape([-1, 200])
    probs_similarity = tf.matmul(probs_users_Embeddings, tf.transpose
(product_matrics))
    sim = (probs_similarity.numpy())
    p = np.argmax(sim, 1)
    print(" People who like the target product also like: ")
    if len(set(p)) < top_k:
       results = set(p)
    else:
       results = set()
       while len(results) != top_k:
          c = p[random.randrange(top_k)]
          results.add(c)
    for val in (results):
       print(products_orig[val])
    return results
```

Get the following predicted results, as shown in Fig. 3.19.

As can be seen, after the model training is completed, we save the embedded vector output of the user tower and product tower to a file (memory database can also be used), and load the recommendation stage directly to the memory for matrix calculation, and the recommendation results can be obtained quickly with only a small calculation cost. The code is detailed in the download file dssm.py.

The predicted rating of customer 8 for product 1387 is: [[4.266861]]
Obtain similar products for product number 1387:
The product you are looking at is: [1401 '1302 New Customer Wealth Management Product - X Bank' No Handling Fee ']
Similar products include:
[1406 '1305 New Customer Wealth Management Product - X Bank' No Handling Fee ']
  New customer wealth management product No. 233227- X bank 'no handling fee']
[1401 '1302 New Customer Wealth Management Product - X Bank' 'No Handling Fee']
Recommend 5 products to User 8:
Here are my recommendations for you:
[318310 New Customer Wealth Management Product - X Bank 'No Handling Fee']
[922'862 New Customer Wealth Management Product - X Bank 'Bond Type']
[2019 '1840 New Customer Wealth Management Product - X Bank' No handling fee for purchase starting from 10000 ']
People who like 1387 products also like:
The target product is: [1387'1289 New Customer Wealth Management Product - X Bank '10000 Starting Purchase Self operated Wealth Management'] The people who like the target product are: [3031'M'184]
[5861 'F' 50 1]]
People who like the target product also like:
[318'310 New Customer Wealth Management Product - X Bank 'no handling fee'] [2019'1840 New Customer Wealth Management Product - X Bank 'no handling fee for purchases starting from 10000']

Fig. 3.19 Prediction results of the twin tower model

3.5 Case Summary

Three recommendation algorithms—collaborative filtering, PersonalRank, and Twin Tower model—can effectively output a tailored set of financial products for a specified customer group. In practice, we analyze the intersection of the outputs from these three methods to derive the product list $P1$, which holds the highest marketing priority. Subsequently, the union set minus the intersection yields the product list $P2$, targeting lower marketing priority. We divided customers with similar characteristics into two groups: Group A and Group B, each comprising 20,000 individuals. Group A followed a strategy prioritizing the marketing of P1 products first and P2 products later, while Group B adopted a random product marketing approach. The AB test results showed that Group A's marketing success rate was six times higher than that of Group B, underscoring the positive impact of the recommendation system in enhancing marketing outcomes.

With the rise of Internet banking, banks increasingly require robust recommendation systems. This case study demonstrates the implementation process of a recommendation system, with a focus on technical execution. By substituting financial product data with online marketing activities, value-added services, or advertisements, an intelligent recommendation system can be developed for these domains. If integrated with the bank's online channels, there is potential for building a comprehensive online ecosystem. Therefore, the recommendation system plays a crucial role in helping banks construct an Internet ecosystem. Readers may use the insights from this chapter to develop their own specialized recommendation systems within their respective fields.

Chapter 4
Assessing the Value of Bank Online Marketing Posts: Reinforcement Learning Techniques

Under the guidance of the "decentralization" business thinking of the Internet, "open banking" and "ecosystem integration" have become frequent keywords. On the one hand, bank products seamlessly integrate into the ecological scenarios of other industries; on the other hand, products from other sectors are consigned to banks for distribution. This includes not only financial products like funds, insurance, and trusts from third-party financial institutions but also rural revitalization and manufacturing products consistent with national policies. The advantages of online sales—such as wide customer reach, 24-h availability, ease of data statistics, and continuous marketing based on customer demand—have led many third-party institutions to opt for online distribution models. As a result, fewer customers visit bank branches, leading more banks to showcase and sell products online.

Selling products online necessitates effective promotional articles. Banks typically hire content advertising companies to write these articles, which are then distributed to target customers through customer service centers, SMS, WeChat public accounts, mini-programs, and apps. Most customers learn about products by reading these articles and then proceed to the transaction page via links provided. Therefore, high-quality promotional articles significantly influence product exposure and sales. Despite this, banks often merely tally the total sales generated from these articles, overlooking the strategic improvements in the marketing process.

This chapter introduces a reinforcement learning-based method for evaluating the value of promotional articles using a single-agent Q-learning algorithm. The method evaluates the value of an article not by examining the content itself, but through feedback from the target customer base. It assesses the expected positive feedback value from customers in different states regarding different articles. This approach offers a novel perspective by focusing on environmental feedback rather than the problem itself, which can be more direct and effective in many scenarios. Implementing this method to identify the best value promotional articles could positively impact transaction volume.

4.1 Introduction to Reinforcement Learning

This section introduces the basic concepts, development background, and design ideas of reinforcement learning techniques.

4.1.1 Development of Artificial Intelligence and Reinforcement Learning

Early AI referred to the technical science of simulating human intelligence, emphasizing "machine learning from people" or "machines simulating human intelligence," focusing on imitation. Human neural networks inherently model external implicit information, also known as "prior knowledge." For instance, when teaching a child to recognize people, the child will know that others are "people" when they see them. Due to the complexity of human neural networks, it is challenging for computers to fully simulate them. Consequently, researchers shifted AI development toward intelligent algorithms better suited to computational characteristics of machines. This shift moved from "biological intelligence" to "algorithmic intelligence." Human neural networks receive stimulation through dendrites, generate excitement, and conduct it to the cell body. The pulse signal then transmits to neighboring neurons via axons. If neighboring neurons also generate excitement, they propagate the signal; if not, they inhibit propagation, as shown in Fig. 4.1.

The neural network in the computer sense receives data and generates forward propagation through stacked neurons and activation function, and backpropagation through loss function, thus adjusting the parameter values of neurons in each layer, as shown in Fig. 4.2. Forward propagation is similar to biological nerve, while back propagation is not possessed by biological nerve. Although the biological brain has a learning mechanism for error feedback, the principle is very different and more complex than that of back propagation. Artificial intelligence technology is inspired by biological mechanisms, but different from biological mechanisms, forming an independent system suitable for computer information extraction, information transmission and information calculation, which is "algorithmic intelligence."

Information technology is evolving from Internet technology to algorithmic intelligence. Internet technology emphasizes process, automation, centralization, and impersonalization, focusing on computing, storage, and information retrieval. Algorithmic intelligence emphasizes intelligence, interaction, decentralization, and personalization, focusing on perception, cognition, and decision-making. The development of algorithmic intelligence has progressed through three stages: symbolism, connectionism, and behaviorism. In the symbolism stage, intelligence was believed to be the expression of knowledge, focusing on symbolic logic, expert systems, and knowledge graphs. This stage was rule-driven with deterministic intelligence. With the advent of big data, the focus shifted to data intelligence, where intelligence is viewed as the expression of data. The emphasis was on extracting valuable

4.1 Introduction to Reinforcement Learning

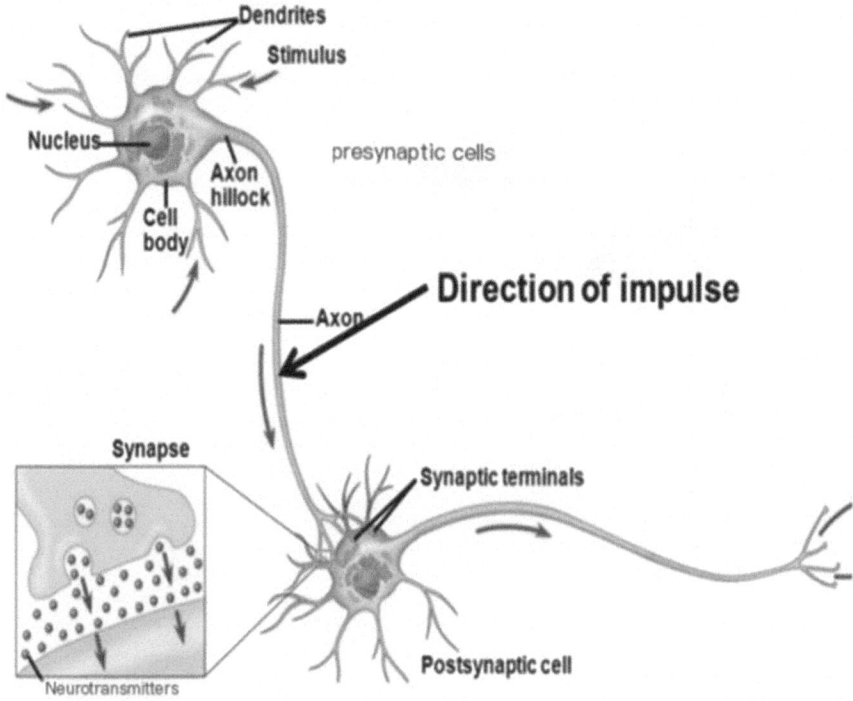

Fig. 4.1 How human neural networks work

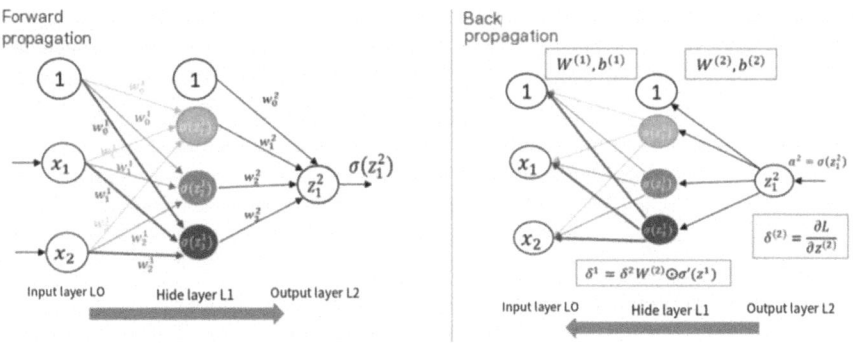

Fig. 4.2 Working mechanism of computer neural network

information through machine learning and deep learning, connecting data to decisions. In this connectionism stage, intelligence became data-driven and uncertain, as data variability directly influenced decisions. In 2016, AlphaGo defeated Lee Sedol, which became a landmark event in the field of AI, and made the world realize the great value of reinforcement learning. And AlphaZero's victory over Ke Jie pushed the self-playing reinforcement learning technique to the fore. Behaviorism

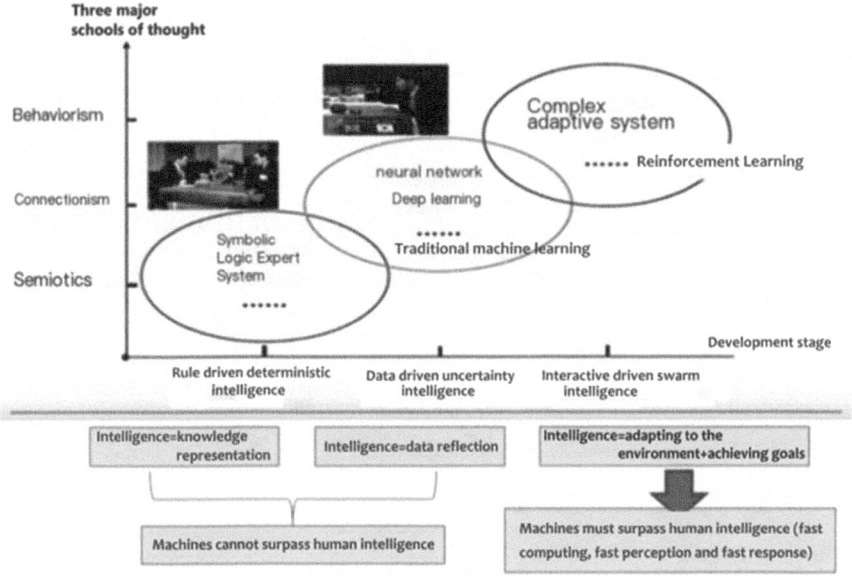

Fig. 4.3 Developmental stages of algorithmic intelligence

represented by reinforcement learning becomes the third stage of the development of algorithmic intelligence, which believes that "intelligence is to adapt to the environment and achieve goals." For example, if a child learns to fall down, he will pick himself up and continue walking. The ability to walk comes from the process of "adapting to a situation and achieving a goal," not from data (books). Later, when people looked at ant colonies, flocks of sheep, and flocks of birds, they found that animals are often more intelligent in groups than they are as individuals, and that their learning ability is derived from adaptation to the environment. Social reinforcement learning based on group intelligence constitutes a complex adaptive system game. In the behaviorism stage, because the computing power of the machine is far superior to that of human beings, it has a stronger perception of the subtle changes in the environment and the agility of its response and decision-making, so the algorithmic intelligence at this stage is likely to surpass human intelligence. In the two stages of symbolic logic and data intelligence, it is human beings that instill knowledge to the machine, so it is difficult for the machine to surpass human beings. The development stage of algorithmic intelligence is shown in Fig. 4.3.

The application of these three stages in banking includes the following. At the symbolic level, knowledge graphs and symbolic rules are primarily used in customer identification, risk management, product recommendation, and similar scenarios. At the connectionism level, neural network-based machine learning is mainly applied in data modeling, image processing, voice processing, and related fields. Business scenarios encompass customer mining, business forecasting, facial recognition, financial review, and intelligent customer service. At the behaviorism level, reinforcement learning-based algorithmic intelligence is primarily used in hedge funds, trading decisions, intelligent investment advisory, and other scenarios.

4.1.2 Basic Concepts of Reinforcement Learning

Reinforcement learning (RL), also known as goal-directed learning, evaluative learning, or associative learning, is used to address the challenge of an intelligent agent maximizing reward objectives through policy learning during interactions with the environment. Reinforcement learning primarily consists of an agent, environment, state, action, reward, policy, cumulative reward, action value function, and state value function.

1. Agent: generally refers to reinforcement learning algorithms. For example, the Q-Learning algorithm in this case.
2. Environment: the data source of the agent, consisting of a set of states. In this case, the target group is used as the environment.
3. State: A data that represents the environment. This case takes the interactive stage of the target customer group's reaction to the post as the state, that is, silence, entering the post page, reading the post (more than 20 s), forwarding or commenting after reading, entering the transaction page from the post link, placing an order on the transaction page and concluding the transaction. There are 6 states in total.
4. Action: the action that the agent can make. The actions in this case are: send six posts numbered 1–6 to the target group, a total of six actions.
5. Reward: the positive/negative feedback signal that the agent receives from the environment after performing an action. In this case, the positive reward is the proportion of the number of people in the target customer group who make a deal.
6. Policy: The one-way mapping from the state to the action becomes the policy, that is, the thinking process of how the agent chooses the action. In this case, the Q matrix is used to represent the policy mapping, that is, the Action corresponding to the maximum value of Q (State,Action) is selected as the policy.
7. Cumulative reward: Considering that the reward of the future step is less valuable than the reward of the current step, so set a attenuation coefficient γ to discount the future reward, the cumulative reward can be written as: $U_t = R_t + \gamma R_{t+1} + \gamma^2 R_{t+2} + \cdots$, where R_t stands for the T-step reward. The cumulative reward in this case is the value of Q (State,Action) in the Q matrix. It is important to note that rewards are immediate, while cumulative rewards are statistical concepts.
8. Action value function: evaluate the value of the action, which is equal to the value of the subsequent state that the action can reach. Its mathematical expression is: $Q_\pi(s_t, a_t) = E[U_t | S_t = s_t, A_t = a_t]$, that is, under the condition of t-step state and

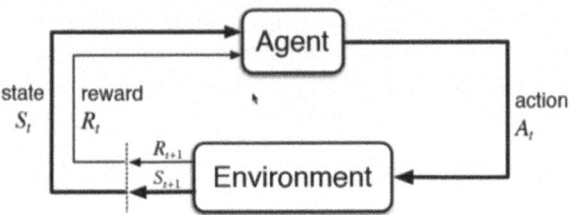

Fig. 4.4 Reinforcement learning process: gain knowledge in action-evaluation (from step t to step $t+1$)

action, the conditional mathematical expectation of cumulative reward. In this case, the action value function is used to guide the updating of the Q matrix.

9. State value function: evaluates the value of a state, which is equal to the value of all actions in that state. Its mathematical expression is: $V_\pi = E_A[Q_t(s_t, A)]$, that is, to find a mathematical expectation of the value of all actions in a specified state.

The reinforcement learning process involves: the agent performs an action, the environment transitions to a new state, and reward signals are received. The agent assesses the target value based on the reward and executes new actions according to strategy. This cycle repeats, aiming to maximize cumulative rewards. If S denotes state, A denotes action, and R denotes reward, the learning process is illustrated by the following sequence data:

$$S_0 \to A_0 \to R_0 \to S_1 \to A_1 \to R_1 \to S_2 \to A_2 \to R_2 \to \ldots\ldots \to S_n$$

Subscripts in the above formula denote specific steps within the iterative process. The agent engages with its environment through a continuous cycle involving states, actions, and rewards. Each action taken by the agent influences subsequent states, actions, and resulting rewards. By employing strategic decision-making, the agent identifies its current state and determines optimal actions to maximize rewards. The reward signal serves as a metric for assessing the effectiveness of the actions taken (typically represented by scalar values). Consequently, reinforcement learning entails acquiring insights from experiential interactions with the environment, continuously refining action strategies to adapt and ultimately achieve the objective of maximizing cumulative rewards, as depicted in Fig. 4.4.

4.1.3 Q-Learning Algorithm

Q-learning, also referred to as Q-Learning, serves as a widely recognized, value-based reinforcement learning algorithm. The fundamental element of this algorithm is the Q matrix, alternatively termed the Q-table or Q Table. This Q matrix comprises column numbers that represent actions and row numbers that denote states, while the values within the matrix are known as Q-values. These Q-values indicate the expected rewards that can be obtained by executing various actions in different states. Expected rewards encompass both immediate rewards and anticipated future

4.1 Introduction to Reinforcement Learning

All possible states / All possible actions	a1	a2	a3
s1	R(s1, a1)	R(s1, a2)	R(s1, a3)
s2	R(s2, a1)	...	...
s3	R(s3, a1)	...	...
s4	R(s4, a1)	...	...

Fig. 4.5 Mapping of the Q matrix

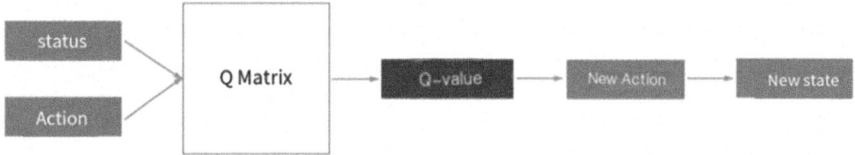

Fig. 4.6 Reasoning process of Q-Learning

Table 4.1 R matrix instant rewards

All possible states / All possible actions	a1	a2	a3
s1	R(s1, a1)	R(s1, a2)	R(s1, a3)
s2	R(s2, a1)	...	...
s3	R(s3, a1)	...	...
s4	R(s4, a1)	...	...

rewards. Due to the fact that the Q matrix correlates states and actions with their respective expected values, we can examine the maximum value within each row to identify the most advantageous action to undertake in a specific state. For a practical illustration, refer to Fig. 4.5, which depicts a basic Q matrix.

Figure 4.6 shows the reasoning process of Q-Learning. Solving the Q matrix is a key step in building a reinforcement learning system.

The premise of solving the Q matrix is to specify the immediate reward for each state and action condition. We use the matrix R to represent this immediate reward, as shown in Table 4.1.

Using both the Q matrix and the R matrix, we can iteratively refine the Q matrix. In scenarios where executing an action transitions to a specific new state s', the Q-Learning algorithm uses the Bellman equation to update the Q matrix: $Q(s,a) = R(s, a) + \gamma(\max(Q(s',a')))$. Where, $Q(s,a)$ represents the value expectation of taking action a in the current state s; s' represents the new state after taking action a; a' represents the action that can be taken in the new state, and $\max(Q(s',a'))$ represents the maximum expected value in the new state, which represents the future value estimation; $R(s,a)$ represents the immediate reward after taking action a in the current state s; γ is the attenuation coefficient, indicating how much to balance the immediate reward with the expected future reward. In many cases, it is possible to jump to several different states after taking an action. In this case, it is only necessary to multiply the transition probability of each new state by the expected value of the corresponding new state when updating $Q(s,a)$. The Bellman equation can be written as:

In practical applications, the transition resulting from an action could lead to multiple potential states. Under such circumstances, updating \($Q(s,a)$\) necessitates adjusting by the transition probability of each possible new state. Each transition probability is multiplied by the expected value of the corresponding state, ensuring an accurate update mechanism. Consequently, the Bellman equation adapts to scenarios with probabilistic state transitions, ensuring comprehensive accounting of potential outcomes in the Q-Learning process.

$$Q_{k+1}^*(s,a) \leftarrow \sum_{s'}(s'|s,a)\left(R(s,a,s') + \gamma \max_{a'} Q_k^*(s',a')\right)$$

where $P(s'|s,a)$ represents the conditional probability of jumping to the new state s' after taking action a on the current state s, while the content in the right parentheses represents the expected value of moving to the new state.

4.2 Case Practice

In this case study, we consider the various stages of marketing for the same product as states, sending different tweets to customers as strategies, the target audience as the environment, and the customer's feedback (purchase or hesitation) as rewards. The algorithm concurrently adapts to the environment (learning from audience feedback) and achieves its goal (maximizing cumulative rewards), thereby determining the optimal tweet strategy for different states. There are six types of tweets in this study (although other quantities could also be used), all promoting the same product. Because we need to track various customer feedback, it is essential to implement data tagging on the tweet pages in advance to perform statistical analysis of customer behaviors across multiple entry points. We acquire a batch of the target audience who have participated in previous marketing activities and gather their responses to previous tweets (not the six tweets in this case), which include six states: "silent, enter post page, read post (for 20 s or more), share or comment after reading, enter transaction page from post link, place an order and complete the transaction." The Q matrix is defined as given in Table 4.2.

In reality, it is quite common for certain products to enjoy prolonged promotional periods within banking institutions. For instance, financial products that provide real-time fund redemption, various types of fund insurance products, and similar offerings often experience extended marketing cycles. Identifying target customer groups across different states becomes straightforward during these cycles. To

4.2 Case Practice

Table 4.2 Q matrix in this case

All possible state / All possible actions	Send tweet number 1	Send tweet number 2	Send tweet number 3	Send tweet number 4	Send tweet number 5	Send tweet number 6
Silent	$Q(s1,a1)$	$Q(s1,a2)$	$Q(s1,a3)$	$Q(s1,a4)$	$Q(s1,a5)$	$Q(s1,a6)$
Enter the tweet page	$Q(s2,a1)$	...	...	...	...	
Reading tweets (over 20 s)	$Q(s3,a1)$	...	...	...	...	
Forward or comment after reading	$Q(s4,a1)$		...			
Enter the transaction page	$Q(s5,a1)$					
Place an order on the	$Q(s6,a1)$		...	...	...	

analyze these customer groups, we undertake a process of random sampling to create six distinct experimental customer groups, ensuring uniformity in group size and equal representation of individuals across states. We then distribute a unique post to each group: group 1 receives post 1, group 2 receives post 2, continuing in this manner for all groups. Given the time lag in customer feedback, we monitor marketing outcomes over a defined observation period to gather insights. Specifically, we assess how many previously inactive customers in each experimental group have purchased the product versus those who remain inactive. Additionally, we examine how many customers, who engaged with the posts, proceed to make purchases compared to those who stay inactive. This methodology allows us to calculate the transaction proportions for each state within each experimental group. These transaction proportions serve as immediate rewards, which we then utilize to construct the reward matrix, referred to as the R matrix. The R matrix, as calculated by the author, is detailed in Table 4.3.

Our actions targeting a customer in one specific state may lead him to transition across multiple other states. For instance, after engaging with the posted content, the customer might exhibit various behaviors: remain unresponsive, initiate a direct transaction, share the post without engaging in a transaction, or visit the transaction page without completing a purchase, which represents a period of indecision. Consequently, for each value within the Q matrix, we delineate a state transition vector encompassing six probability values, each corresponding to the likelihood of transitioning among six defined states. We implement divergent actions for each experimental customer cohort and, upon the conclusion of the marketing observation period, compute the state transition probabilities for each group. This analysis culminates in the formation of a state transition matrix T, structured as a $6 \times 6 \times 6$ array. The T matrix values are derived from the actual customer behavior feedback received during our marketing activities. An illustrative example is provided in Table 4.4.

Table 4.3 The R matrix of this case

All possible actions\\All possible states	Send tweet number 1	Send tweet number 2	Send tweet number 3	Send tweet number 4	Send tweet number 5	Send tweet number 6
silent	0.01	0.01	0.02	0.05	0.03	0.01
Enter the tweet page	0.12	0.03	0.02	0.01	0.02	0.02
reading tweets(over 20 seconds)	0.08	0.02	0.03	0.07	0.09	0.01
Forward or comment after reading	0.06	0.03	0.07	0.02	0.04	0.01
Enter the transaction page	0.01	0	0.02	0.01	0	0
Place an order on the	0.03	0.01	0	0	0.05	0

The next step is code writing. We use Python to implement the case natively, first import the relevant library package:

```
import numpy as np
import random
```

In the initial state, the Q matrix is an all-zero matrix of 6x6:

```
Q = np.zeros((6, 6))
Q = np.matrix(Q)
```

Based on the aforementioned statistics, define the immediate reward matrix R and the state transition matrix:

```
R = np.matrix([[0.01,0.01,0.02,0.05,0.03,0.01],
[0.12,0.03,0.02,0.01,0.02,0.02],[0.08,0.02,0.03,0.07,0.09,0.01],
[0.06,0.03,0.07,0.02,0.04,0.01],[0.01,0,0.02,0.01,0,0],
[0.03,0.01,0,0,0.05,0]])
T = np.array([[[0.8,0.05,0.03,0.08,0.02,0.02],
[0.7,0.03,0.05,0.03,0.04,0.15],[0.75,0.05,0.01,0.01,0.1,0.06],
[0.92,0.02,0.03,0.01,0.01,0.01],[0.82,0.03,0.05,0.07,0.01,0.02],
[0.88,0.03,0.01,0.02,0.03,0.03]],
 [[0,0.77,0.06,0.05,0.08,0.04],[0,0.9,0.03,0.02,0.03,0.02],
[0,0.85,0.03,0.07,0.3,0.02],[0,0.88,0.02,0.04,0.03,0.03],
[0,0.93,0.03,0.02,0.01,0.01],[0,0.87,0.06,0.03,0.03,0.01]],
 [[0,0,0.73,0.08,0.09,0.1],[0,0,0.78,0.04,0.07,0.11],
[0,0,0.82,0.06,0.5,0.07],[0,0,0.76,0.1,0.07,0.07],
[0,0,0.85,0.07,0.06,0.02],[0,0,0.92,0.05,0.02,0.01]],
 [[0,0,0,0.84,0.06,0.1],[0,0,0,0.89,0.03,0.08],
```

4.2 Case Practice

Table 4.4 State transition matrix T of this case

All possible states / All possible actions	Send tweet number 1	Send tweet number 2	Send tweet number 3	Send tweet number 4	Send tweet number 5	Send tweet number 6
silent	[t111,t112.t113.t114,t115,t116]	[t121.t122,t122,t124,t125,t126]	[t131,t132,t133.t134,t135,t136]	[t141.t142,t143,t144.t145.t146]	[t151,t152,t153,t154,t135,t156]	[t161,tL162,t163,t164,t165.t166]
Enter the tweet page	[:211.t212.t213.t214.+215,t216]	...	...	...	...	...
Reading tweets(over 20 seconds)	[+311.t312.t313.t314.+315.t316]	...	...	...	...	...
Forward or comment after reading	[t411.t412.t413.t414.t415,t416]	...	...	...		
Enter the transaction page from the tweet link	[+511.t512.t513.t514.t515,t516]	...			...	...
Place an order and close the transaction on the trading page	[+611.t612.t613.t614.+615,t616]	...			...	...

[0,0,0,0.72,0.16,0.12],[0,0,0,0.64,0.16,0.2],
[0,0,0,0.7,0.13,0.17],[0,0,0,0.8,0.09,0.11]],
 [[0,0,0,0,0.77,0.23],[0,0,0,0,0.86,0.14],[0,0,0,0,0.78,0.22],
[0,0,0,0,0.91,0.09],[0,0,0,0,0.81,0.19],[0,0,0,0,0.92,0.08]],
 [[0,0,0,0,0,1],[0,0,0,0,0,1],[0,0,0,0,0,1],[0,0,0,0,0,1],
[0,0,0,0,0,1],[0,0,0,0,0,1]]])

The attenuation parameter determines the extent to which future value estimates are adopted. This case is set to 0.8:

$$\gamma = 0.8$$

Here comes the iteration of the Q matrix, that is, for every action in every state, update its value according to the Bellman equation:

```
[[0. 23337923  0. 22846     0. 23029909  0. 27019923  0. 25305268  0. 22807604]
 [0. 43603222  0. 36520735  0. 38545534  0. 34166359  0. 36094872  0. 35373695]
 [0. 34410237  0. 28932316  0. 36599197  0. 33855801  0. 36707146  0. 29581132]
 [0. 27168206  0. 24452027  0. 2730831   0. 22160578  0. 24462866  0. 2190285 ]
 [0. 16085106  0. 14510638  0. 17021277  0. 15191489  0. 14829787  0. 1412766 ]
 [0. 23        0. 21        0. 2         0. 2         0. 25        0. 2       ]]
```

Fig. 4.7 Convergent Q matrix obtained in this case

```
for j in range(1000): # Iterate 1000 rounds
  for state in range(6): # iterate through 6 states
    r_pos_action = [0,1,2,3,4,5] #6 actions
    # Select one of the 6 actions at random to iterate over
    action = r_pos_action[random.randint(0, len(r_pos_action) - 1)]
    # Gets the state transition vector for the corresponding location
    probability = T[state, action]
    p_Q = 0
    # Calculate future value expectations
    for i in range(len(probability)):
      p_Q += probability[i] * γ * (Q[i]).max()
    # Update the Q matrix by the Bellman equation
    Q[state, action] = R[state, action] + p_Q
```

End up with the updated Q matrix:

```
print(Q)
```

It is worth noting that when the number of iterations reaches a certain level, the Q matrix will converge to a fixed value, and the value of the Q matrix cannot be changed by continuing iterations. In this case, after 1000 iterations, the Q matrix has converged, and the result is shown in Fig. 4.7.

At this point, we query the Q matrix to get the best post. For example, the best post for the silent customer is Post number 4 (0.27019923), and the best post for the customer who entered the post page is Post number 1 (0.43603222). We can push the best post to customers in different states first, and thus get a higher transaction rate. After sending the post, if the customer status has migrated and there is no transaction, we send the best value post in the new status; If the status of the customer has not changed, we can send the posts in order of the Q matrix value from high to low (within the range of disturbance tolerated by the customer), and wait for a marketing observation period to obtain new feedback results to update the R and T matrix, and then recalculate the Q matrix. We iterate and learn from reality until the Q matrix stabilizes. See the download QL.py for the code.

4.3 Case Summary

In this case, an AB test was carried out in actual work. Users in various states were randomly divided into equal numbers of group A and Group B. Group A adopted reinforcement learning marketing strategy and group B sent marketing posts at random. As A result, the actual turnover of group A was 30% higher than that of group B, which was a considerable improvement.

The case study applies reinforcement learning techniques to demonstrate a method for evaluating the value of tweets in the context of online marketing of financial products, thereby promoting increased transaction volume. Given the frequent and similar marketing efforts for financial products, this method is particularly suitable. The book primarily focuses on discussions regarding the applications of artificial intelligence in the banking sector, rather than delving deeply into algorithmic details. The six customer states and six marketing actions presented in this chapter serve merely as examples, allowing readers to design according to their specific needs. This chapter aims to spark interest, introducing Q-Learning, which is a relatively simple reinforcement learning algorithm limited to finite states and finite actions. Readers are encouraged to experiment with deep Q networks (DQNs) to handle more complex scenarios. Nonetheless, this case illustrates a crucial point: once artificial intelligence algorithms penetrate the operational details of a specific industry, they can significantly improve those details. In this instance, we established evaluation standards for bank online marketing tweets by learning from the feedback information of the target customer group. This marks an attempt to improve the traditionally lacking digital content marketing evaluation methods in banks, leveraging the reinforcement learning concept of "evaluating by engaging with the evaluator," which is a direct and impactful approach.

Reinforcement learning is a method of sequential data analysis that incorporates elements of game theory, particularly the interaction between an intelligent agent and its environment. Banks, during their operational processes, generate extensive sequential game data, thereby making reinforcement learning widely applicable in banking scenarios. For instance, within the realm of intelligent financial advisory, it facilitates decision-making on the buying and selling of stocks, funds, and bonds. It also optimizes pricing strategies for loans and deposits from the perspective of financial market trends and customer needs, thereby maximizing bank profits. Currently, reinforcement learning not only finds wide application in fields like gaming, robotics control, computer vision, natural language processing, and recommendation systems but is also poised to further integrate into the daily operations across various industries, creating substantial additional value.

Chapter 5
Modeling Binary Causal Effects of Related Repayments: Causal Inference Techniques

The long-tail customers in retail banking represent a vast demographic with a relatively low individual value. This segment, however, holds considerable potential and can be a significant asset. Nevertheless, leveraging this potential necessitates the use of technological advancements rather than relying solely on manual efforts. Enhancing long-tail customer engagement translates into increasing the reliance of the general populace on a banking institution, decreasing customer attrition, and bolstering assets under management (AUM) along with the number of monthly active users (MAU). The term "dual-card customers" refers to retail clients who possess both a credit card and a debit card. Typically, banks encourage valuable customers to link their credit and debit cards, allowing the system to automatically deduct repayment amounts from the debit card's current account on the credit card's due date. The business rationale for this practice is multifaceted: it requires customers to maintain sufficient funds in their debit card current accounts, thereby increasing deposits and the usage rate of debit cards. It also consolidates idle funds within the designated bank, simultaneously enhancing AUM and MAU while mitigating the risk of inadvertent credit card payment delays. Moreover, the linkage of credit and debit cards fosters greater customer loyalty. For customers, failing to link the cards necessitates either transferring repayment funds from another bank or using services like Alipay or WeChat, both of which involve additional transaction fees and a higher likelihood of incurring overdue interest due to missed payments, making this approach less advantageous.

This case seeks to precisely identify the most valuable marketing customer lists through specific technical means. In conventional machine learning, the typical approach for such predictive tasks involves creating a supervised learning model. This model maps customer feature sets to a pre-prepared set of Y-value labels, and then uses this model to predict outcomes for the test set. This data-fitting method maximizes the ability to find high-value customer groups within unbound populations that exhibit characteristics of bound customers. However, it fails to address the extent of the marketing's impact on the bank's overall value contribution. Specifically, precision marketing needs to account for individual heterogeneity, meaning that even for similar customers receiving the same marketing action, the actual marketing effect varies. This variation is similar to the differing efficacy of a single drug on various patients based on factors like genetic makeup and mental state. Therefore, traditional machine learning methods that identify potentially marketable customer groups fall short as they do not account for the individual heterogeneity of marketing effects. In the realm of precision marketing, further steps are required to predict the marketing effect on each unique customer.

Causal inference is a breakthrough technology that addresses the issue of marketing individual heterogeneity. It draws causal links based on specific conditions occurring at particular times. Using mathematical, statistical, and probabilistic methods, it analyzes how changes in causal variables influence the outcome variable. It is a branch of the science of complex systems causality. The author has applied this technique to data modeling in scenarios involving dual-card customer binding and repayment associations, achieving successful implementation results. Given that causal inference is part of the broader field of causal science, a brief introduction to the background of causal science is necessary before delving into the intricacies of causal inference.

5.1 Introduction to Causal Science

Let us first examine the attitudes of statistical luminaries toward the concept of causality. Galton stated, "There is no causality in the scientific world," while Fisher advocated that "statistics is the reduction of data." Russell criticized causality as an "outdated philosophical concept." Evidently, since the inception of statistics, the concept of causality has been continuously dismissed. This is because scientists believe that scientific research subjects must be objectively existing entities, whereas the concept of causality is not; it is merely a construct within the human brain and, therefore, not a subject of scientific inquiry. Consequently, causal science research has long been neglected.

Next, let us review the current state of artificial intelligence (AI). Machine learning and deep learning primarily involve using function curves to fit data, which remains at the level of discovering correlations between data rather than understanding it. To truly comprehend the data, algorithms must uncover the underlying causes behind the data. Simply put, algorithms must be capable of deciphering both "what" the data phenomena are and "why" they occur. Although causality is a construct of the human brain, it serves as a perspective and tool for understanding the world. We can deduce causal variables, outcome variables, and their strength of association through mathematical expressions from observational data, thus providing highly promising new methods for AI modeling in the era of big data. Moreover, the significant change that causal science offers to AI is interpretability. We know that neural networks are a crucial AI technology, but one major flaw is their poor interpretability. For instance, the meaning each neuron represents is unstable, and there is no clear business explanation for network structures and

5.1 Introduction to Causal Science

hyperparameters. The consequence of poor interpretability is that a well-performing model might suddenly fail in some scenarios, or changing the value of a single input feature could lead to entirely different model predictions. Models with poor interpretability exhibit uncertainty and instability. By integrating causal science with neural networks, we can embed causal explanations of events into the model, thereby enhancing the model's interpretability and increasing its certainty and stability. This is the profound significance of causal science.

Causal science represents an interdisciplinary field bridging statistics, probability theory, advanced mathematics, machine learning, sociology, and psychology. It encompasses areas such as causal structure learning, causal representation learning, and causal emergence. Causal structure learning derives causal variables and their effects from observational data, creating a structural diagram of the causal system, known as a causal graph—this process is also referred to as causal discovery. Causal representation learning combines causal reasoning with machine learning techniques to infer the strength of causal relationships and hypothesize about causal variables. Causal emergence integrates the concept of causality with complex systems, exploring causal connections at micro-, meso-, and macro-levels, delineating the links and distinctions between macroscopic systems and microscopic states. Causal inference, dedicated to identifying causal relationships between variables and using these relationships for reasoning, includes both causal structure learning and causal representation learning. Causal effect estimation, a subset of causal inference, models the intensity of causal associations between variables, commonly referred to as causal effects. Recently, causal science has merged with traditional artificial intelligence technologies, leading to new domains such as causal reinforcement learning, causal machine vision, causal natural language understanding, causal transfer learning, and emergent causal behaviors within complex systems. This case exemplifies the practical implementation of causal effect estimation within the field of causal inference.

Judea Pearl, renowned recipient of the Turing Award, father of Bayesian networks, National Academy of Sciences member, and author of pivotal works like "Causality: Models, Reasoning, and Inference" and "The Book of Why," posits that the first scientific revolution anchored around data has paved the way for the era of big data. In contrast, the second revolution centers around causal science, guiding data toward policy formulation, explanation, and generalized mechanisms. Pearl asserts that humanity's dominance on Earth hinges on the ability to ask "why" and comprehend underlying causes, which is fundamental to establishing accurate epistemology and worldview. This capacity differentiates humans from machines. If machines could autonomously determine causes, human supremacy would diminish. Thus, causal science marks a significant pivot from AI 1.0 to AI 2.0. Judea Pearl contends that the world embodies uncertainty and causal relationships, and mathematical models that reflect real-world entities most accurately hold promising potential. Consequently, probabilistic methods capable of handling uncertainty and performing causal reasoning will steer the future trajectory of AI development. Pearl's scholarly work "The Book of Why" has redefined causal science, elevating it to a burgeoning field of scientific inquiry.

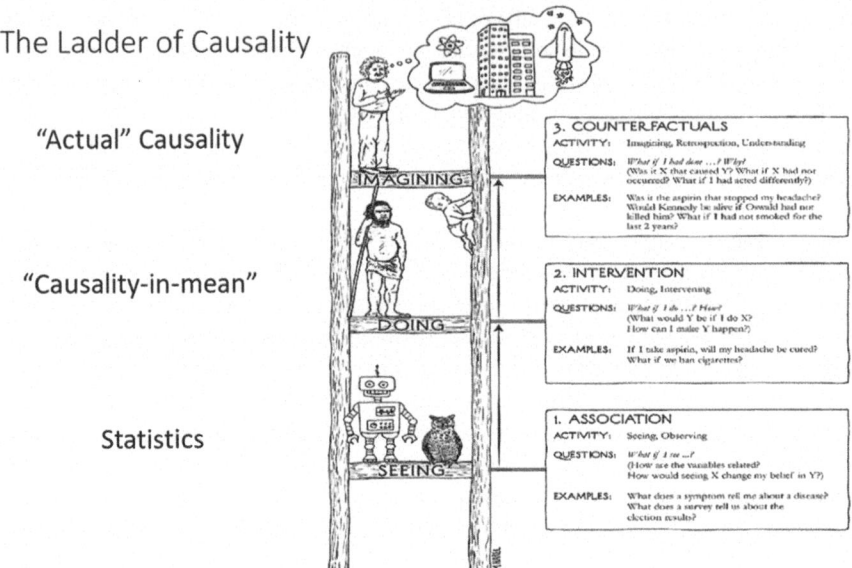

Fig. 5.1 The three-tier causal ladder

In "The Book of Why," Pearl introduces a theoretical framework called the "ladder of causation" comprised of three tiers, serving as the foundational structure for the entire causal science discipline, as illustrated in Fig. 5.1.

The Ladder of Causation comprises three distinct levels: the Association Level, the Intervention Level, and the Counterfactual Level. At the first level, known as the Association Level (Seeing), the focus is on analyzing correlations within observational data. Techniques such as machine learning and deep learning operate at this level, which Judea Pearl critiques as "curve fitting." In mathematical terms, this approach involves learning a function to model the conditional probability distribution, $p(y|x)$, which describes the probability distribution of y given the observed variable x. The second level, referred to as the Intervention Level (Doing), is concerned with understanding causation by analyzing the effects of interventions. The fundamental approach involves perturbing a potential cause variable while holding other conditions constant, to compare the probability distributions of outcomes with and without the intervention. Pearl introduced the concept of do-calculus, denoted as $p(y|do(x))$, to represent the distribution of y under an intervention where x is manipulated. For example, $p(Sunrise|do(RoosterCrows))$ illustrates the probability of sunrise given the artificial intervention of a rooster crowing. This probability is evidently different from $p(Sunrise|RoosterCrows)$, which is simply the observed correlation between sunrise and rooster crowing. This distinction reveals that rooster crowing is not a cause of sunrise. Conversely, taking medication could be identified as a cause for the amelioration of an illness. In the second level, reasoning transitions from cause to effect, exemplified by causal

Directed Acyclic Graph (DAG) models. This particular case falls within the second level. The third level, known as the Counterfactual Level (Imagining), involves hypothetical reasoning–speculating what would have occurred if a certain action had not been taken. If A results in B and the absence of A results in the absence of B, then A can be considered the cause of B. This level of reasoning involves deducing causes from observed effects, a backward transmission of inference, with Structural Causal Models standing as the representative method. The Ladder of Causation's tripartite structure has illuminated pathways for the advancement of artificial intelligence, marking the emergence of causal science as a comprehensive theoretical and methodological framework encompassing current AI developments, and signifying a landmark achievement in the field.

5.2 Causal Forest Algorithm

To achieve precise marketing, we need to understand the impact of marketing actions on individual returns, rather than predicting the overall returns for the marketing customer group. This involves quantifying individual heterogeneous treatment effects. Beyond marketing, personalized treatment plans are critical in healthcare. Even if patients have similar medical histories, ages, and immune levels, they can respond differently to the same treatment. Hence, doctors must evaluate the consequences of treatments on different individuals before taking action. This is a differential assessment of individual outcomes, not an estimate for the patient group as a whole.

In the traditional machine learning context, models typically fit the probability of events given certain conditions. Let X, y, t represent customer characteristics, marketing returns, and the presence or absence of marketing actions, respectively. The objective function in traditional machine learning is $P(\text{return } y \mid \text{features } X, \text{action } t)$, where the vertical bar reads as "seen," indicating data observed in natural conditions. In causal learning, models fit the probability of events under intervention states. Interventions are not natural states but imposed conditions. For instance, forcibly applying a marketing action to a customer during the natural marketing process involves calculating returns as $P(\text{return } y \mid \text{features } X, do(\text{action } t))$. The "do" operator indicates forced intervention, representing the effect of perturbing data for variable t. Traditional machine learning focuses on the association between random variables, while causal learning calculates the gain from interventions, assessing causal effects. In this case study, only one type of intervention exists, producing "marketing" and "non-marketing" outcomes, forming a "binary causal effect model." Multiple interventions yield various outcomes, known as a "multiple causal effect model."

Based on marketing effectiveness, customers are categorized into four quadrants: "marketing-sensitive," "natural converters," "indifferent," and "adverse responders," as shown in Fig. 5.2.

Fig. 5.2 Marketing effectiveness group classification

The problem we want to solve is: how to target marketing resources to sensitive groups as much as possible (i.e., give gifts when completing debit card binding credit card repayment), so as to avoid wasting resources? If each customer sample is given marketing resources and their contribution to the bank's earnings is $y1$, but no marketing resources are given and their contribution to the bank's earnings is $y0$, then $y1$–$y0$ is the causal effect of marketing, or Uplift Score. Obviously, the Uplift Score for the market-sensitive population is positive, the Uplift Score for the converted and indifferent population approaches 0, and the Uplift Score for the reactive population is negative. Now here's the problem: it is impossible for a customer to observe both $y1$ and $y0$ at the same time, because marketing and non-marketing are mutually exclusive states. Uplift Score calculations require both $y1$ and $y0$ for each customer in an observational dataset. At this point, the algorithm needs to solve the problem of approximating the unobservable value as closely as possible for each customer sample.

Classification and Regression Tree is a common decision tree algorithm, referred to as CART. The algorithm can be used for both classification and regression. Where the output of the classification tree is the class of the sample, and the output of the regression tree is continuous real numbers. Figure 5.3 shows the general process of the algorithm: traversing all features and segmentation points of the data set, feature splitting in turn, after reaching the preset conditions, all samples are assigned to the leaf node, while minimizing the mean square error MSE, its calculation formula is: $MSE = \frac{1}{N}\sum_{j}\sum_{i}(y_i - c_j)^2$.

Causal tree is an algorithm that leverages classification and regression trees (CART) to compute the Uplift Score. It inputs the dataset into a CART tree for feature splitting and minimizes the mean square error, aiming to cluster similar samples into the same leaf node. Its advantages include immunity to the dimensionality curse, suitability for multi-covariate causal prediction, and strong interpretability. Proposed by Athey and Imbens in 2015, the algorithm posits that the marketing

5.2 Causal Forest Algorithm

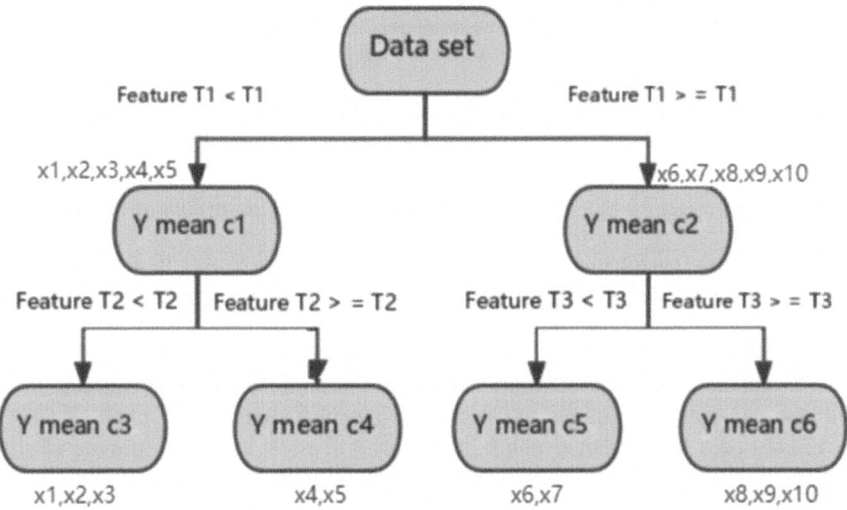

Fig. 5.3 Schematic diagram of CART algorithm for classification regression tree

gain for two similar customers is also similar. Therefore, one can approximate the marketing gain of each customer using the outcomes of their similar counterparts. The approach involves building a classification regression tree to identify customer groups with analogous feature vectors (X). Within each group, $y1$ and $y0$ can be observed from certain samples. The difference between the mean $y1$ and mean $y0$ in each group serves as the Uplift Score for that group and for each customer within it.

Wager and Athey proposed the Causal Forest algorithm. It integrates random forest Bagging with causal decision trees. Essentially, it involves random sampling in both sample and feature dimensions to build B causal trees and determine the Uplift Score of the predictions. Bagging these predicted results provides the final Uplift Score. In detail, the algorithm uses a training set from sampling without replacement to construct B decision trees. It trains these trees using feature variables X and outcome variables y based on recursive partitioning. The process involves top-down feature splitting from the root node, aiming to maximize heterogeneity. This ensures that the causal effects within the same node are as consistent as possible, while differences between nodes are maximized. During splitting, the causal effect stability is preserved. This means minimizing the MSE loss of sample causal effects and the mean square error within each leaf node group until every sample lands on a leaf node. The individual causal effect for a sample at a leaf node is then determined.

$$\hat{\tau}_i(x) = \frac{1}{|i: W_i = 1, X_i \in L|} \sum_{|i:W_i=1, X_i \in L|} Y_i$$
$$- \frac{1}{|i: W_i = 0, X_i \in L|} \sum_{|i:W_i=0, X_i \in L|} Y_i(i \in E)$$

where $\frac{1}{|i:W_i=1, X_i \in L|} \sum_{|i:W_i=1, X_i \in L|} Y_i$ represents the mean of the revenue observation y1 of all marketing customers of the leaf node (i.e., the sample that implemented the marketing action), $\frac{1}{|i:W_i=0, X_i \in L|} \sum_{|i:W_i=0, X_i \in L|} Y_i(i \in E)$ represents the mean of the revenue observation y0 of all unmarketed customers of the leaf node. It is important to note that customers without marketing do not mean that there is no revenue, they will naturally grow or lose naturally, and this revenue is observable.

Finally take the predicted mean of B causal trees as the final estimate of the individual causal effect: $\hat{\tau}_i(x) = \frac{1}{B} \sum_{b=1}^{B} \hat{\tau}_{i,b}(x)$.

5.3 Developing the Library

This case study represents a typical Uplift Modeling scenario. Pylift is a Python library package for constructing uplift models. Building upon Sklearn, Pylift has implemented a series of optimizations specifically for uplift modeling, along with an integrated set of evaluation metrics. Therefore, Pylift's core remains Sklearn, with an additional API wrapper for user-friendly operation. The code repository for Pylift is hosted at https://github.com/wayfair/pylift, following the BSD-2-Clause open-source license. In this case study, Pylift library is utilized to calculate evaluation metrics for the causal effects of marketing actions generated by the model.

Cforest is a Python algorithm wrapper for the causal forest, which can be used to estimate the heterogeneity of treatment effects in a potential outcomes framework. The algorithms implemented in CForest largely replicate the causal tree and causal forest algorithms proposed by Athey and Imbens (2016), and Athey and Wager (2019). The CForest library provides the CausalForest class to implement all algorithms of the causal forest. For similar functionality in R, refer to the grf library (https://github.com/grf-labs/grf). The code repository for Cforest is https://github.com/timmens/causal-forest, released under the MIT license. This case utilizes this library to build a causal forest model.

5.4 Case Practice

This section describes the project development process from the aspects of data preparation, development and operation environment construction, code writing and model evaluation.

5.4.1 Data Preparation

The goal of this case study is to identify credit card customers with a positive Uplift Score for debit card automatic repayment marketing and create a marketing list ranked by Uplift Score. Therefore, when selecting customer characteristics, focus on business explanatory variables strongly related to marketing. Customers with both credit and debit cards who are not enrolled in automatic repayment must manually repay before the due date, which may lead to missed payments and short-term overdue. If linked to WeChat Pay, they incur repayment fees proportional to the amount. Forgetting to repay or incurring fees burdens the customer, making automatic repayment marketing likely to succeed for such customers. Additionally, customers with higher credit card usage and stronger ties to the bank's financial products are more likely to respond positively to marketing. Basic customer characteristics also impact marketing success. Thus, the following data labels as of the marketing date are extracted as the customer feature matrix X for the model:

1. Credit card usage intention: the number of credit card transactions in the last 6 months, and the maximum single overdraft amount of credit card in the last 6 months.
2. Repayment fees: the amount of WeChat repayment in the last 6 months, the amount of Alipay repayment in the last 6 months, the number of WeChat repayment in the last 6 months, the number of Alipay repayment in the last 6 months.
3. Viscosity with bank products: the number of products held, the number of electronic payment transactions in the last 6 months, the number of debit card transactions in the last 6 months, the number of days of credit card holding, the nine assets of debit card, the number of days since the most recent login of mobile bank, the amount of debit card transactions in the last 6 months, the amount of electronic payment transactions in the last 6 months.
4. Customer base characteristics: credit card credit line, gender, gender, age.

After a marketing date, use the asset rating of bound repayment customers to gauge marketing income. This income equals the product of three variables: $y =$ Binding automatic repayment within one observation period post-marketing date * Binding automatic repayment to our bank * Customer asset rating on the observation date. Here, the value for "whether to bind automatic repayment to our bank" is 0 if not bound, 1 if bound to our bank, and −1 if bound to other banks, indicating customer loss.

Extract the list of customers who carried out marketing actions to customers on the marketing day as the intervention variable t, with the value of 1 for implemented marketing and 0 for non-implemented marketing.

The task to be completed by the model is to calculate the personalized marketing revenue for each customer:

$$y1 = P(\text{revenue } y|\text{ feature } X, do(\text{marketing } t = \text{True}))$$

$$y0 = P(\text{revenue } y|\text{ feature } X, do(\text{marketing } t = \text{False}))$$

$$\text{Uplift Score} = y1 - y0$$

5.4.2 Environment Setup

In the conda environment, run the conda install-c timmens cforest command to install the causal forest algorithm package.

On the CLI, run pip install pylift to install the Uplift model computation package.

Run pip install matplotlib pandas numpy to install the dependency package.

5.4.3 Code Practices

First import the package
  from cforest.forest import CausalForest
  from pylift import TransformedOutcome

Next, load the prepared data. It is important to note that the range of values of the feature values is not indicative of the importance of the feature, so we need to scale all feature columns to the range [0,1]. Use the numpy array to represent X, y, t.

```
df = pd.read_csv('./data/data.csv', encoding='gb2312')
df = df[(df['0424 Whether to bind bank card automatic repayment (0 no 1 our bank -1 other line) ']==0) | df[' Whether to bind automatic repayment after 4.1 ']==1]
df = df.reset_index()
print(df)
df2 = pd.DataFrame()
df2[' Number of credit card transactions in the last 6 months '] = df[' Number of credit card transactions in the last 6 months ']*100/max(df[' Number of credit card transactions in the last 6 months '])
```

5.4 Case Practice

```
df2[' Maximum single overdraft amount in the last 6 months '] = df[' Maximum single overdraft amount in the last 6 months ']*100/max(df[' Maximum single overdraft amount in the last 6 months '])
df2[' Repayment fee '] = (df[' most recent 6 months WeChat repayment amount ']+df[' most recent 6 months Alipay repayment amount '])*100/max(df[' Most recent 6 months WeChat repayment amount ']+df[' Most recent 6 months Alipay repayment amount '])
# df2 [' overdue number 1-29] = df [' overdue number 1-29] * 100 / Max (df [' overdue number 1-29])
df2['0424 products held '] = df['0424 products held ']*100/max(df['0424 products held '])
df2[' Number of e-payment transactions in the last 6 months '] = df[' Number of e-payment transactions in the last 6 months ']*100/max(df[' Number of e-payment transactions in the last 6 months '])
df2[' Number of debit card transactions in the last 6 months '] = df[' Number of debit card transactions in the last 6 months ']*100/max(df[' Number of debit card transactions in the last 6 months '])
#df2[' age '] = df[' age ']*100/max(df[' age '])
df2 [' credit card holding days'] = df [' credit card holding days'] * 100 / Max (df) [' credit card holding days']
df2[' Debit card 9 days '] = df[' debit card 9 days ']*100/max(df[' debit card 9 days '])
df2[' academic qualification '] = df[' academic qualification ']*100/max(df[' academic qualification '])
df2[' credit line '] = df[' credit line ']*100/max(df[' credit line '])
df2[' gender '] = df[' gender ']*100/max(df[' gender '])
df2[' Days from last login of mobile banking '] = df[' days from last login of mobile banking ']*100/max(df[' days from last login of mobile banking '])
df2[' Debit card transactions in last 6 months '] = df[' Debit card transactions in last 6 months ']*100/max(df[' Debit card transactions in last 6 months '])
df2[' Last 6 months electronic payment transaction amount '] = df[' Last 6 months electronic payment transaction amount ']*100/max(df[' Last 6 months electronic payment transaction amount '])
X = df2.values
df['y'] = df[' Whether to bind automatic repayment after 4.1 '] * df['0424 Whether to bind automatic repayment of bank card (0 no 1 our bank -1 other bank) '] * df['0424 Customer asset rating ']
y = df['y'].values
t = (df[' Customer marketing authorization '].values==1)
```

We then use the CausalForest class encapsulated by cforest to construct a causal forest with ten causal fruit trees. Of course, more causaltrees can be specified to fit the data, depending on the machine's computing power and project needs. The use of the CausalForest class is very simple, you just need to define the parameters of the causal forest.

```
cfparams = {
  'num_trees': 10, #10 cause trees
  'split_ratio': 1, # feature splitting strategy
  'num_workers': 10, # number of concurrent cores
  'min_leaf': 5, # minimum number of leaves
  'max_depth': 10, # maximum tree depth
  'seed_counter': 1,
}
cf = CausalForest(**cfparams) # Builds a causal forest with specified parameters
```

Next, we call the fit method of the CausalForest class to pass the data to the model. The algorithm will automatically train the causal forest and decide the feature splitting strategy of each fruit tree, place the appropriate samples into the appropriate leaf nodes, and complete the data-fitting work.

```
cf = cf.fit(X, t, y) # Fits the model with the observed data
```

Next, use the trained model to make predictions on the test set to get the Uplift Score for each client, saved as a result column.

```
result = cf.predict(X_test, num_workers=10) #10 concurrent predictions
df_result = pd.DataFrame({'result':result})
df_result['custid'] = df['customer number']
```

We sorted the Uplift Score from highest to lowest to find sections of customers greater than zero that can be marketed on this list.

```
df_result = df_result.sort_values(by='result', ascending=False)
df_result.to_csv('predict.csv', index=False)
```

Cause and effect 3D visualizations are drawn using the plot_predicted_treatment_effect function, which plots the degree to which a pair of X feature columns are associated with the Uplift Score.

```
plot_predicted_treatment_effect(cf, figsize=(14, 10), npoints=75, num_workers=10)
plt.show()
```

5.4 Case Practice

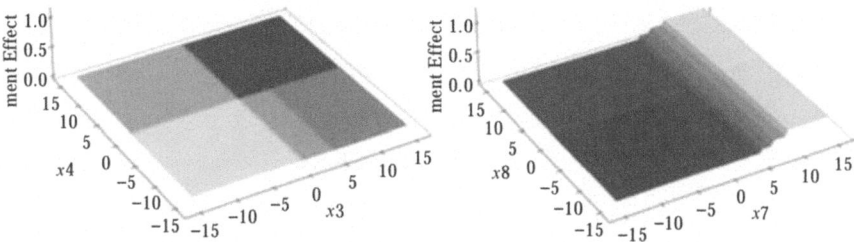

Fig. 5.4 Visualization of uplift score in the subcolumn of feature X (example)

Table 5.1 Model parameters in this case

Causal tree number	Tree node number	Left subtree	Right subtree	Node hierarchy	Splitting features	Split value	Causal effect
0	0	1	2	0	7	0.244539	
0	1	3	4	1	5	0.052896	
0	2	101	102	1	5	0.581857	
0	3	5	6	2	6	2.749351	
0	4	55	56	2	1	0	
0	5			3			0.290323
0	6	7	8	3	4	0.245499	
……	……	……	……	……	……	……	……

You can see that each subcolumn of the X feature has a stepped effect on the Uplift Score, as shown in Fig. 5.4.

Save the causal forest model parameters to a model.csv file.

```
cf.save(filename="model.csv", overwrite=True)
```

Table 5.1 Model parameters saved in the model.csv file.

At this point, we have successfully constructed a causal forest model. The next question is, how do you evaluate the performance of this model? In general, there are two ways to evaluate causal models: online evaluation and offline evaluation. Online evaluation is to conduct marketing according to the customer list predicted by the model, and then compare the marketing results with the marketing results of random sampling control group, so as to obtain the actual performance of the model. Offline evaluation is based on the observed data, through certain mathematical operations to calculate the indicators of the prediction group and the control group, so as to evaluate the predictive performance of the model. The offline evaluation method is described below.

Rank the uplift score predicted by the model from large to small and define the Uplift Curve function for t samples up to the end: $f(t) = \left(\frac{Y_t^T}{N_t^T} - \frac{Y_t^C}{N_t^C}\right)(N_t^T + N_t^C)$. Where, Y_t^T represents the number of positive instances (where the uplift score is positive) of the implementation group (i.e., marketing occurred) in the first

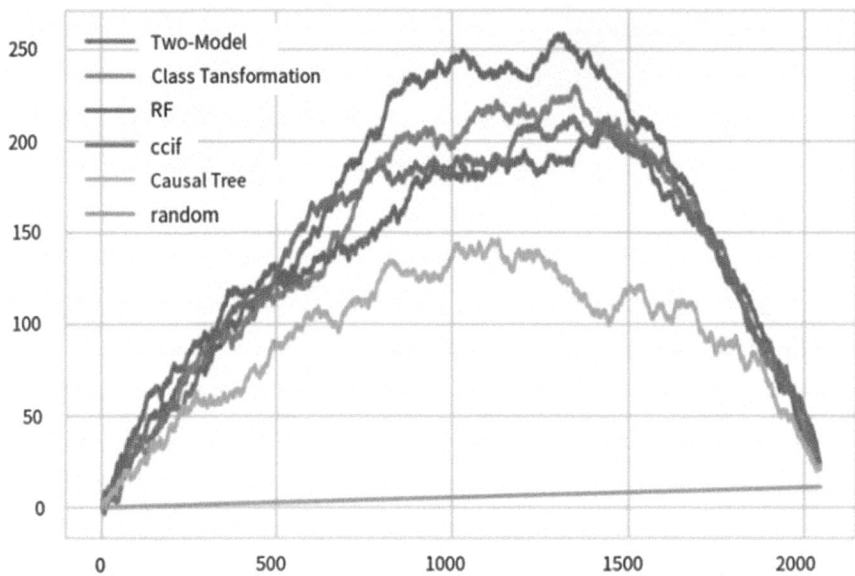

Fig. 5.5 Uplift curves

t samples; Y_t^C represents the number of positive cases (uplift score positive) in the control group (no marketing occurred) in the first t samples; N_t^T represents the number of samples from the implementation group in the first t samples; N_t^C represents the number of samples in the control group in the first t samples. From this, $f(t)$ can be seen that it reflects the total amount of improvement in the uplift score positive cases obtained by the model in the first t customer samples.

In Fig. 5.5, the Uplift Curve function is visualized with the X-axis being the sample order and the Y-axis being the cumulative causal effect. Each curve at the top represents the cumulative causal effect if ranked in descending order by the Uplift Score; The lines at the bottom represent the causal effects of randomly selecting a sample to perform a marketing intervention. The end points of each curve intersect, representing the causal effect after the full sample marketing intervention. In the curve, the higher the arch, the better the effect of the model, indicating that the model can mine more causal effect Uplift, that is, the larger the Area Under the curve, the better. This is similar to the Area Under ROC Curve (AUC) in binary classification evaluation, called Area Under Uplift Curv (AUUC).

The problem with the uplift curve above is that there is a bias in the increment of expression when the intervention and control samples are not consistent. Therefore, a scaling modification of the above equation is equivalent to scaling the control group based on the sample size of the intervention group, and the cumulative curve drawn is called the Qini curve. $g(t) = Y_t^T - \frac{Y_t^T N_t^T}{N_t^C}$, if it reflects the positive increment of Uplift Score in the first t customer samples after scaling the control group, the Qini curve expression is: $f(t) = g(t) \frac{N_t^T + N_t^C}{N_t^T}$, it reflects the positive uplift of the first

5.4 Case Practice

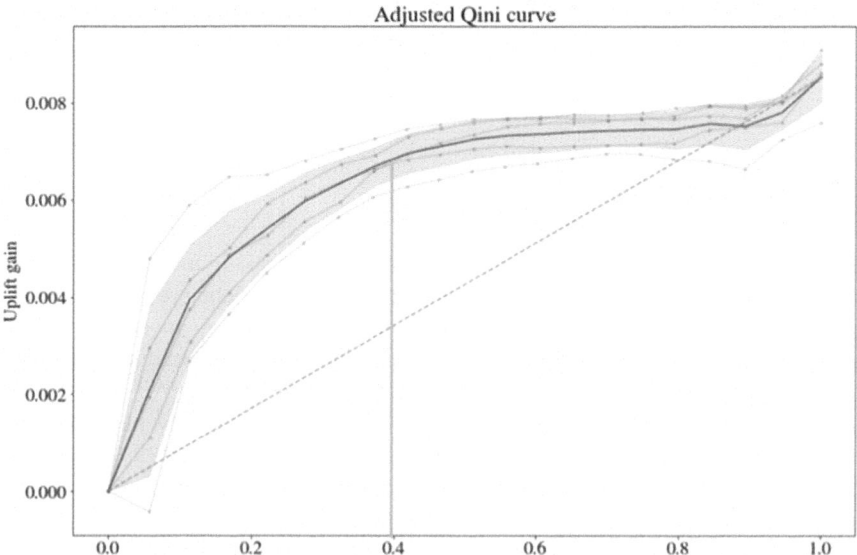

Fig. 5.6 Qini curve for this case

t customer samples. Visualize the Qini expression with the X-axis being the sample bit-order and the Y-axis being $f(t)$, get the Qini curve. This model uses the pylift module to draw the Qini curve:

```
up = TransformedOutcome(df, col_treatment='t', col_outcome='y',
stratify=df['t'])
    up.randomized_search(n_iter=20, n_jobs=10, random_state=1)
    up.fit(**up.rand_search_.best_params_)
    up.shuffle_fit(params=up.rand_search_.best_params_, nthread=30,
iterations=5);
    up.noise_fit()
    up.plot(plot_type='aqini',
        show_shuffle_fits=True,
        show_random_selection=True,
        shuffle_band_kwargs={'color':[0.7,0.7,0.3], 'alpha':0.3},
        shuffle_lines_kwargs={'color':[0.3,0.3,0.3], 'alpha':0.2},
        shuffle_avg_line_kwargs={'color':[1,0,0]}
    )
```

Run the above code and you will get the Qini curve as shown in Fig. 5.6.

The range of light yellow lines shows the performance under different strategies, the dark curves are the Qini curves of this model, and the straight dashed lines are the causal effects obtained by random sampling. It can be seen that when the X-axis is 0.4, the dark curve is farthest from the straight dashed line, indicating that in the case of limited marketing resources, we can take the top 40% of the causal effect values output by the model from large to small, and the marketing benefits will be the largest at this time.

5.5 Case Summary

This case explores causal inference in the field of retail banking marketing, demonstrating a complete methodology from data preparation, modeling, to model evaluation, aiming to inspire readers to think about new technology modeling. In the author's practical work, this case achieved noticeable results: we conducted marketing activities based on Uplift Score from high to low, and through AB testing with randomly selected customer samples, we found that the marketing revenue of the top 20% of the predicted customer samples was nearly three times that of the random samples. The top 40% also reached twice the level, which indicates that the model identified marketing-sensitive groups, allocating limited marketing resources to the customer groups most likely to drive marketing performance indicators. The X features in this case can utilize derived variables, and the Qini curve can be used to group the top t customer samples, calculating the causal effect uplift rate for each group. These subsequent refinements remain for readers to further enrich and enhance.

Causal inference is a tool to explain the world and a framework for intelligent modeling with big data, making it highly suitable for explaining and analyzing sequentially related events, many of which are present in banking operations. In addition to marketing scenarios, there are use cases such as fraud detection, control, evaluation, customer quality assessment, loan pricing adjustment, and risk-return management. Therefore, causal inference technology has positive significance in improving the quality and effectiveness of bank operations. It can quantify the gains from interventions (causal effect model) and identify potential causal relationships among seemingly unrelated variables in complex systems (causal discovery model). Moreover, causal inference can further enhance the actual work efficiency of AI in tasks such as natural language processing, image and video processing, and multimodal processing, improving model interpretability and advancing AI from "data fitting" to "data understanding," thereby occupying the high ground in the future development of AI.

Part II
Intelligent Risk Control

Chapter 6
Telecom Fraud Money Laundering Account Recognition Case: Multiple Machine Learning Techniques

On July 28, 2020, the Ministry of Public Security held a press conference where Liu Zhongyi, Director of the Criminal Investigation Bureau, reported that with the rapid development of the information society, the crime structure has undergone profound changes. Traditional contact crimes continue to decline, while new crimes represented by telecom network fraud are on the rise, becoming the fastest-growing and most strongly reported crimes by the public. These crimes show characteristics such as continuous high occurrence rates, rapid growth in online fraud, quick relocation of fraud nests, gradual generalization of criminal groups, and rampant black and gray industries. In recent years, both the "practitioners" of such cases and the number of cases have grown explosively. The report from the Ministry of Public Security pointed out that under the influence of multiple factors such as the pandemic, economic downturn, and increased employment pressure, production and life have accelerated their shift online, leading to various types of fraud targeting different groups—loan fraud targeting netizens with financial shortages, part-time job fraud targeting unemployed netizens, online shopping scam targeting frequent shoppers, fake investment platforms targeting those with investment intentions, and impersonation scams of public security and legal officials. In 2020, there were 256,000 telecom fraud cases solved, 263,000 criminal suspects arrested, 140 million scam calls intercepted, 870 million scam texts intercepted, and 120 billion yuan in direct economic losses recovered.

Telecom fraud generally consists of three parts: fraud, money laundering, and technical operations. Fraud methods include online part-time jobs, forex scams, online gambling and drugs, loan fraud, investment scams, illegal fundraising, order brushing fraud, and impersonation of customer service. Money laundering methods include underground banks, virtual assets, point-run platforms, and large cash withdrawals. Technical methods include dark web transactions, fake base stations, Trojan horse links in SMS, AI-generated fake voices and videos, and network identity forgery. In recent years, numerous overseas fraud companies and specialized money laundering platforms have emerged, forming a large-scale, highly specialized socialized industry chain for fraud, making telecom fraud highly covert, transnational, organized, professional, anonymous, diverse in form, technology-driven, and complex in scenarios. Coupled with the weak self-protection awareness of vulnerable groups and untimely detection and handling, this makes the prevention and control of such cases difficult and the recovery of losses challenging. Telecom fraud causes significant harm to society and the nation: it damages the country's image, leaves a negative impression on the international community, affects the business investment environment and national development; a significant portion of the illicit funds flows overseas, exerting pressure on the internationalization of the RMB, affecting national strategy; and it damages social trust, affects social stability, and increases social management costs.

Efforts to combat telecom fraud have escalated to a national level with the cooperation of multiple ministries. In early 2016, the "Great Wall Operation" was launched to crack down on overseas fraud pretending to be law enforcement. By February 2017, the National Anti-Fraud Center was established in Beijing. From June to September 2019, the Supreme People's Court and the Supreme People's Procuratorate issued interpretations on handling criminal cases related to the illegal use of information networks. In October 2019, measures were taken to freeze QQ, WeChat, and Alipay accounts associated with fraudulent activities in heavily affected areas in Northern Myanmar, and operations against cash smuggling were initiated. In October 2020, the State Council held an inter-ministerial meeting to address new types of telecom network crimes and launched the "Card Breaking" operation nationwide starting October 10. In May 2021, the Ministry of Industry and Information Technology held a conference on preventing telecom network fraud within the telecommunications industry. By June 2021, both the Ministry of Industry and Information Technology and the Ministry of Public Security issued a notice to clean up and rectify phone cards, IoT cards, and linked online accounts used for fraud. Commercial banks formed special task forces and took proactive measures to safeguard citizens' lawful rights and curb the proliferation of telecom fraud cases. Currently, anti-telecom fraud efforts have become a societal focal point, and the identification and management of accounts involved in telecom fraud have become critical tasks demonstrating the "political and people-oriented" nature of commercial banks.

6.1 Case Pain Point: Limitations of Anti-Telecom Fraud Risk Control Rules in the Banking Industry

For a long time, the banking industry has the following three limitations in the identification of telecom fraud cases:

1. **Risk control rules are difficult to adjust dynamically in real time, leading to missing checks**: When novel criminal activities emerge, existing risk control protocols struggle to detect them, resulting in investigative oversights. These protocols typically rely on the collective expertise of specialists, based on

previously observed cases, encapsulating historical patterns. Despite intensified enforcement efforts, criminal methodologies continuously evolve. Consequently, for robust prevention and control, risk control protocols often lag, making near-real-time adaptation challenging. For example, initial investigations might target newly issued cards, prompting risk control measures to focus on new cardholders. However, once the risk associated with new cards is mitigated, offenses involving existing cards become more prevalent. This necessitates the expert group to devise updated risk control protocols. These adaptations are frequently delayed due to the inherent latency in revising risk control measures, thus failing to provide real-time dynamic adjustments.

2. **The risk control rules are not objective and comprehensive, leading to missing checks and wrong checks**: Risk management protocols often exhibit a high degree of subjectivity, failing to impartially and comprehensively encapsulate all case characteristics. The subjective nature arises from the manual summarization process that forms these rules. Thus, these protocols do not offer an objective and thorough judgment of the entire data set. Typically, risk management focuses on explicit feature descriptions, adhering to a "what you see is what you get" approach. This methodology neglects the intricate, latent features inherent in the case data set. To address these shortcomings, it is imperative to devise a comprehensive evaluation mechanism that amalgamates multidimensional features. This integrative assessment should leverage machine learning techniques, utilizing full-sample feature fitting to ensure accuracy and inclusiveness. For instance, conventional risk management primarily identifies account and transaction attributes, disregarding the interconnected characteristics of associated parties. Moreover, these protocols emphasize the observable logic at face value, ignoring derivative and cryptic features within the data set. Consequently, the current rules fall short of providing an objective and holistic evaluation of risk.

3. **Vague risk control rules are not clearly expressed, leading to missing and wrong checking**: The compressive outline of risk control protocols often contains ambiguous language. For instance, the terms "frequent transaction" and "large transaction" are not clearly defined. The interpretation of these terms can vary by region and individual managers, resulting in inconsistent application. This discrepancy causes the same vague rule to have different triggering thresholds and leads to divergent outcomes in similar investigations, rendering the standardization of investigation procedures difficult. As an illustration, historically for the vague rule of "large transactions," an account involving amounts exceeding 100,000 yuan would often trigger risk controls, leading to a 100,000 yuan threshold. However, as crime detection has become more sophisticated, the typical amount involved in such transactions has decreased to less than 50,000 yuan. If the threshold is not adjusted promptly, it will result in missed or erroneous investigations.

What this case should do is to use a series of clear mathematical models to describe the identification process of the account involved in the above three limitations, and then convert it into computer code, so as to efficiently and accurately complete the investigation process.

6.2 Modeling Techniques and Scenario Analysis

The banking sector must innovate technically to effectively combat telecom fraud, especially given the technological advancements in fraudulent schemes and the limitations of traditional counter-fraud methods. To address the challenges posed by "unscientific risk control rules" and "simulated manual investigation," this case study introduces three pivotal measures: "continuous real number deep feature synthesis (DFS)," "unsupervised adversarial machine learning," and "fuzzy mathematical control." These measures significantly enhance the identification rate of fraudulent accounts, substantially reduce the need for manual investigations, and markedly improve the efficiency of risk control mechanisms. It is important to highlight that, due to the unique nature of anti-telecom fraud initiatives and the practical efficacy of data-driven approaches, this case study presents a novel solution at the research and testing stages only; it does not imply that this case study offers the ultimate or definitive choice for implementing such measures.

6.2.1 A Solution for "Real-Time Dynamic Adjustment of Risk Control Rules": Continuous Real Depth Feature Synthesis Technology

This case scenario exhibits the characteristics of "reconnaissance and counter-reconnaissance." From a holistic data perspective, regardless of the variations in transaction behaviors of the involved samples, certain features or their derived features will statistically diverge from those of normal samples. One common algorithm to address such issues is anomaly detection. Given that this case subsequently adopts machine learning methods, the aim in the feature engineering stage is to identify the statistical characteristics of the anomalous samples within the fluctuating dataset. Consequently, a "continuous real number DFS" algorithm was devised. The core idea of this algorithm is to express all feature dimensions as continuous real numbers and to perform various mathematical operations between any two continuous real number features to obtain new derived features. By calculating the contribution of these derived features to predicting the target variable (y-value), the significant features for modeling are determined. This case collects features across multiple dimensions, such as "transaction," "account," and "behavior," each containing several sub-features. For sequence data features, variance, mean, median, mode, skewness, extreme values, and change points of the sequence are used to represent the sequence itself. Similarly, the first-order difference sequence (the new sequence obtained by subtracting the preceding values) is represented by its variance, mean, median, mode, skewness, extreme values, and change points. Thus, sequence features are converted into numerical features expressed as continuous real numbers. The DFS method involves performing pairwise operations—addition, subtraction, multiplication, and division—on these

6.2 Modeling Techniques and Scenario Analysis

continuous real number features to derive new features. For instance, while height and weight alone may not strongly correlate with health, their ratio significantly correlates with health. By computing the correlation between these new derived features and the known anomalous sample labels (y-values), highly correlated derived features are identified, which then become essential features to be included in the model. It is crucial to note that the original features must comprehensively cover all aspects involved in transaction behavior and customer attributes. The advantage of continuous real number DFS lies in its ability to automatically compute features from the complete dataset without human intervention, thereby eliminating incomplete expert knowledge. When perpetrators employ counter-reconnaissance tactics to alter their behaviors, the comprehensive coverage of features ensures that the algorithm consistently identifies characteristics distinguishing involved samples from uninvolved ones on a statistical level, achieving the risk control effect of "as you change, I adapt."

In feature engineering, the new features obtained through feature learning using raw data are referred to as derived features. Deep feature synthesis (DFS) is a method for generating derived features that utilizes two main techniques: "aggregations" and "transformations." The "aggregations" are similar to performing variable aggregation calculations through multiple table joins in relational databases. For example, in a scenario where a transactions table is linked to a products table via the product_id field, linked to a sessions table via the session_id field, and the sessions table is linked to a customer table via the customer_id field, the original features consist of 8 variables: transaction time, amount, brand, device, session start time, join date, date of birth, and zip code. In featuretools, tables are referred to as entities, and the collection of multiple tables along with their relationships is known as an entity set, as shown in Fig. 6.1.

Featuretools is an open-source framework for DFS. It was launched by Feature Labs, a U.S. startup that has been acquired by data analytics provider Alteryx Inc. Feature Labs' technology has been integrated into Alteryx, a one-stop data analytics platform. Website address is https://www.alteryx.com/open-source. The Featuretools framework comes with the above data set. It can easily be loaded with the following code:

```
import featuretools as ft
es = ft.demo.load_mock_customer(return_entityset=True)
```

Print the entity set information, you can see that this entity set is named transactions, where the data shape of each entity (number of column and column fields) and its associated fields are as follows:

```
print(es)
```

Run the above code and you will get the entity set shown in Fig. 6.2.

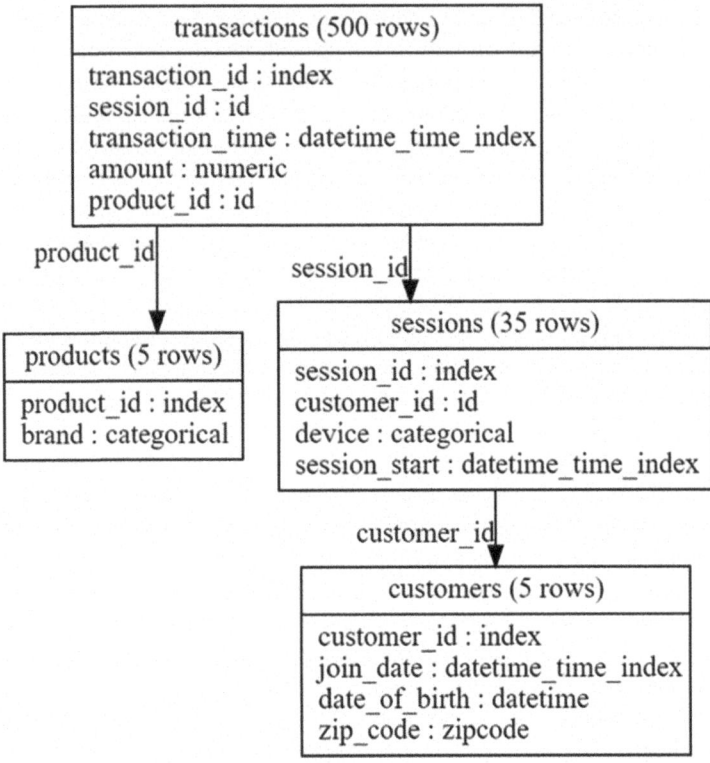

Fig. 6.1 Example of an association of multiple tables

Fig. 6.2 Featuretools entity set

```
Entityset: transactions
Entities:
    transactions [Rows: 500, Columns: 5]
    products [Rows: 5, Columns: 2]
    sessions [Rows: 35, Columns: 4]
    customers [Rows: 5, Columns: 4]
Relationships:
    transactions.product_id -> products.product_id
    transactions.session_id -> sessions.session_id
    sessions.customer_id -> customers.customer_id
```

Featuretools uses the dfs() function to perform DFS. It simply passes in the entity set and the target entity. The framework traverses each associated path in the entity set and generates derived features through various aggregation operations, including summing, averaging, and counting. Execute the following statement to get a feature matrix with 5 rows and 77 columns, as shown in Fig. 6.3.

6.2 Modeling Techniques and Scenario Analysis

```
             zip_code  COUNT(sessions)  ...  MODE(transactions.sessions.device)  MODE(transactions.sessions.customer_id)
customer_id
5               60091                6  ...                              mobile                                        5
4               60091                8  ...                              mobile                                        4
1               60091                8  ...                              mobile                                        1
3               13244                6  ...                             desktop                                        3
2               13244                7  ...                             desktop                                        2

[5 rows x 77 columns]
```

Fig. 6.3 Featuretools feature matrix

```
feature_matrix, features_defs = ft.dfs(entityset=es,
target_entity="customers")
print(feature_matrix.head(5))
```

In this example, as the target entity is the customer table, the algorithm will aggregate all fields based on the unique index (customer_id) of the customer table. It will be joined with the session table, transaction table, and product table separately to calculate derived features for each customer_id, such as session count COUNT (sessions), mode of device in session table MODE(transactions.sessions.device), and mode of customer_id in session table MODE(transactions.sessions. customer_id), and so on. This process results in 77 derived features from the original 8 features. It is important to note that this is a deep operation, where the algorithm can traverse even deeper features. In the concept of DFS, the term "deep" refers to the degrees of association between entity tables, for instance, if A is associated with B, B is associated with C, and A is not associated with C, then the aggregation depth of A and B is 1, and the aggregation depth of A and C is 2, and so forth.

In addition to aggregation, Featuretools also provides a "transformation" method to generate derived features. The difference between aggregation and transformation is that the former requires statistical processing from a group of data, while the latter processes only a single piece of data. Execute the following code:

```
feature_matrix, feature_defs = ft.dfs(entityset=es,
                        target_entity="customers",
                        agg_primitives=["count","mode"],
                        trans_primitives=["month"],
                        max_depth=1)
```

The agg_primitives parameter specifies the aggregation primitives, the trans_primitives parameter specifies the transformation primitives, they set the aggregation method and the transformation method respectively, and the max_depth parameter specifies the depth. When the depth is specified as 1, the algorithm only computes the target entity table and its directly associated session table fields. Since the conversion method is month, meaning the month in which the original feature is taken, it only takes effect for the date format field, and non-date format fields are ignored. The dfs() function returns the derived feature matrix and derived feature names, and we end up with the result shown in Fig. 6.4.

```
           zip_code  COUNT(sessions)  MODE(sessions.device)  MONTH(date_of_birth)  MONTH(join_date)
customer_id
5             60091                6                 mobile                     7                 7
4             60091                8                 mobile                     8                 4
1             60091                8                 mobile                     7                 4
3             13244                6                desktop                    11                 8
2             13244                7                desktop                     8                 4
[<Feature: zip_code>, <Feature: COUNT(sessions)>, <Feature: MODE(sessions.device)>, <Feature: MONTH(date_of_birth)>
, <Feature: MONTH(join_date)>]
```

Fig. 6.4 Featuretools feature matrix

name	type	description
num_true	aggregation	Finds the number of 'True' values in a boolean.
percent_true	aggregation	Finds the percent of 'True' values in a boolean feature.
time_since_last	aggregation	Time since last related instance.
num_unique	aggregation	Returns the number of unique categorical variables.
avg_time_between	aggregation	Computes the average time between consecutive events.
all	aggregation	Test if all values are 'True'.
min	aggregation	Finds the minimum non-null value of a numeric feature.
mean	aggregation	Computes the average value of a numeric feature.
seconds	transform	Transform a Timedelta feature into the number of seconds.
second	transform	Transform a Datetime feature into the second.
and	transform	For two boolean values, determine if both values are 'True'.
month	transform	Transform a Datetime feature into the month.
cum_sum	transform	Calculates the sum of previous values of an instance for each value in a time-dependent entity.
percentile	transform	For each value of the base feature, determines the percentile in relation
time_since_previous	transform	Compute the time since the previous instance.
cum_min	transform	Calculates the min of previous values of an instance for each value in a time-dependent entity.

Fig. 6.5 Partial feature primitives

```
print(feature_matrix)
print(feature_defs)
```

The mathematical operations of aggregation and transformation in Featuretools are called feature primitives. The algorithm computes variables by stacking primitives. Figure 6.5 shows the primitives.

The biggest strength of Featuretools is its reliability and ability to handle information leaks. In addition, the framework can also work with time series data. Because this part not involved in this case, interested readers can refer to Featuretools official documentation at https://featuretools.alteryx.com/en/stable/. Featuretools code hosting for https://github.com/alteryx/featuretools, the highest version for 1.3.0, the BSD - 3 - Clause License License agreement, heat is 5.9 K star.

6.2 Modeling Techniques and Scenario Analysis

Critics have scrutinized the methodology of DFS, primarily focusing on two critical issues. First, the synthesized features often lack interpretability, resulting in newly constructed features derived from operations on unrelated data points. If these features show high correlation with predicted labels, it represents "meaningless" data fitting. Consequently, training models with such features hinders the ability to determine if the extracted information aligns with the required insights. Second, the computational demand for applying feature primitives across all potential feature combinations is substantial. This process leads to the creation of numerous new features, causing inefficiencies, and high-resource consumption. We recognize that innovative methods invariably spark debate. This particular case study emphasizes the practical application outcomes of the model, specifically the accuracy and recall metrics pertaining to the involved account. Additionally, the exploratory nature of the case allows for experimenting with new methodologies.

6.2.2 A Solution for "Non-objective and Incomplete Risk Control Rules": Unsupervised Adversarial Machine Learning Techniques

To enhance the reader's comprehension, we will introduce some foundational concepts before delving into the technical solution. We will first examine the functioning of a BP neural network. A BP neural network refers to a neural network based on the Back Propagation algorithm. As illustrated in Fig. 6.6, the initial layer, termed Layer1, is the neural network's input layer. It receives the input vector X, where each component of X is multiplied by the corresponding weight of each neuron in Layer1 and, after adding a bias term b, undergoes a nonlinear transformation via an activation function before being passed to Layer2. Layer2, known as the hidden layer, takes the output from Layer1, multiplies it by the weights of its neurons, adds its bias term b, applies another nonlinear transformation via an activation function, and passes the result to Layer3. Layer3, the final layer, or the output layer, processes the output from Layer2 by multiplying it by the weights of its neurons and applying an activation function to yield the maximum value as the network's final output. This describes the forward propagation process of a feedforward neural network.

Figure 6.7 illustrates the use of the activation function. In forward propagation, each layer corresponds to an activation function, and each neuron is activated one by one.

Without activation functions, the functions that the network fits in the data set can only be linear and can only slice the data set with straight lines or hyperplanes, as shown in Fig. 6.8.

After the addition of the activation function, the function that the network fits in the data set can be nonlinear, and the data set can be shred with curves or hypersurfaces, as shown in Fig. 6.9.

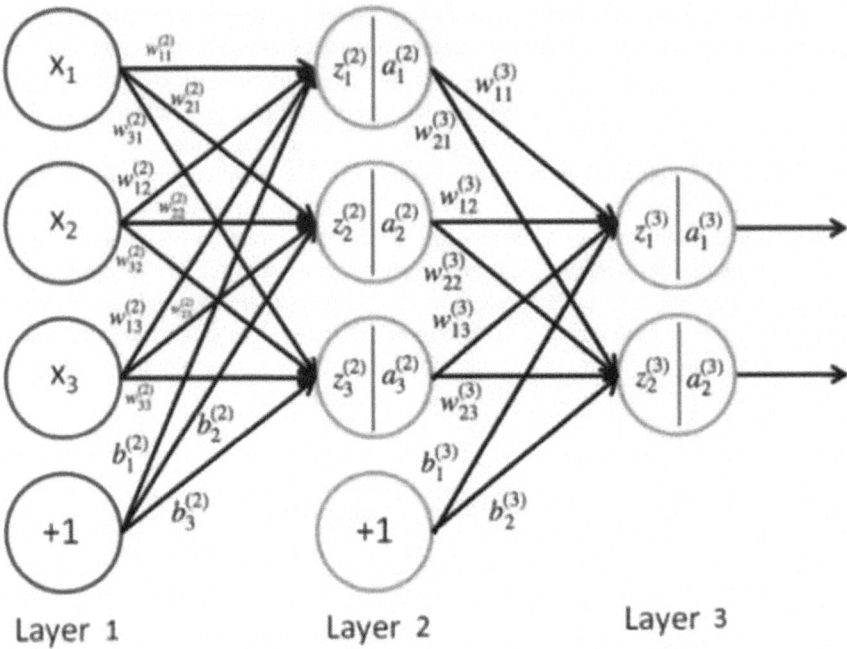

Fig. 6.6 Forward propagation process of BP neural network

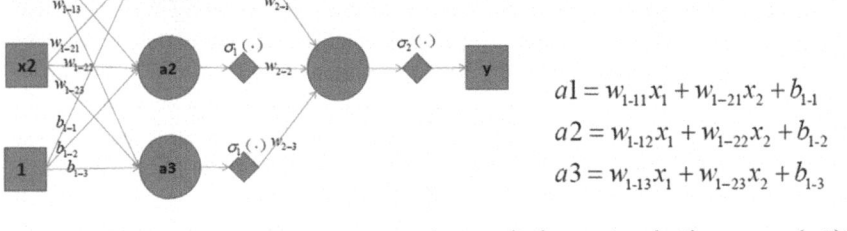

$$a1 = w_{1-11}x_1 + w_{1-21}x_2 + b_{1-1}$$
$$a2 = w_{1-12}x_1 + w_{1-22}x_2 + b_{1-2}$$
$$a3 = w_{1-13}x_1 + w_{1-23}x_2 + b_{1-3}$$

$$y = \sigma_2(w_{2-1}\sigma_1(a1) + w_{2-2}\sigma_1(a2) + w_{2-3}\sigma_1(a3))$$

Fig. 6.7 Activation function in forward propagation

There are several common activation functions, as shown in Fig. 6.10. Limited to the space, the specific meaning of the reader can be Baidu.

The forward propagation in a backpropagation (BP) neural network starts with the input vector X. As X passes through the multiple interconnected layers of the network, the output generated is the predicted value y. However, this predicted value y often deviates from the actual value y' that should correspond to X. To correct this, BP calculates the loss value using a predefined loss function. This loss value subsequently adjusts the weights in each layer, aiming to make future predictions

6.2 Modeling Techniques and Scenario Analysis

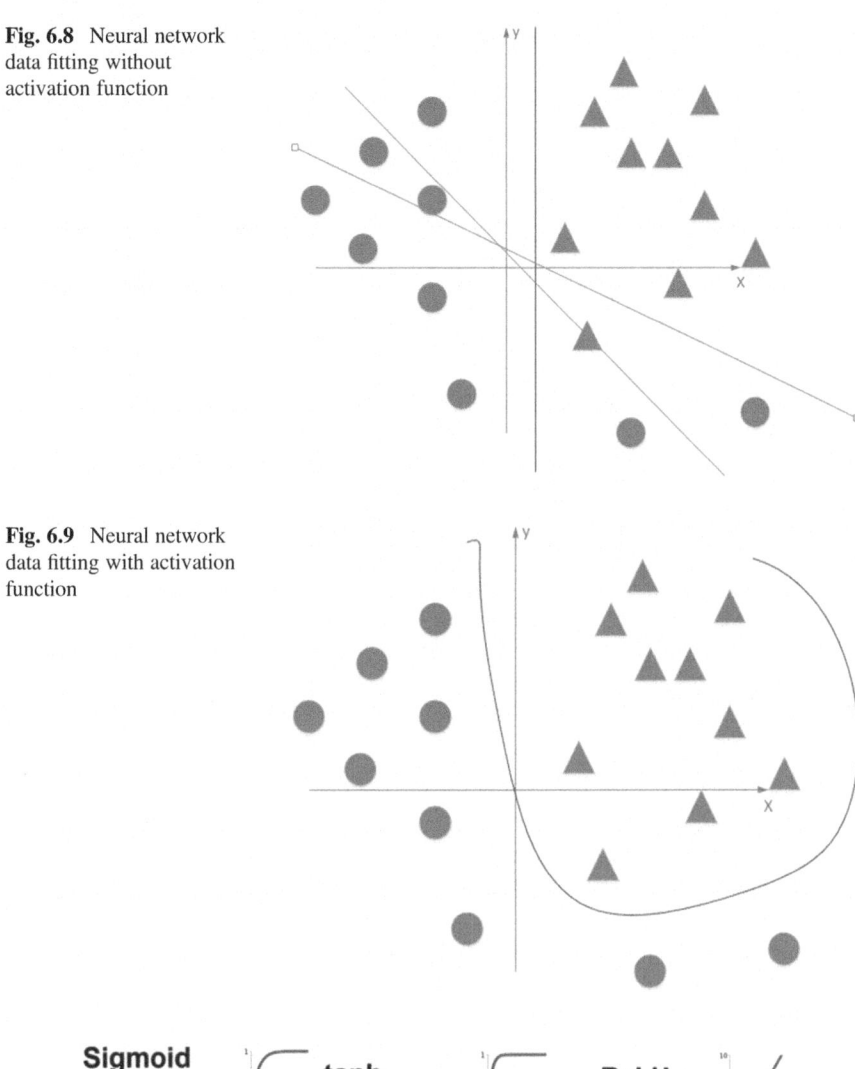

Fig. 6.8 Neural network data fitting without activation function

Fig. 6.9 Neural network data fitting with activation function

Fig. 6.10 Common activation functions

closer to the real value y'. This iterative adjustment process is referred to as network optimization, training, fitting, or learning. The essence of the optimization procedure is to minimize the loss function, directing the network toward improved accuracy. Thus, the loss function fundamentally guides the optimization trajectory of the neural network. For a more detailed understanding, refer to the schematic representation in Fig. 6.11, which illustrates the operational mechanism of the BP neural network.

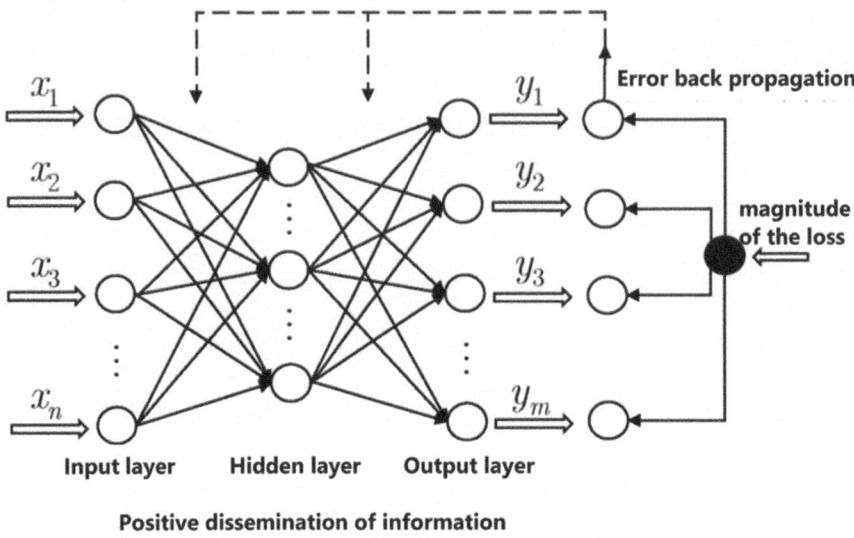

Fig. 6.11 Forward and reverse propagation of BP neural network

$$MSE = \frac{1}{N}\sum_{i=1}^{N}(y_i - Y_i)^2 \quad cost = -\frac{1}{N}\sum_{i=1}^{N}[y_i \ln a_i + (1-y_i)\ln(1-a_i)] \quad Loss\frac{1}{m}\sum_{i=1}^{m}|\hat{y}_i - y_i|^2$$

Fig. 6.12 Common loss functions

For different tasks, different loss functions can be chosen. Several common loss functions are shown in Fig. 6.12. Due to space limitations, readers can search for the specific meanings on their own.

It is worth noting that both forward and back propagation are transmitted layer by layer. The layer-by-layer transmission of BP is called the "chain rule," as shown in Fig. 6.13.

BP involves the use of various optimizers that align with distinct optimization algorithms. In essence, BP transmits data concerning the gradient of the loss function to a specific weight parameter. Here, the gradient signifies a vector composed of partial derivatives, providing critical information for adjusting the parameters to minimize the loss function in neural network training. Since the equation $\frac{\partial L}{\partial x} = \frac{\partial L}{\partial z}\frac{\partial z}{\partial x}$ holds, that is, the partial derivative of L with respect to x is equal to the partial derivative of L with respect to z multiplied by the partial derivative of z with respect to x, which is similar to the divisor operation, and $\frac{\partial L}{\partial z}$ can be regarded as the gradient of the Loss value loss to the reverse first layer weight, and $\frac{\partial z}{\partial x}$ can be regarded as the gradient of the reverse first layer weight to the reverse second layer weight. In this manner, adjusting the weights of each layer becomes straightforward through the sequential computation of gradients across layers. During the training phase of a BP

Fig. 6.13 Backpropagation based on the chain rule

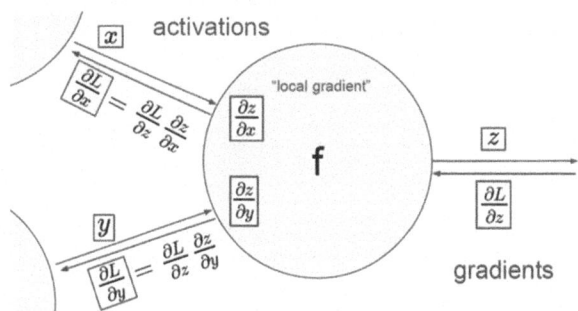

neural network, the optimal adjustment of network weights occurs by modifying the input dataset (X) and iterating through both forward and backward propagation processes. When there exists a definite correlation between the feature set (X) and the label vector (y), the network weights tend to converge to a value that mirrors this correlation. Consequently, the loss function approaches a minimal value, indicating high accuracy. Upon achieving this state of convergence, the refined network weights are preserved in a model file. This model can subsequently be utilized to predict outcomes for new input data (X').

Now let's look at autoencoder (AE). Also known as the Diabolo Network or Autoassociator, the AE was originally proposed by Hinton and the PDP group in the 1980s. AEs also use BP algorithms for learning, and unlike BP neural nets, AEs aim to reconstruct the same input. The network structure of the AE is shown in Fig. 6.14.

The AE can be thought of as consisting of two cascaded BP neural networks, the first BP network is the encoder, which is responsible for receiving the input vector x and transforming it into the implicit representation signal Code of a lower dimension, which is equivalent to nonlinear dimensionality reduction. The second BP network is the decoder, which takes the Code signal output by the encoder as its input to obtain a reconstructed signal with dimensions consistent with x, named $\tilde{x}$, which is equivalent to nonlinear dimensionality reduction. Define the error e as the difference between the original input x and the reconstructed signal $\tilde{x}$, $e = \|x - \tilde{x}\|$. The error is backpropagated to the hidden layers, and both the encoder and decoder networks are optimized in the direction of minimizing the error. Through repeated iterations, the output of the network $\tilde{x}$ will eventually be as consistent as possible with x. From this it can be seen that the essence of the AE is that the network weights reflect the probability distribution of the input data x. Since the AE reduces the dimension of x and then raises it, and can transform from low-dimensional Code back to x, it means that low-dimensional Code is the information compression of high-dimensional x, and its spatial structure is consistent with that of x space. This is an important property of AEs.

Now look at generative adversarial networks or GANs. They are a very hot technology in deep learning in recent years. It consists of two neural networks, one called a generator, which regenerates noise data into data; and a discriminator, which learns features from real data while identifying whether the generator's data is

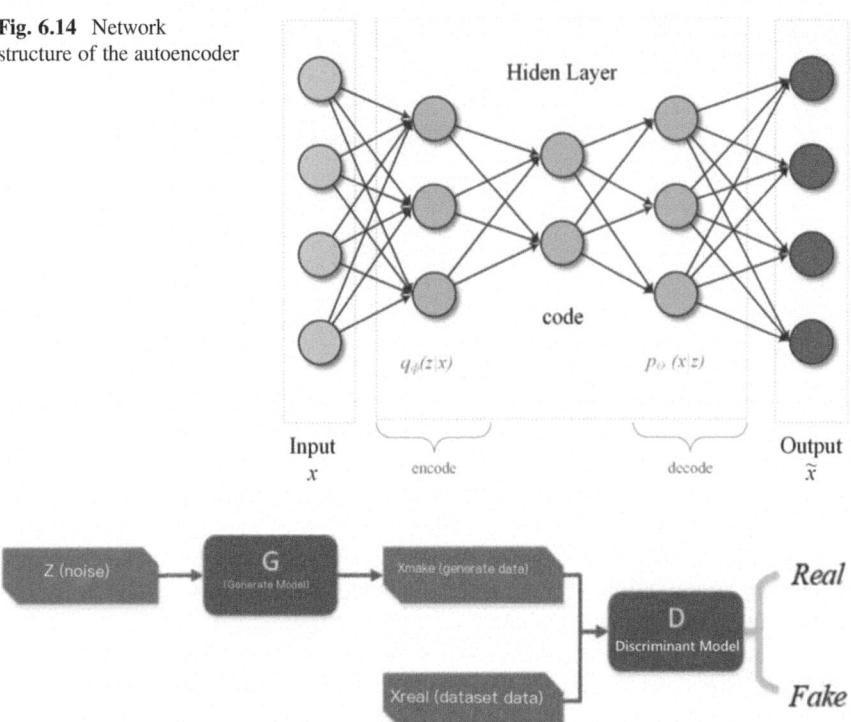

Fig. 6.14 Network structure of the autoencoder

Fig. 6.15 Generate adversarial network structure

real or generated. The data can be either an image or a set of vectors, the network structure of which is shown in Fig. 6.15.

In a generative adversarial network (GAN), the discriminator undertakes the binary classification task, aiming to differentiate between true and false data. The model prioritizes optimizing toward accurately classifying false data as false and true data as true. This process is driven by iterative optimization of the loss function. Conversely, the generator constructs its loss function based on feedback received from the discriminator, striving to craft synthetic data that the discriminator deems as true. Throughout this optimization process, the GAN progressively refines its capability to discern real data from generated data. The training phase concludes when the generator and discriminator reach a Nash equilibrium. At this juncture, the discriminator cannot distinguish between the authentic data and the data synthesized by the generator, signifying that the generated data has achieved a level of deceptive realism.

After gaining a preliminary understanding of the aforementioned concepts, let's return to this case. This case is characterized by the fact that the black samples consist of accounts reported for fraud or bank-confirmed fraudulent accounts, resulting in a very low number of black samples. In contrast, the white samples (normal accounts) are extremely large in quantity, leading to a severely imbalanced dataset. Typically, supervised learning can employ the SMOTE algorithm to achieve dataset balance through undersampling (discarding some white samples) or oversampling (generating synthetic black samples within the known feature distribution boundaries of black samples). However, a significant issue arises because the essence of supervised learning is to learn the correlation between the feature matrix X and the label vector y, restricting it to learning sample features only within the labeled sample feature distribution boundaries. In supervised learning tasks, the severe imbalance in the number of black and white samples causes the model to predominantly learn the features of white samples (the majority class), resulting in a strong bias toward white samples in prediction outcomes, thereby missing a significant number of fraudulent accounts. This case exemplifies such a scenario. Therefore, unsupervised learning becomes essential at this point. Unsupervised learning uses input data solely consisting of X (composite feature matrix) without y (label vector) and analyzes the structural features of all samples within the complete dataset, thereby clustering similar samples together. Compared to supervised learning, unsupervised learning offers greater comprehensiveness. Its fundamental objective is to learn the spatial structure of the dataset, leading to the formation of new data views (such as new spaces formed after dimensionality reduction and clusters formed through clustering). Typical approaches include dimensionality reduction, clustering, and restricted Boltzmann machines. Furthermore, supervised learning usually fits a model to the multidimensional features of labeled samples, which is not beneficial for revealing the various feature tendencies of numerous labeled samples. Conversely, the clustering method in unsupervised learning can subdivide the various feature tendencies of numerous labeled samples into subsets. As the father of deep learning, Geoffrey Hinton, stated, unsupervised learning represents the future direction of artificial intelligence. Given the advantages of unsupervised learning outlined above, this case adopts this approach for modeling.

Addressing the issue of "subjectivity in risk control rules" is quite challenging. The key to resolving this lies in designing an algorithm that identifies appropriate latent semantic feature representations within the entire dataset. This facilitates the creation of new, objective, comprehensive, and dynamic risk control rules. This case prioritizes the effectiveness of risk control rules—specifically the accuracy of predictive results—while not focusing on the interpretability of the algorithm-generated risk control rules due to the inherent lack of explainability in the machine learning process. The technical approach employs an unsupervised algorithm combining "Adversarial Autoencoders (AAE) + Gaussian Mixture Clustering (GMM)." The overall strategy involves representing each white sample (normal account) and black sample (known fraudulent account) as a high-dimensional (M-dimensional) vector, thereby forming a high-dimensional feature matrix of $N*M$ for N samples. If we plot each vector in this matrix into an M-dimensional space, we get N sample points, forming an original high-dimensional feature space. We use AEs to compute embedded representations, mapping this original high-dimensional feature space to a lower-dimensional embedded space. Then, using GAN, we constrain the low-dimensional embedded space vectors to a single distribution (the case here employs the standard normal distribution). Subsequently, we employ Gaussian

Mixture Models for clustering, enabling the grouping of accounts with similar latent features into a single class. It is important to note that the latent features here are derived from neural network training on the entire dataset, meaning they evolve with changes in the dataset. Thus, provided the dataset comprehensively covers the characteristics of criminal activities, the appearance of new criminal methods will alter the sample distribution structure in the low-dimensional embedded space, consequently changing the model's output of suspicious samples. This enables the dynamic adjustment of AI models to accommodate changes in criminal methods. The essence of the low-dimensional embedded space is to generalize the latent features behind each original sample. These features are learned by the neural network from the inter-sample relationships within the dataset. Usually, due to the consistency in transactional purposes, the latent feature representations of fraudulent accounts are relatively similar. Therefore, after clustering via Gaussian Mixture Models, known fraudulent accounts typically cluster into a limited number of groups. Since all samples within the same cluster share similar latent features, scrutinizing the white samples within clusters heavily populated with known black samples will accurately identify suspicious samples, achieving precise and dynamic risk control. This case reframes the problem of "discovering black samples within mixed black-and-white samples" into one of spatial structural relationships among samples, using algorithms to delineate new spatial views.

Figure 6.16 shows the model workflow in this case, with dark dots representing known scam accounts (black sample) and light dots representing normal accounts (white sample). In this case, the original cluttered and unevenly distributed high-dimensional feature space of the original sample is transformed into a new low-dimensional feature space after nonlinear dimensionality reduction against the AE.

In the realm of neural networks, an AE exhibits two fundamental characteristics: the structure of the latent space remains analogous to that of the original data space, and data points sharing similar latent features cluster more tightly. When visualized within a reduced-dimensional space, samples displaying identical features aggregate more cohesively than in the original space. Within a multi-feature dimension landscape, fraudulent account groupings might exhibit diverse feature patterns, subsequently forming multiple distinct clusters. These clusters encapsulate both

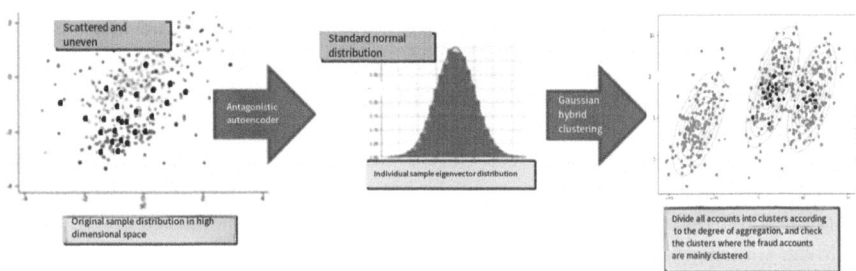

Fig. 6.16 Workflow of this case model

6.2 Modeling Techniques and Scenario Analysis

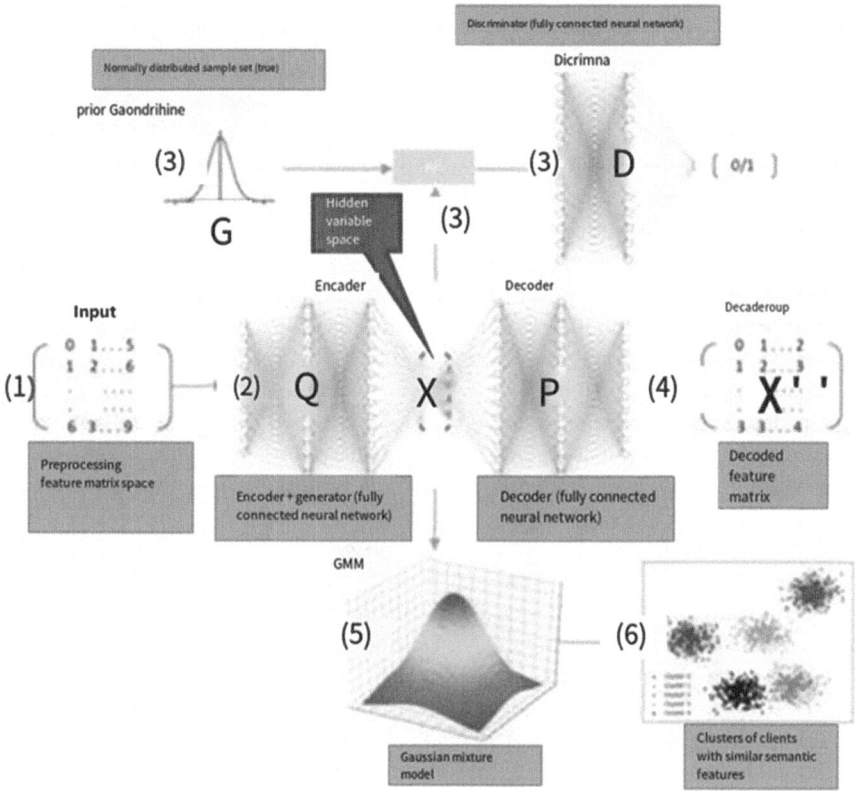

Fig. 6.17 Adversarial autoencoder network model structure

known deceptive samples and unidentified suspicious samples (i.e., white samples with anomalous traits). Predictably, white samples with analogous features also cluster together more closely, resulting in additional discrete clusters. Given that in subsequent clustering processes, it's often imperative to assess the similarity of sample vectors across each dimension, maintaining the original feature distribution within the reduced dimension space is crucial. Implementing a GAN ensures that each sample vector preserves a standard normal distribution. Following this, fitting the low-dimensional space to a multidimensional Gaussian distribution allows for the computation of the probability of each sample's affiliation with the various Gaussian distributions, thereby delineating distinct clusters.

The adversarial AE used in this case is superimposed to generate adversarial neural network (GAN) based on theAE. The network structure is shown in Fig. 6.17.

The network workflow is as follows:

1. The high-dimensional feature matrix X of continuous real depth feature synthesis output is fed into the encoder Q.

2. Encoder Q maps high-dimensional features to low-dimensional features, generating the hidden variable space X' (that is, an embedded representation of X), where X corresponds one-to-one to X'. Here, an incomplete AE is usually used, so that nonlinear dimensionality reduction can be achieved, and the advantage is that the low-dimensional space better reflects the aggregation structure of the high-dimensional space, creating conditions for downstream clustering tasks.
3. The data generator G generates data with a fixed distribution, usually in the standard normal distribution (the distribution commonly used in machine learning). X' is sent to discriminator D together with the generated standard normal distribution data set, and the discriminator D judges whether X' meets the standard normal distribution (binary classification) given by G. The discriminator keeps learning iteration through loss function, BP optimizes itself, and constantly improves the discrimination ability.
4. The decoder P decodes the reduced matrix X'' obtained from X', and it optimizes in the direction of making X' as consistent as possible with X'' by back-propagating the optimization of P and Q through a loss function metric difference.
5. Encoder Q builds a loss function according to the discriminant output of discriminator D, and the optimization goal is to generate the data itself, so that the discriminator D discriminates as true (standard normal distribution).
6. For AEs, Q is the encoder; For GANs, Q is the generator. Q thus assumes both the encoder and generator roles. For each round of training, Q is a neural network with two loss functions that simultaneously optimize its weight in both directions. The encoder Q is optimized in both directions to generate a normal distribution of the hidden variable X'. Then X' has two properties at the same time: first, it is spatially consistent with the original feature space X, and second, every single vector sample of X' remains normally distributed (probabilistic mass function).
7. The Gaussian mixture model is used to fit X' with multiple Gaussian distributions, and the probability that each embedded sample in X' belongs to a certain Gaussian distribution is found, so as to achieve the clustering effect in the embedded space.

It should be noted here that the discriminator identifies X' vector one by one, and makes the X' generated by the generator more prone to standard normal distribution after each round of training. What are the benefits of embedding X' into the spatial sample with a consistent distribution strategy? If different X' samples have different data distributions, then they are not comparable to each other in clustering, because clustering measures the proximity of samples in various dimensions. Therefore, the work of the discriminator provides conditions for downstream clustering tasks. In addition, as mentioned above, the AE has the property of "the hidden space structure is consistent with the original space structure," and in the embedded space after dimensionality reduction, the samples with similar implicit characteristics have a closer connection, which also improves the downstream clustering effect.

As for "the distribution of samples in high-dimensional space is sparser than that in low-dimensional space," it can be simply explained as follows: As shown in

6.2 Modeling Techniques and Scenario Analysis

Fig. 6.18 The sample distribution in a high-dimensional space diverges to the edge

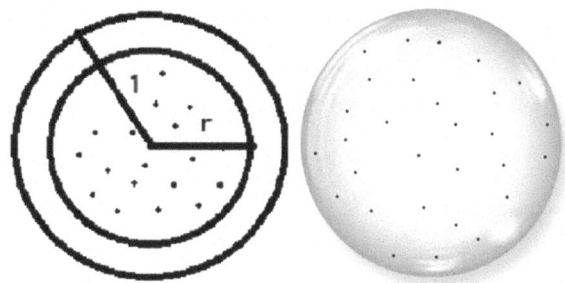

Fig. 6.18, suppose that in 2-dimensional space, the samples are distributed in a ring, the radius of the outer ring is 1, the radius of the inner ring is r, and the volume of the sample distribution is $s = \pi r^2$. If it becomes a 3-dimensional space, then the volume of the sample distribution inside the 3D inner ring is $s = 4/3\pi r^3$. Considering the n-dimensional space case, then the sample distribution volume is $s = K\pi r^N$, where K is a constant coefficient. Since $r < 1$, s must approach 0 as N approaches infinity. If we calculate the ratio of the inner ring volume of the sample distribution to the outer ring volume of the entire sphere, we find that this ratio also tends to zero as N approaches infinity. That is to say, as the spatial dimension continues to increase, the samples originally distributed inside the inner circle will gradually be distributed to the outer circle, that is, the samples will become more sparse. This explains that the dimensionality reduction operation will make the sample distribution more compact while maintaining the original spatial data structure, as shown in Fig. 6.18.

If the sample space distribution of variable X' is uniform, clustering yields no valuable results. Conversely, the more heterogeneous the distribution of X', the more effective the clustering outcome. Thus, evaluating the clustering feasibility of X' before initiating the clustering process is imperative. For this evaluation, the Hopkins statistic H is utilized. Generally, an H value less than 0.5 signifies an absence of clustering propensity in X', while values closer to 1 suggest a strong clustering tendency. In the given scenario, the calculated H value is 0.98, indicating a significant clustering tendency.

Following these preliminary evaluations, one can proceed to the clustering phase. This step is a standard procedure in machine learning, where a complex, non-standard function or distribution is approximated using multiple standard functions or distributions with varying parameters. For instance, the Fourier transform achieves this by employing multiple sinusoidal functions, each with distinct amplitudes and phases, to approximate any continuous function. This function can be narrow, meaning a continuous periodic function, or broad, encompassing continuous aperiodic functions, thereby standardizing them across multiple components, as illustrated in Fig. 6.19.

The Gaussian mixture model used in this case is also of this kind of thinking. It uses multiple Gaussian distributions (i.e., normal distributions) to fit any continuous function, which is the case in 2-dimensional space (i.e., X is 1-dimensional and Y is

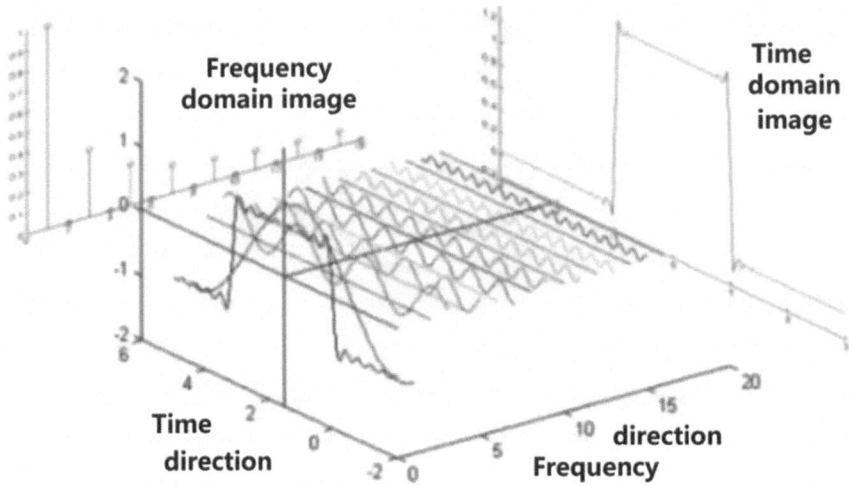

Fig. 6.19 Schematic diagram of the Fourier transform

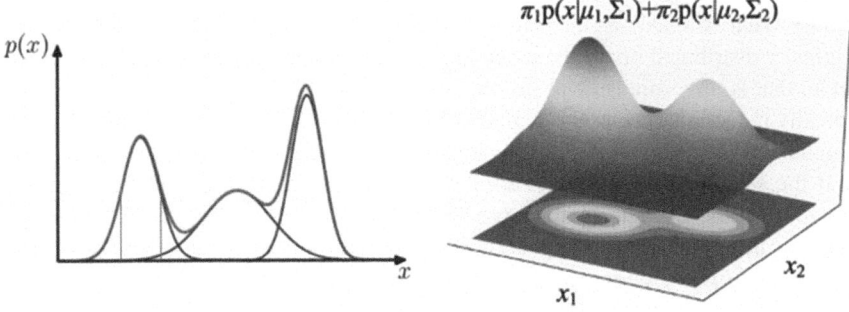

Fig. 6.20 Schematic diagram of Gaussian mixture model in 2D and 3D space

the probability of this random variable). For example, the figure on the left uses three Gaussian distributions to fit this function curve (the gray line represents the contours). In the 3-dimensional space, as shown in the figure on the right (i.e., X is 2-dimensional and Y is the probability of this random variable), circles represent contours, as shown in Fig. 6.20.

In N-dimensional space, it is also possible to fit the data distribution of a multidimensional matrix with a high-dimensional Gaussian distribution. The Gaussian mixture model will give the probability that each sample belongs to each Gaussian distribution, which can achieve the effect of clustering. Moreover, compared with K-Means clustering algorithm, Gaussian distribution can fit data according to ellipses (that is, adjust the variance of Gaussian distribution in each dimension), while K-Means can only fit data according to the exact circle. Therefore,

6.2 Modeling Techniques and Scenario Analysis

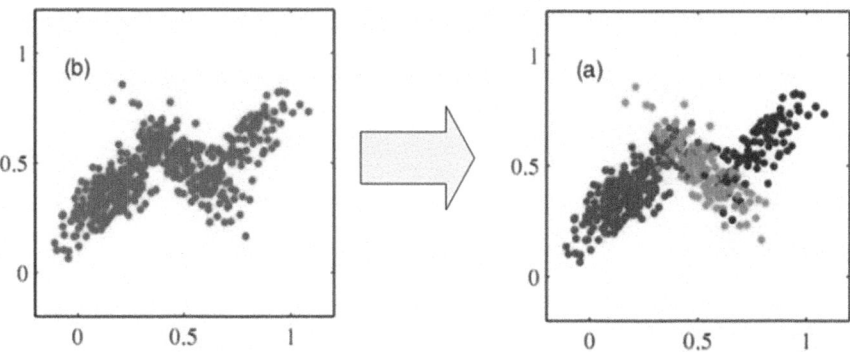

Fig. 6.21 Clustering effect of Gaussian mixture model

Gaussian mixture model can adapt to irregular data sets better than K-Means in clustering effect. Finally, we can cluster the X' hidden space matrix against the AE output to obtain sample clusters with similar embedding features, as shown in Fig. 6.21.

How many clusters should be gathered? This example uses the Calinski-Harabaz (CH) index to search for the best number of clusters. The property of CH index is: the smaller the covariance of the data within the cluster and the larger the covariance of the data between clusters, the higher the CH index, and the smaller the vice versa. Therefore, a cluster number interval is set up in this case. The CH value is calculated for each cluster number in the interval, and the cluster number with the highest CH value is taken as the final cluster number.

In this case, the number and proportion of known fraudulent accounts in each cluster are calculated and sorted out as given in Table 6.1.

At this point, the model can output a list of predicted scam accounts. Manual checks focus on the white sample accounts in the cluster of known fraudulent accounts because the characteristics of the white sample are similar to those of the known black sample. With this result table, when we want to check more accounts, we can select more samples in the cluster of known fraud accounts. When we want to pursue the investigation accuracy, we can select the white samples with a higher proportion and a smaller European distance from the known fraud accounts in the cluster for investigation. In this way, you can choose a balance between the "amount of investigation" and "accuracy" according to your needs.

To summarize, the advantages of this method are:

1. The unsupervised algorithm does not need to monitor labels at all, and is suitable for business scenarios without labels;
2. Reasoning on the full data set, regardless of the size of the data set, is valid, and can be arbitrarily balanced between "precision" and "coverage";
3. Having the ability to learn cryptic representation, and can calculate the "recessive feature" of each sample in a large number of "dominant feature" datasets, which is the overall expression of the dataset.

Table 6.1 Clustering results of this case

Cluster number	Number of known fraudulent accounts	Known proportion of fraudulent accounts (%)
Unsupervised model final result: The fraudulent accounts are basically clustered in these two clusters, and all accounts in these two clusters are manually screened	703	3.12
36		
94	362	0.98
82	54	0.56
98	42	0.32
17	14	0.13
67	11	0.12
10	15	0.09
27	4	0.01
12	3	0
15	8	0

6.2.3 A Solution for "Unclear Expression of Fuzzy Risk Control Rules": Fuzzy Control Techniques

We recognize that the essence of a data model lies in "digitization + mathematization," meaning it involves digitizing real-world entities and their relationships through feature engineering, then using mathematical language to establish data models based on these digitized features, and finally solving mathematical problems through algorithms to map them back to real-world issues, completing the model application. As such, mathematical language constitutes the soul of data models. However, typically, concepts described in mathematical language are precise, clear, and well-defined, embodying an absolute nature—much like a number cannot be simultaneously equal to 1 and 2. In this particular case, many risk control concepts lack clear boundaries, necessitating the integration of these vague notions to accurately determine whether an account is suspicious. For instance, the rule of "rapid multiple transactions" in risk control cannot precisely define the terms "rapid" and "multiple" due to varying individual perceptions of speed and quantity. Consequently, these notions are challenging to delineate mathematically. Moreover, the combination of speed and quantity concepts mimics the process of manual scrutiny of suspicious accounts. Therefore, we are confronted with a paradox: modeling requires clear and well-defined conceptual boundaries, yet the concepts to be

6.2 Modeling Techniques and Scenario Analysis

modeled are ambiguous and lack definitive boundaries. How do we resolve this issue?

The solution is fuzzy mathematics. In 1965, American cybernetics expert Professor Lotfi A. Zadeh published a paper titled "Fuzzy Sets" in the journal Information and Control. Since then, fuzzy mathematics has emerged as a new branch of mathematical theory. This theory uses "fuzzy sets" to describe collections of ambiguous concepts, "membership functions" to depict the transition of concept intensity, and "fuzzy rules" to delineate decision rules among multiple fuzzy concepts under different linguistic variables. Fuzzy mathematics transcends the absolute relationships of "equal or not equal" and "belongs or does not belong" found in classical mathematics, allowing it to describe "the degree to which a concept belongs to both fuzzy set A and fuzzy set B." Subsequently, this framework has given rise to subfields such as fuzzy control, fuzzy inference, fuzzy clustering, and fuzzy pattern recognition, which find extensive applications in areas like engineering control and machine learning. To facilitate readers' understanding, I will briefly introduce some fundamental concepts of this theory.

Domain of discourse: a nonempty set of all characteristic concepts. For example, if the feature concept is the age of a person, then the domain of discourse is the set from 0 to 100.

Fuzzy set: Given a domain U, then a mapping from U to the unit interval $[0,1]$ is called a fuzzy set on U. As an example, in this case, "number of transactions" is a characteristic concept. "Few transactions," "moderate transactions," and "frequent transactions" are the three fuzzy sets of "number of transactions." If max (trade_count) is the maximum number of transactions in all accounts, then each element of these three fuzzy sets has different values within the range of the discourse domain $[0,\max(\text{trade\_count})]$. Like the classical sets, two fuzzy sets can be found intersection, union, complement and other operations. But the biggest difference between fuzzy set and classical set is: one element can belong to multiple fuzzy sets, which is described by membership function; In a classical set, an element can only belong to one set.

Membership function: If there is A function $A(x) \in [0,1]$ corresponding to any element x in the domain U (the scope studied), then $A(x)$ is called the membership function of x with respect to A. In this case, we use the triangular membership function to describe the three fuzzy sets of "few transactions," "moderate transactions," and "frequent transactions," as shown in Fig. 6.22.

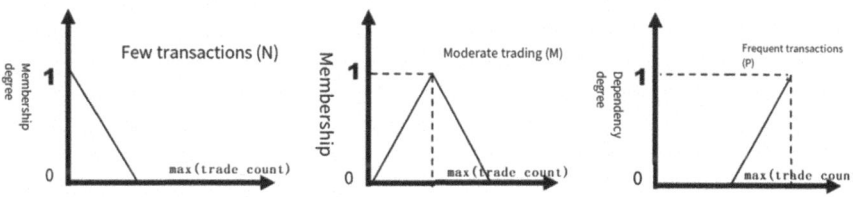

Fig. 6.22 Membership function of number of transactions, discussion domain and fuzzy set

Table 6.2 Fuzzy Rule determination table of this case (examples)

		Ranking by number of transactions		
		N	N	P
Fast forward the block rank	N	N	N	M
	M	N	M	M
	P	M	M	P

Fuzzy rule: A rule table for the goal determination of multiple fuzzy concepts. As given in Table 6.2, the two concepts of the ranking of the number of transactions and the ranking of the time interval (fast forward and fast out) of two adjacent transactions in this case constitute a fuzzy rule decision table, where "N," "M," and "P," respectively, represent the weak, medium, and strong concepts, which are called linguistic variables. After checking this table, we can determine the degree of fraud tendency of an account (the green part) according to the ranking of the number of transactions and the ranking of the fast forward and fast out, as given in Table 6.2.

Fuzzy rules are formulated based on the expertise of specialists. In this case, an iterative approach is employed to cyclically determine the fraud propensity of N feature concepts in combination. Initially, the fuzzy calculation results of feature concept 1 and feature concept 2 are used as new inputs. These inputs are then recalculated with feature concept 3 according to the fuzzy rule table. The resultant calculation is further processed with feature concept 4, and the process continues accordingly. Detailed procedures can be found in the attachment "fuzzy_rules.xlsx." It is important to ensure that the values of all linguistic variables align consistently with the final determination direction. Specifically, if the final result approaches the meaning of linguistic variable N, creating a higher likelihood of fraud, all feature concepts must be combined in a manner that maintains a uniform direction. Upon completing these tasks, we initiate the fuzzy computation framework. Each feature value of the sample is input into the framework, and the model outputs a fraud propensity score for the sample. Compared to the aforementioned machine learning scheme, the benefits of using fuzzy control include strong interpretability, comprehensibility, and the absence of a training requirement. Additionally, this method offers fast operational speed and comparable computation results.

6.3 Case Practice

This section introduces the development and operation environment construction, modeling code implementation process and operation results of this case. After reading this section, you will understand the specific implementation process of this case in detail.

6.3 Case Practice

6.3.1 Environment Setup

This case training requires the use of an Nvidia GPU, so first verify that the computer has an Nvidia graphics card (N card) installed. The author's hardware environment is 8GB RTX 2080 Super, 128GB RAM, CPU i9-10900K, and 1T SSD.

Next, install Anaconda, a machine learning suite. See Sect. 1.3.1 of this book for installation instructions. Once installed, create a virtual environment named PyTorch on the command line and install Python 3.6.7 (the reader can specify the virtual environment name and Python version as needed) and execute the following statement:

```
conda create -n pytorch python=3.6.7 -y.
```

To activate the virtual environment, execute the following statement:

```
conda activate pytorch.
```

At this point, a "(pytorch)" prompt appears on the command line, indicating that you have entered the virtual environment.

In fact, there is an easier way to install CUDA and CUDNN. You can install support for CUDA and cudnn in the virtual environment by executing the command conda install cudatoolkit=10.0 cudnn= 7.6.4-y after activating the virtual environment. It is worth noting that the CUDA and CUDNN installed in this installation mode need to activate the virtual environment to be effective, and only the minimum volume version required for running is installed, and the complete version is not installed. The CUDA and CUDNN installed in the download and installation mode mentioned above is globally valid and the complete version. After PyTorch 1.2.0 is installed, run the conda list|findstr cuda command to view CUDA and CUDNN installed in the virtual environment, as shown in Fig. 6.23.

The following software is used in this example. Install CUDA in sequence by following the following commands:

Featuretools is a DFS development framework that provides automated derived feature generation. Feature_selector is a feature importance selection library that evaluates the relative importance of each feature among multiple features. They are used for feature engineering in this case. Run the command: pip install featuretools feature_selector to complete the installation.

```
(pytorch) E:\>conda list|findstr cuda
cudatoolkit            10.0.130                 0    defaults
cudnn                  7.6.4           cuda10.0_0    defaults
pytorch                1.2.0    py3.6_cuda100_cudnn7_1    pytorch
```

Fig. 6.23 CUDA installation result

TensorboardX is a deep learning aid that allows you to visualize, track, compare, and interpret the experimental process of deep learning projects. install it with the following command: pip install tensorboardX.

Pandas is a well-known data processing library that provides both DataFrame and Series data structures. This case uses it to process the feature data of a sample set. Do pip install pandas installation.

Matplotlib is a data visualization library that is used in this case to draw a visualization of Gaussian mixture model clustering. Execute the command pip install matplotlib to complete the installation.

Sklearn is a highly packaged machine learning library that mainly implements frequency-oriented statistical machine learning methods and provides tools such as classification, clustering, regression, dimensionality reduction, and data preprocessing. Execute the command pip install scikit-learn to complete the installation.

Skfuzzy is a fuzzy mathematics library that provides method encapsulation of fuzzy clustering, fuzzy control, fuzzy and defuzzy, fuzzy filtering, fuzzy mathematics, and fuzzy image. This case uses SKfuzzy to build a fuzzy control model. Run the command pip install scikit-fuzzy to complete the installation. At this point, the runtime environment is set up.

6.3.2 Code Practice

Feature Engineering Practice

In the DFS stage, we carry out the following processing of sequence features and basic features.

```
# Import library package
import featuretools as ft
# Load the account dataset
zhanghs = pd.read_csv('zhangh.csv')
# New account entity set
es = ft.EntitySet(id = 'zhangh')
# Add account entities to entity set
es = es.entity_from_dataframe(entity_id = 'zhanghs', dataframe = zhanghs, index = 'zhangh')
# Specify derived feature generation methods as: addition, subtraction, multiplication, division
 trans_primitives=['add_numeric', 'subtract_numeric', 'multiply_numeric', 'divide_numeric']
```

6.3 Case Practice

```
# Perform a binary operation on all input features of each sample to
generate new derived features, max_depth specifies the depth of the
feature synthesis, a value of 1 means that only the original feature is
operated on to generate new features
  features, feature_names = ft.dfs(entityset = es, target_entity =
'zhanghs', max_depth=1, verbose=1,
trans_primitives=trans_primitives)
  # Save the generated derivative features
  writer = pd.ExcelWriter('zhangh.xlsx')
  features.to_excel(writer,'Sheet1')
  writer.save()
```

After adding the label data to zhangh.xlsx, the code for the feature selection phase is as follows.

```
  from feature_selector import FeatureSelector
  # Read data
  data = pd.read_excel('zhangh.xlsx')
  # Specifies a label data column with 0 for white samples and 1 for black
samples
  train_labels = data.flag
  # Remove label columns in feature data
  train_features = data.drop(columns='flag')
  # Send feature columns and label columns to the feature selection frame
  fs = FeatureSelector(data=train_features, labels=train_labels)
  # First evaluate zero importance feature, parameters specify
classification task, evaluation index as precision, number of iterations
and early end flag respectively
  fs.identify_zero_importance(task='classification',
eval_metric='auc', n_iterations=10, early_stopping=True)
  # Reevaluate the relative importance of each feature. Note that the sum
of the relative importance of all features is 1. The Threshold parameter
specifies the cumulative importance, and the plot_n parameter specifies
the top N important features to be displayed
  fs.plot_feature_importances(threshold=0.3, plot_n=20)
```

The program outputs the following derived feature in order of relative importance from high to low, as shown in Fig. 6.24. This feature is fed into the adversarial AE model.

Due to the sensitivities involved in this case, the business characteristics and their data cannot be released at this time. The characteristic data of this case is the data of all aspects related to the case.

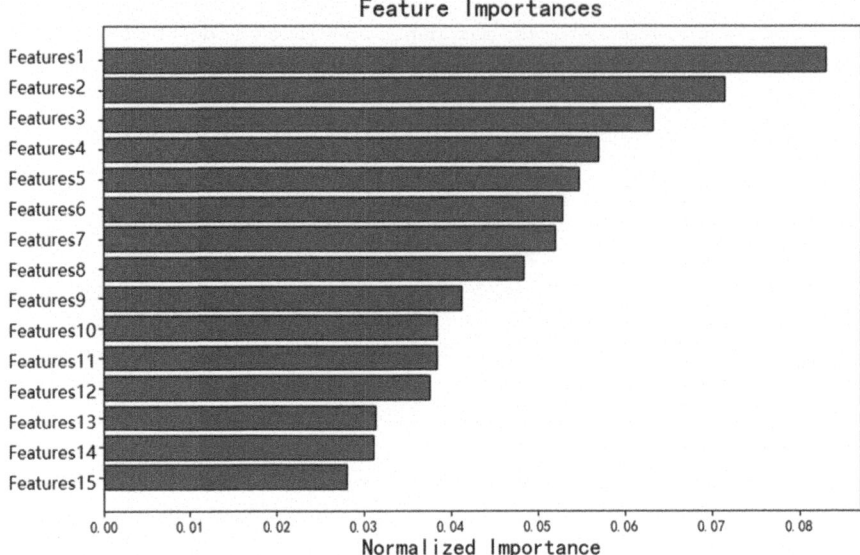

Fig. 6.24 Selected important derived features

Unsupervised Adversarial Learning Practice

The key codes for adversarial AEs are as follows.

```
# Build encoder + generator network Q, a three-layer fully connected BP network
  class Q_net(nn.Module):
    def __init__(self,X_dim,N,z_dim):
      super(Q_net, self).__init__()
      self.lin1 = nn.Linear(X_dim, N)
      self.lin2 = nn.Linear(N, N)
      self.lin3_gauss = nn.Linear(N, z_dim)
    def forward(self, x):
      x = F.dropout(self.lin1(x), p=0.25, training=self.training)
      x = F.relu(x)
      x = f.ropout(self.lin2(x), p=0.25, training=self.training)
      x = F.relu(x)
      z_gauss = self.lin3_gauss(x)
      return z_gauss
# Build decoder P, a 3-layer fully connected BP network
  class P_net(nn.Module):
    def __init__(self,X_dim,N,z_dim):
      super(P_net, self).__init__()
      self.lin1 = nn.Linear(z_dim, N)
```

6.3 Case Practice

```python
        self.lin2 = nn.Linear(N, N)
        self.lin3 = nn.Linear(N, X_dim)
    def forward(self, x):
        x = F.dropout(self.lin1(x), p=0.25, training=self.training)
        x = F.relu(x)
        x = f.ropout(self.lin2(x), p=0.25, training=self.training)
        x = self.lin3(x)
        return F.sigmoid(x)

# Build discriminator D, a 3-layer fully connected BP network
class D_net_gauss(nn.Module):
    def __init__(self, N, z_dim):
        super(D_net_gauss, self).__init__()
        self.lin1 = nn.Linear(z_dim, N)
        self.lin2 = nn.Linear(N, N)
        self.lin3 = nn.Linear(N, 1)
    def forward(self, x):
        x = F.dropout(self.lin1(x), p=0.2, training=self.training)
        x = F.relu(x)
        x = f.ropout(self.lin2(x), p=0.2, training=self.training)
        x = F.relu(x)
        return F.sigmoid(self.lin3(x))
# Define the data generator
class Dataset(torch.utils.data.Dataset):
    def __init__(self, csv_file):
        self.features_frame = pd.read_csv(csv_file, encoding='utf-8')
...

    def __len__(self):
        return len(self.features_frame)

    def __getitem__(self, idx):
        try:
            features = [list(next(self.features_it)[1])]
        except:
            self.features_it = iter(self.features_frame.iterrows())
            features = [list(next(self.features_it)[1])]
        return torch.Tensor(features)
# Then, use iterators to speed up data reading. Iterators call
__getitem__() each time to read data the length of batch_size
dataset = Dataset('features.txt')
data_loader = torch.utils.data.DataLoader(dataset=dataset,
batch_size=512, shuffle=True)
EPS = 1e-15
gen_lr = 0.0001 # Specify the AE learning rate
```

```python
reg_lr = 0.00005 # Specifies the GAN learning rate
z_red_dims = 100 # specifies the dimension of the hidden variable
# Initialize three networks: encoder/generator, decoder,
discriminator. If training is interrupted, you can continue learning
from the retained model file breakpoints. Note that the cuda() function
puts the network model into video memory
Q = Q_net(152,200,z_red_dims).cuda()
if Path("./model/Q_encoder_weights.pt").is_file():
    print('load model... ')
    Q.load_state_dict(torch.load('./model/Q_encoder_weights.pt',
map_location=lambda storage, loc: storage.cuda(0)))
# decoder
P = P_net(152,200,z_red_dims).cuda()
if Path("./model/P_decoder_weights.pt").is_file():
    print('load model... ')
    P.load_state_dict(torch.load('./model/P_decoder_weights.pt',
map_location=lambda storage, loc: storage.cuda(0)))
# discriminator
D_gauss = D_net_gauss(200,z_red_dims).cuda()
if Path("./model/D_gauss_weights.pt").is_file():
    print('load model... ')
    D_gauss.load_state_dict(torch.load('./model/D_gauss_weights.
pt', map_location=lambda storage, loc: storage.cuda(0)))
# Specify the optimizer used by the encoder/decoder in
backpropagation. Use the Adam optimizer to achieve an adaptive learning
rate, fast convergence, and speed up the model optimization process
optim_P = torch.optim.Adam(P.parameters(), lr=gen_lr)
optim_Q_enc = torch.optim.Adam(Q.parameters(), lr=gen_lr)
# GAN partial optimizer
optim_Q_gen = torch.optim.Adam(Q.parameters(), lr=reg_lr)
optim_D = torch.optim.Adam(D_gauss.parameters(), lr=reg_lr)
# data iterator
data_iter = iter(data_loader)
iter_per_epoch = len(data_loader)
print('iter_per_epoch=',iter_per_epoch)
total_step = 30000
writer = SummaryWriter('./Result') # Data is stored in this folder
for step in range(total_step):
    if (step+1) % iter_per_epoch == 0:
        data_iter = iter(data_loader)

    # Save encoder's arguments
    if step > 0 and step % 5000 == 0:
        print('save model... ')
        torch.save(Q.state_dict(), './model/Q_encoder_weights.pt')
```

6.3 Case Practice

```python
    torch.save(P.state_dict(), './model/P_decoder_weights.pt')
    torch.save(D_gauss.state_dict(), './model/D_gauss_weights.pt')
# Take the sample from the iterator
features = next(data_iter)
#features = features.float()
features = to_var(features)
# Empty the cumulative gradients of these three models
P.zero_grad()
Q.zero_grad()
D_gauss.zero_grad()
############### autoencoder section #####################
# encoder encodes x and generates z
z_sample = Q(features)
# decoder decodes z and generates x'
X_sample = P(z_sample)
# Here we compute the reconstruction error of the autoencoder |x' -x |
recon_loss = F.binary_cross_entropy(X_sample + EPS, features + EPS)
# Optimize autoencoder
recon_loss.backward()
optim_P.step()
optim_Q_enc.step()
############### GAN part #############################
# From the normal distribution, sample real gauss (true-Gaussian
distribution sample points)
z_real_gauss = V(torch.randn(features.size()[0], z_red_dims) *
5.).cuda()
# The discriminator checks the true sample and gets loss
D_real_gauss = D_gauss(z_real_gauss)

# Generate fake sample with encoder
Q.eval() # Cut to test form where Q(i.e. encoder) is not involved in
optimization
z_fake_gauss = Q(features)
# Discriminator is used to identify false samples and loss is obtained
D_fake_gauss = D_gauss(z_fake_gauss)

# Total error of the discriminator
D_loss = -mean(log(D_real_gauss + EPS) + log(1 - D_fake_gauss + EPS))

# Optimize discriminator
D_loss.backward()
optim_D.step()
# Encoder acts as a generator
Q.train() # switch training modality, Q(i.e. encoder) takes part in
optimization
```

```
z_fake_gauss = Q(features)
D_fake_gauss = D_gauss(z_fake_gauss)

G_loss = -mean(log(D_fake_gauss + EPS))

G_loss.backward()
# Optimize Q only
optim_Q_gen.step()

 print(step, 'recon_loss=', round(recon_loss.item(),2),
'D_loss=', round(D_loss.item(),2), 'G_loss=', round(G_loss.item
(),2))
    if step > 0 and step % 100 == 0:
      writer.add_scalar('recon_loss', recon_loss, step)
      writer.add_scalar('D_loss', D_loss, step)
      writer.add_scalar('G_loss', G_loss, step)
```

For details of the antagonistic AE, see: Antagonistic Autoencoder.py. In the low-dimensional hidden variable space generated by the adcoder, whether the sample has clustering (that is, the non-uniformity of the distribution) is measured by Hopkins statistics in this case. The key codes are as follows.

```
# Input :DataFrame 2-D data, output: float Hopkins statistic
# The default scale sampled from the dataset is 0.3
Def hopkins_statistic (data: pd. DataFrame sampling_ratio: float =
0.3) - > float:
   # If the sampling ratio exceeds 0.1 to 0.5, either end of the range is
replaced by the endpoint value
   sampling_ratio = min(max(sampling_ratio,0.1),0.5)
   # Number of samples
   n_samples = int(data.shape[0] * sampling_ratio)
   # Sample data drawn in raw data
   sample_data = data.sample(n_samples)
   # Data remaining after sampling the original data
   data = data.drop(index = sample_data.index) #, inplace = True)
   # Sum of distances from the sample drawn in the raw data to the nearest
neighbor
   data_dist = cdist(data,sample_data).min(axis = 0).sum()
   # artificial generate samples, sampled from average distribution
   ags_data = pd.DataFrame({col:uniform(data[col].min(),data[col].
max(),n_samples)\
             for col in data})
   # Sum of the distance of the artificial sample to its nearest neighbor
   ags_dist = cdist(data,ags_data).min(axis = 0).sum()
```

6.3 Case Practice

```
# Compute the Hopkins statistic H
H_value = ags_dist / (data_dist + ags_dist)
return H_value

df = pd.read_csv('dataF.txt')
df = df.sample(frac=0.04,axis=0)
print(hopkins_statistic(df))
```

See Hopkins Statistics.py for the above code. In the clustering phase of Gaussian mixture model, the key code is as follows.

```
X = np.array(x)
df_kehhao = pd.read_csv('features.txt.fx', usecols=
['khh','yh_flag'], encoding='utf-8')

# Search the optimal number of clusters for Gaussian mixtures
score_all=[]
List1 = range(80120).
```

#Calinski-Harabaz (CH index) is used to select the best number of clusters, and the operation speed is much higher than the contour coefficient. When the intra-cluster covariance is smaller and the inter-cluster covariance is larger, the CH index is higher. The number of clusters with the highest CH value is taken as the final cluster number, which in this case is 110.

```
for i in range(80,120):

    print('gmm, i=', i)
    gmm = GMM(n_components=i).fit(X) # specifies the number of clustering centers as i
    y_pred = gmm.predict(X) # To obtain a classification label for each vector
    # Draw a scatter plot of the results
    #plt.scatter(X[:, 0], X[:, 1], c=y_pred)
    #plt.show()
    score=metrics.calinski_harabasz_score(X, y_pred)
    score_all.append(score)
# Draw the clustering effect corresponding to different k values
plt.plot(list1,score_all)
plt.show()

# Gaussian mixture model clustering, using high-dimensional Gaussian distributions to fit the data set
gmm = GMM(n_components=110).fit(X) # specifies the cluster number to be 110
labels = gmm.predict(X) # To obtain a categorical label for each vector
```

The multi-dimensional data derived from clustering is reduced to three dimensions and plotted on three dimensional coordinates for

observation
```
X_tsne3d = TSNE(n_components=3, random_state=33).fit_transform(X)
from mpl_toolkits.mplot3d import Axes3D
fig = plt.figure()
ax = Axes3D(fig)
ax = plt.subplot(111, projection='3d') # Create a 3D mapping project
# Divide the data points into three parts, with a distinction in color
ax.scatter(X_tsne3d[:,0], X_tsne3d[:,1], X_tsne3d[:,2], c=labels,
s=30, cmap='coolwarm') # Draw the data points

ax.set_zlabel('Z') # Coordinate axes
ax.set_ylabel('Y')
ax.set_xlabel('X')
plt.show()
```

See Gaussian mixed cluster.py for the above code. Finally, we get the 3D visual data representation after clustering, as shown in Fig. 6.25. Among them, clusters with the same color represent account groups with the same semantic features, and different color shades represent different semantic features of samples between clusters.

We find that known fraudulent accounts mainly cluster in two clusters, indicating that other samples of these two clusters have similar semantic characteristics to known fraudulent accounts, and they have become important objects for manual investigation, as given in Table 6.3.

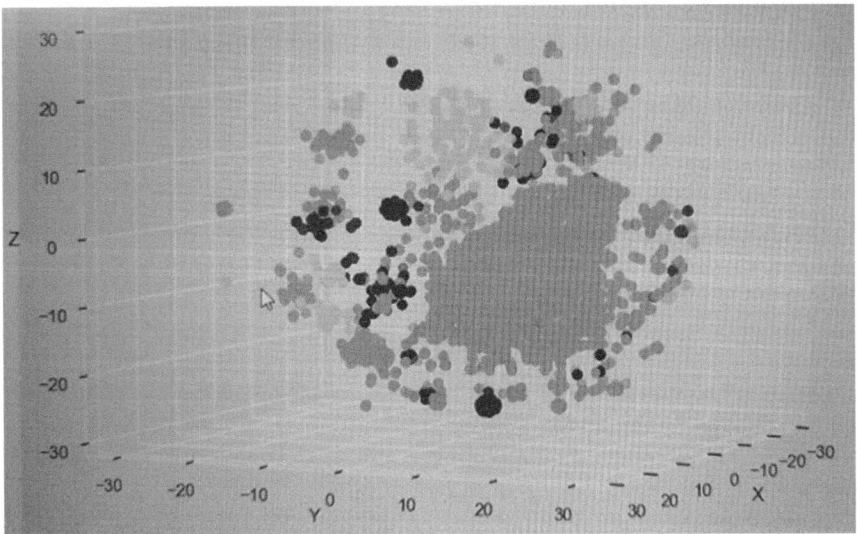

Fig. 6.25 Three-dimensional data visualization effect of Gaussian mixture model clustering

6.3 Case Practice

Table 6.3 Known fraudulent accounts are mainly clustered in two clusters

Cluster number	Number of known fraudulent accounts	Known proportion of fraudulent accounts (%)
36	703	3.12
94	362	0.98
82	54	0.56
98	42	0.32
17	14	0.13
67	11	0.12
10	15	0.09
27	4	0.01
12	3	0
15	8	0

Fuzzy Control Scoring Practice

Fuzzy control provides a complex scoring mechanism under fuzzy rules. Its key codes are as follows:

```
import skfuzzy as fuzz
import skfuzzy.control as ctrl
# Fuzzy control model: Input variable values for x1 and x2 and output y
according to fuzzy rules
x1_max = 1000
x2_max = 1000
y_max = 1000
x1_range = np.arange(0,x1_max,1,np.float32)
x2_range = np.arange(0,x2_max,1,np.float32)
y_range = np.arange(0,y_max,1,np.float32)

def fuzzy_model(x1_value, x2_value):
  # Create fuzzy control variables
  x1 = ctrl.Antecedent(x1_range, 'x1')
  x2 = ctrl.Antecedent(x2_range, 'x2')
  y = ctrl.Consequent(y_range, 'y')

  # Define fuzzy sets and triangular membership functions
  X1 = [' N '] fuzz. Trimf (x1_range, [0, 0, int (x1_max / 2)])
  x1['M']=fuzz.trimf(x1_range,[0,int(x1_max/2),x1_max])
  x1['P']=fuzz.trimf(x1_range,[int(x1_max/2),x1_max,x1_max])
  X2 = [' N '] fuzz. Trimf (x2_range, [0, 0, int (x2_max / 2)])
  x2['M']=fuzz.trimf(x2_range,[0,int(x2_max/2),x2_max])
  x2['P']=fuzz.trimf(x2_range,[int(x2_max/2),x2_max,x2_max])
```

```python
Y = ['N'] fuzz.Trimf(y_range, [0, 0, int(y_max / 2)])
y['M']=fuzz.trimf(y_range,[0,int(y_max/2),y_max])
y['P']=fuzz.trimf(y_range,[int(y_max/2),y_max,y_max])

# centroid unfuzz way
y.defuzzify_method='centroid'
# Rule that outputs N
rule0 = ctrl.Rule(antecedent=((x1['N'] & x2['N']) |
                              (x1['M'] & x2['N']) |
                              (x1['N'] & x2['M'])),
                  consequent=y['N'], label='rule N')
# Output rule with M
rule1 = ctrl.Rule(antecedent=((x1['M'] & x2['M']) |
                              (x1['N'] & x2['P']) |
                              (x1['M'] & x2['P']) |
                              (x1['P'] & x2['N']) |
                              (x1['P'] & x2['M'])),
                  consequent=y['M'], label='rule M')
# Rule that outputs P
rule2 = ctrl.Rule(antecedent=((x1['P'] & x2['P'])),
            consequent=y['P'], label='rule P')
# System and runtime environment initialization
system = ctrl.ControlSystem(rules=[rule0, rule1, rule2])
sim = ctrl.ControlSystemSimulation(system)
sim.input['x1'] = x1_value
sim.input['x2'] = x2_value
Sim.pute() # Run the system
return sim.output['y']
# child process code
def run(j, max_x1, max_x2, max_x3, max_x4, max_x5, max_x6, max_x7,
df_zh, df_x1, df_x2, df_x3, df_x4, df_x5, df_x6, df_x7):
    f = open('result/result.csv'+str(j-1), "w")
    f.write('zhangh,score\n')
    f.close()
    
    for i in range(len(df_x1)):
       # Compute the fuzzy rule by pin-two variables, scaling the variable value to 1000 first
        y = fuzzy_model(df_x1[i]*1000/max_x1, df_x2[i]*1000/max_x2)
        y = fuzzy_model(y, df_x3[i]*1000/max_x3)
        y = fuzzy_model(y, df_x4[i]*1000/max_x4)
        y = fuzzy_model(y, df_x5[i]*1000/max_x5)
        y = fuzzy_model(y, df_x6[i]*1000/max_x6)
        y = fuzzy_model(y, df_x7[i]*1000/max_x7)
        if i%100==0:
```

6.3 Case Practice

```
            print(j, i, len(df_x1))
        f = open('result/result.csv'+str(j-1), "a")
        f.write(str(df_zh[i])+','+str(round(y,2))+'\n')
        f.close()
if __name__ == '__main__':
    freeze_support()
    df_x1 = pd.DataFrame()

    for i in range(20):
        df = pd.read_csv('out/df2.csv.'+str(i))
        df_x1 = pd.concat([df_x1,df])
    df_x1 = df_x1.reset_index(drop = True)
    # The following are the multiple features that are fed into the fuzzy rule
    df_x1['rank_trans'] = df_x1['trans'].rank(method='min')
    df_x1['rank_delay'] = df_x1['delay'].rank(method='min')
    df_x1['rank_probe'] = df_x1['probe'].rank(method='min')
    df_x1['rank_ye'] = df_x1['ZHHUYE'].rank(method='min')
    df_x1['rank_hkcs'] = df_x1['hkcs'].rank(method='min')
    df_x1['rank_sus_counts'] = df_x1['sus_counts'].rank(method='min')
    df_x1['rank_ast9'] = df_x1['ast9'].rank(method='min')

    # To speed up the computation, start 20 processes in parallel
    length = int(len(df_x1)/20)
    for k in range(1,21):
        if k<20:
            endid = k*length
        else:
            endid = len(df_x1)-1
        p = Process(target=run, args=(k, max(df_x1['rank_trans']), \
            max(df_x1['rank_delay']), max(df_x1['rank_probe']), \
            max(df_x1['rank_ye']), max(df_x1['rank_hkcs']), max(df_x1['rank_sus_counts']), max(df_x1['rank_ast9']), \
            df_x1['zhangh'][(k-1)*length:endid].reset_index(drop = True), \
            df_x1['rank_trans'][(k-1)*length:endid].reset_index(drop = True), \
            df_x1['rank_delay'][(k-1)*length:endid].reset_index(drop = True), \
            df_x1['rank_probe'][(k-1)*length:endid].reset_index(drop = True), \
            df_x1['rank_ye'][(k-1)*length:endid].reset_index(drop = True), \
```

```
        df_x1['rank_hkcs'][(k-1)*length:endid].reset_index(drop =
True), \
        df_x1['rank_sus_counts'][(k-1)*length:endid].reset_index
(drop = True), \
        df_x1['rank_ast9'][(k-1)*length:endid].reset_index(drop =
True), ))
      p.start()
```

The final score of the fuzzy rule is in the file under the result directory, the code is in the download file "Fuzzy Control.py."

6.4 Case Summary

In this scenario, the researcher identified for the first time 2270 suspicious accounts in the actual operational environment and confirmed 910 fraudulent accounts through manual verification. The model exhibited an average accuracy rate of approximately 40%. In contrast, traditional expert rule methods demonstrated an accuracy range between 9% and 15%. The implementation of the new technology thus improved identification accuracy by 200–300%. This heightened accuracy has halved the workload for manual investigations, expedited risk detection lead times, and bolstered risk control efficiency by an average factor of three to five times. The model has the capability to operate on a daily basis, thereby supporting sustained investigation and operational activities.

Methodologically, the use of a variational autoencoder (VAE) serves as an alternative to the adversarial autoencoder (AAE) in this context. Both approaches aim to regulate the latent space of the AE to conform to a predetermined distribution. The key distinction lies in their mechanisms: the VAE employs the Kullback-Leibler (KL) divergence technique, whereas the AAE utilizes a discriminator component. Despite these differences, both methods accomplish parallel outcomes. Additionally, fuzzy control represents a pattern recognition methodology, enabling machines to emulate human decision-making rules. When the dimensionality of input variables is high, fuzzy control proves advantageous by providing faster and more precise evaluations compared to manual checks.

In conclusion, this case incorporates three advanced technologies: continuous real DFS, unsupervised adversarial machine learning, and fuzzy control. The practical outcomes are evident. It exemplifies the transformative potential of big data and artificial intelligence technologies, showcasing their capacity for replication and generalization. This case holds significant political and practical value by aligning with The State Council's anti-telecom fraud directives, safeguarding consumers' legitimate rights and interests, promoting societal harmony, and enabling financial institutions to uphold social responsibilities. It is imperative to highlight that this study focuses on technical aspects, serving as an inspiring exploratory endeavor

6.4 Case Summary

rather than an optimal solution. Due to the unique nature of the case, the original data characteristics remain unpublished. Readers are asked for their understanding in this regard. Future research could explore enhancements such as integrating behavioral time series anomaly detection algorithms, transaction text feature extraction at the historical transaction level, or extracting heterogeneous features using graph technology. Interested readers are encouraged to follow the book's WeChat public account to engage in discussions.

Chapter 7
Developing a Dialectal Speech Phone Collection Bimodal Robot from Scratch: Intelligent Voice Q&A Technology

The Bible states, "In the beginning was the Word, and the Word was with God, and the Word was God." Language constructs the world, establishing communication, trust, and commerce. It is the most critical mode of client interaction. Once artificial intelligence algorithms surmount linguistic barriers, they can generate immense commercial value. We already observe smart voice assistants, voice-controlled automobiles, and intelligent home devices emerging as novel customer engagement platforms. Such smart voice interaction technologies yield sustained, substantial profits for tech companies. For instance, a smart voice assistant enables users to simply name a song, and the device searches the internet to play it. Additionally, it can facilitate online transactions, social interactions, voice chatting, reminder setting, and other interactive functions. A basic speaker costing 200 RMB, when equipped with language and interactive capabilities, can transform the customer experience and retail for 800 RMB, quadrupling its price. However, product enhancement is not the sole value of intelligent voice interaction technology. The crucial aspect lies in establishing a new commercial ecosystem, internet traffic entry points, and customer acquisition channels for technology companies. Mature artificial intelligence language technologies have the potential to rebuild societal norms and commercial models, without exaggeration.

The AI industry mirrors this trend, as does the commercial banking sector. During operations, commercial banks engage with clients frequently, creating a substantial demand for intelligent language technologies. Various applications exist for intelligent voice interaction within commercial banking, such as post-loan recovery, telemarketing financial products, customer communication and service, voice knowledge databases, and many more. These applications span customer marketing, internal management, and risk control, involving technologies like automatic speech recognition, natural language understanding, and automatic speech synthesis. Among these, post-loan recovery offers a comprehensive and representative implementation. The author's development models have yielded favorable outcomes in practical work. Thus, this chapter focuses on the application of these technologies within the context of post-loan recovery.

Commercial banks traditionally rely on human operators to conduct post-loan collection calls. This approach faces several challenges, including workspace constraints, training requirements, personnel management, and labor costs. Additionally, the number of calls a person can make daily is limited, making it difficult to manage communication with a large volume of clients. Supply arises from demand. In response, the artificial intelligence sector has developed intelligent voice outbound robots that offer a comprehensive solution. This includes concurrent and efficient call handling, understanding customer intent, smart conversational interactions, and generating outbound call reports. Delegating labor-intensive, fixed-process, and repetitive tasks to AI systems has become a significant trend in commercial banks.

In analyzing post-loan collection scenarios, we observe that current market products generally adopt a predefined customer interaction strategy. This involves scripting responses to various customer intents, forming a strategic selection mechanism wherein the NLP module maps customer input to response strategies during model inference. The TTS module or prerecorded modules then execute voice responses. Traditional implementation methods involve "response systems based on structured data" and "question answering systems based on knowledge graphs."

Figure 7.1 illustrates the processing workflow of a structured data-based response system. It analyzes the syntax and semantics of input data and queries the database for the most appropriate answer. The system's simplicity and ease of setup are clear advantages, but its scalability and intelligence are limited.

Figure 7.2 illustrates the operational workflow of the question answering system that leverages knowledge graph technology. This system pre-constructs an entity-relation semantic knowledge graph, which serves as its foundational data structure. The system employs advanced techniques such as named entity recognition, triplet extraction, and entity-relation extraction to delineate the "subject-verb-object" configurations within the dataset. This approach ensures the identification of the most accurate answer within the graph, marking a significant research focus within the

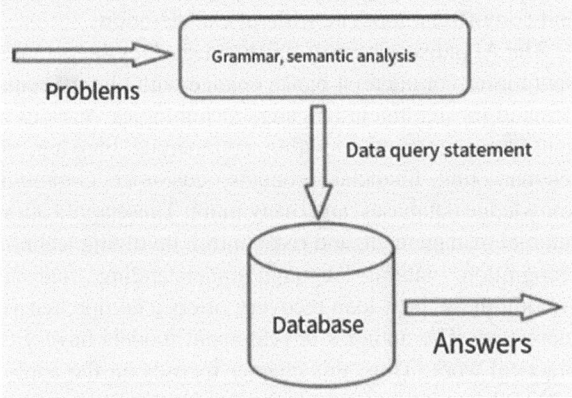

Fig. 7.1 Processing flow of the response system based on structured data

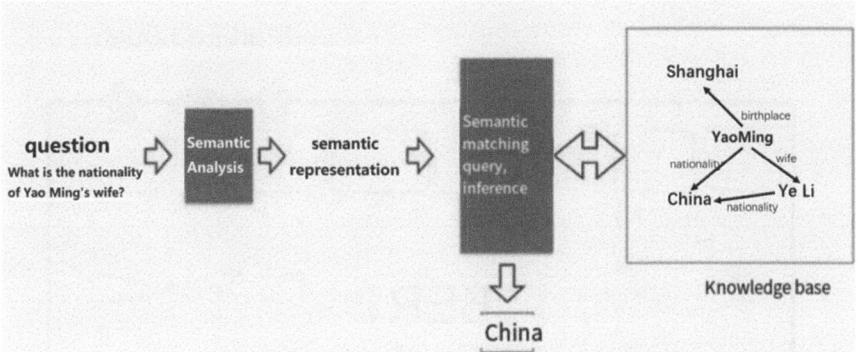

Fig. 7.2 Process flow of the question answering system based on knowledge graph

domains of graph databases and graph neural networks. The system's strengths include its precise semantic clarity and robust interpretability. However, it also presents challenges, primarily due to the formidable technical expertise required to construct and maintain the system.

Both aforementioned practices necessitate aligning the customer's intention to a predefined node, which invariably results in a suboptimal customer experience. This misalignment occurs because the customer's articulated intent may not correspond to any existing database entry or pre-configured graph node, thereby rendering the machine's response inadequate. Consequently, we advocate for an enhanced alternative: the "question answering system based on free text reading comprehension." Illustrated in Fig. 7.3, this system processes the customer's input as a query, enabling the algorithm to perform a reading comprehension task on the pre-existing free text to intelligently identify the optimal response to the customer's intent. Notably, the algorithm interprets both the customer's input and the free text corpus at a semantic level, ensuring that inquiries with analogous semantics receive appropriate and consistent responses. This approach boasts a high degree of algorithmic sophistication, considerable scalability, and significantly improved user experience.

Through rigorous experimentation, the author concludes that this approach not only maintains the robustness of the model but also enhances the accuracy of target interactions and customer satisfaction to a notable degree. Specifically, many customer inquiries fall outside the realm of open text format, such as statements like "I didn't hear clearly" or "It's not convenient now." For these particular queries, predefined semantic nodes are established, and a semantic matching response mechanism that leverages structured data is implemented. Conversely, for other types of questions, a question-and-answer framework centered on the comprehension and analysis of free text is utilized. The workflow for this system is illustrated in Fig. 7.4. In practice, the integration of both methodologies yields optimal results. Due to space constraints, this chapter focuses exclusively on the approach that involves reading comprehension of free text.

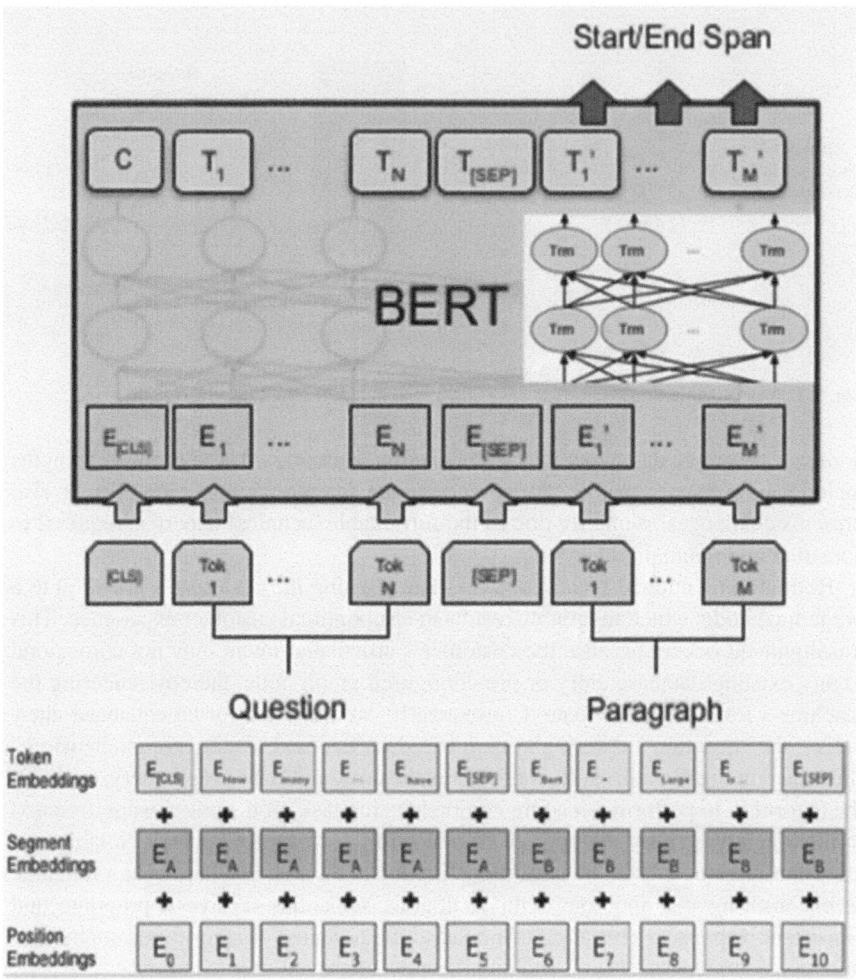

Fig. 7.3 Processing flow of a question-and-answer system based on free text reading comprehension

7.1 Scheme Design

The term "dual-mode robot" represents a system that recognizes both Mandarin and various dialects through dedicated models. This robot interprets customer speech, either in Putonghua or a dialect, based on singular model compatibility per sentence. Thus, during client interaction, the robot must switch between languages seamlessly, depending on whether the spoken input is Mandarin or dialect.

The project's objective is to develop an intelligent voice response system from the ground up. This entails managing the complete spectrum of data activities: data

7.1 Scheme Design

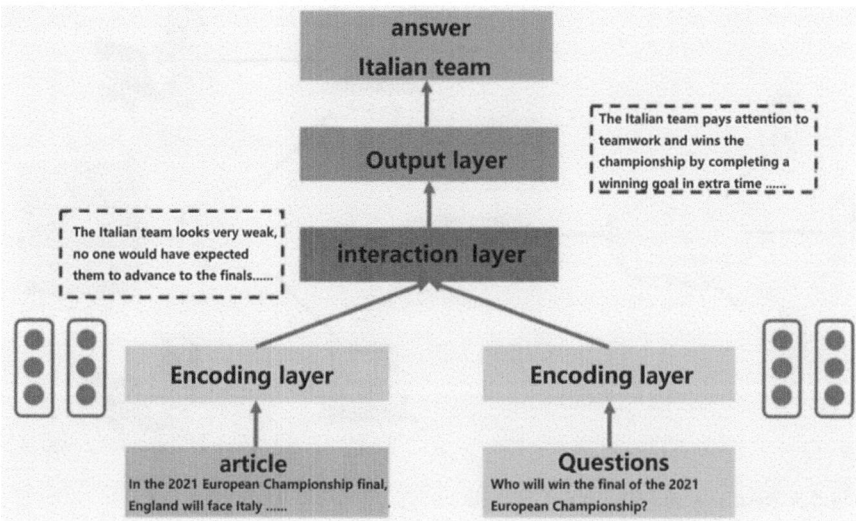

Fig. 7.4 Workflow of Q&A system based on free text reading comprehension

collection, processing, and customer interaction within voice queries. The system undertakes multiple tasks such as retrieving unstructured text data about clients, initiating automated phone calls, recognizing customer intents via voice analysis, matching responses to these intents, creating synthesized voice outputs, and conducting dialogues. Furthermore, it records both dialogue and voice data, collecting information about clients' repayment intentions, which are stored in a database for subsequent analysis. The entire deployment of this solution remains localized, ensuring that all voice engines, models, and codebases are executed on local servers. Given the necessity to accommodate dialect speech, specialized dialect datasets have also been curated to enhance speech recognition accuracy. This case involves intricate integrations such as session initiation protocol (SIP), voice gateway hardware, and database operations. Due to the breadth of content and complexity, this chapter introduces only the primary five components: custom corpus transfer learning, automatic speech recognition, free text reading comprehension, text-to-speech synthesis, and SIP protocol integration.

The system architecture, depicted in Fig. 7.5, utilizes SIP to facilitate the IP telephony system. The dialect voice phone collector interconnects with the SIP server via the SIP client, enabling user registration, call initiation, and voice transmission. Internally, the robot communicates through a local area network (LAN) with the voice gateway. The gateway then converts digital signals into analog signals, allowing voice interaction through the carrier's telephone network.

For the voice interactive robot, the logical architecture shown in Fig. 7.6 is used in this case.

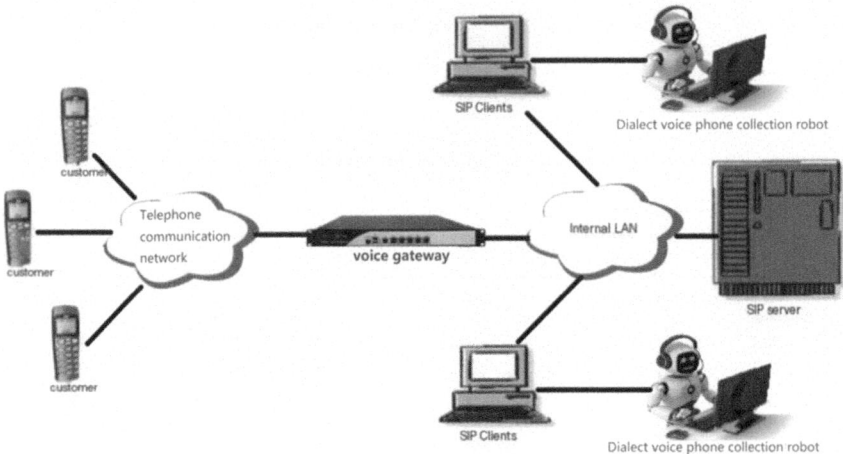

Fig. 7.5 System architecture of this example

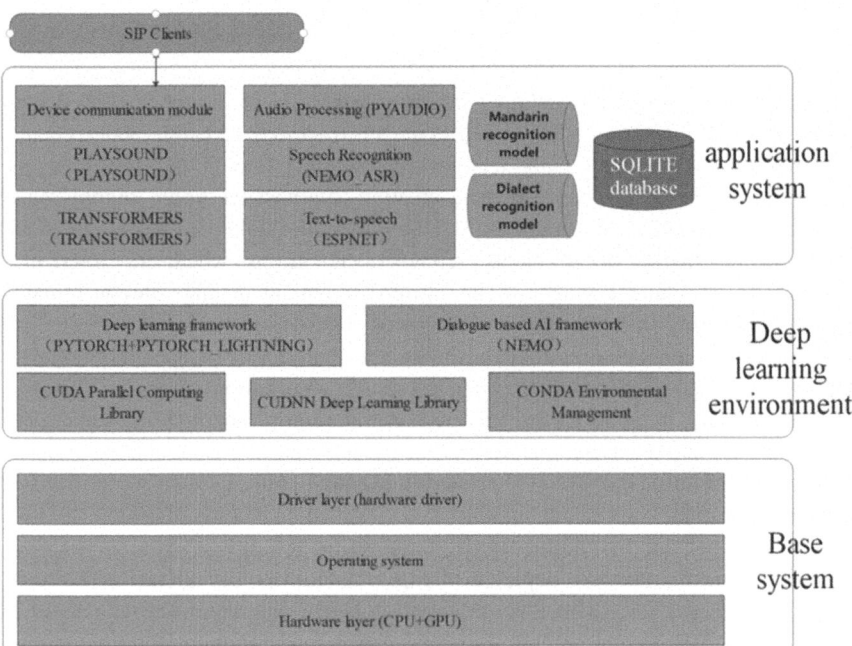

Fig. 7.6 Logical architecture of dialect-based voice phone collection dual-mode robot

The prevalent solution for voice interaction robots in the industry utilizes the open-source Kaldi framework. However, this case study introduces an alternative open-source approach using the Nemo conversational AI framework for a multi-model voice interaction solution. This innovative solution is founded on Nvidia's deep learning architecture and leverages GPU hardware for high-performance computing. It integrates Compute Unified Device Architecture (CUDA) for parallel processing, CUDNN (Deep Neural Network Acceleration Library) for neural network optimization, and CONDA (an environment management tool) as the foundational support. Building upon the PyTorch deep learning framework, this solution establishes a sophisticated intelligent question answering application. At the application layer, the Nemo_asr module facilitates accurate dialect speech recognition. Additionally, the transformers library is employed for natural language processing (NLP) to ensure robust semantic understanding. For text-to-speech (TTS) conversion, the Espnet network demonstrates efficacy in producing natural-sounding speech. Moreover, the Pyaudio and Playsound modules enable seamless voice interaction capabilities. The communication module further enhances functionality by integrating with a session initiation protocol (SIP) client. Once the SIP client connects with the SIP server, it allows for natural voice interactions with customers via the voice gateway and telephone communication network. The detailed workflow, as demonstrated in Fig. 7.7, showcases the comprehensive integration and operational efficiency of this innovative voice interaction solution.

7.2 Intelligent Q&A Technology

This section introduces the basic tasks and processes of intelligent voice question answering system and elaborates the implementation principle and development framework of the three key technologies of ASR, NLP, and TTS in detail. At the same time, it also introduces the transfer learning technology in dialect ASR task, the related open-source community model and SIP communication protocol.

7.2.1 Basic Tasks of Intelligent Voice Q&A System

The intelligent voice Q&A system in this case is based on deep neural network to realize natural language communication between human and machine, so that the machine can understand the words spoken by human, read the sentences written by human, write the sentences understood by human, and speak the words understood by human, and maintain the barrier-free semantic communication between human and machine during this process. The tasks of human–computer interaction are shown in Fig. 7.8.

An intelligent voice question answering system usually consists of three subsystems as shown in Fig. 7.9.

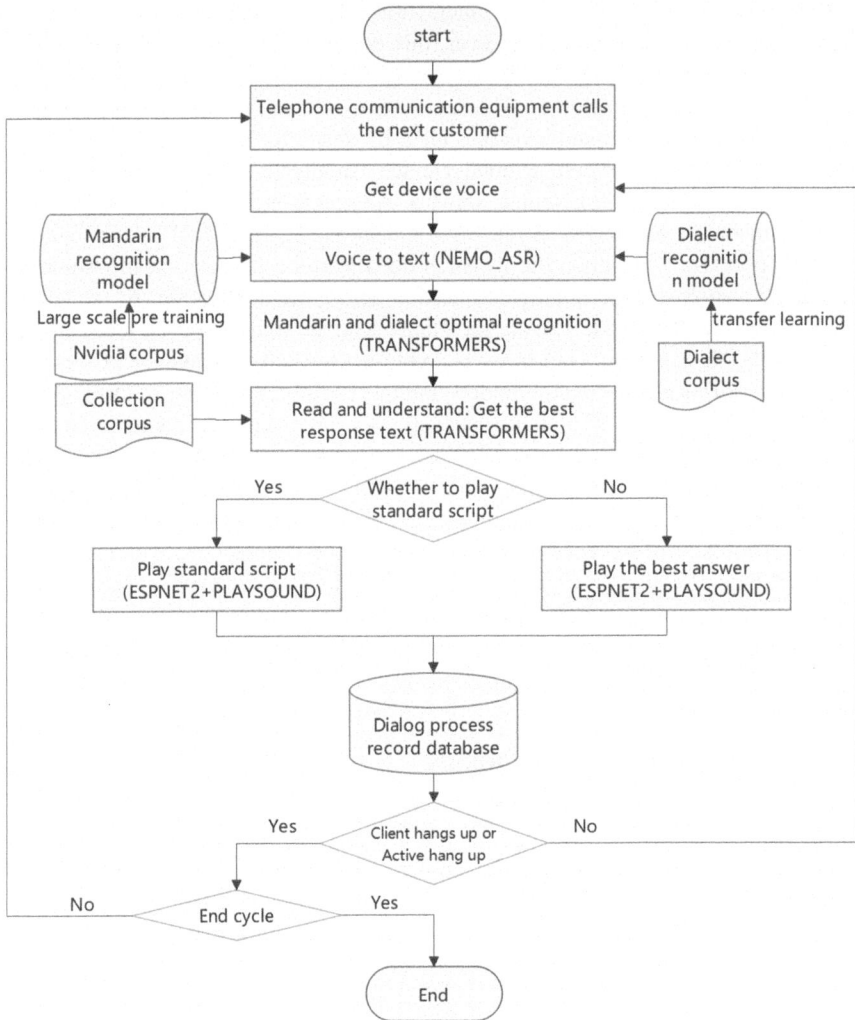

Fig. 7.7 Work flow of dialect voice phone collection dual-mode robot

Automatic speech recognition (ASR) technology transforms a person's spoken language into written text. This field spans multiple disciplines, including acoustics, phonetics, linguistics, digital signal processing theory, information theory, and computer science. Essentially, ASR functions as the auditory interface for computers. In this instance, ASR facilitates the recognition of both Mandarin and various dialects. Natural language processing (NLP) plays a pivotal role by processing the text recognized by ASR to determine the most accurate response from free text data. This process relies on several technologies, such as text vectorization, semantic understanding, and question answering systems, effectively serving as the cognitive component of the computer. TTS (Text to Speech) technology converts written text

7.2 Intelligent Q&A Technology

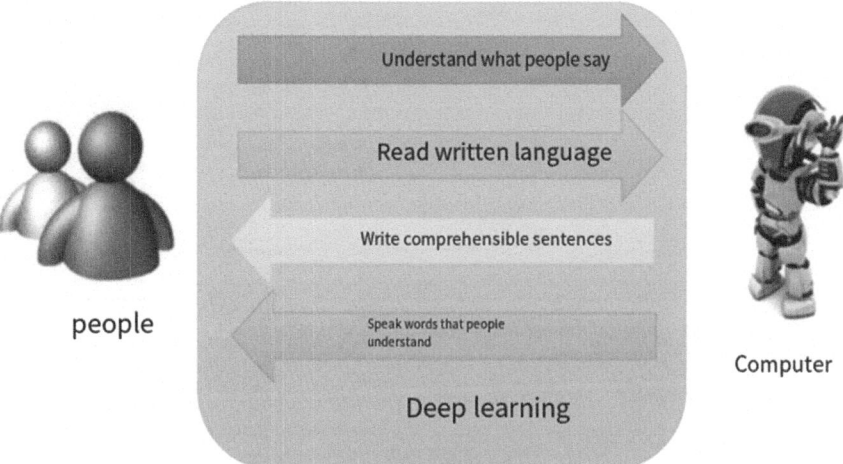

Fig. 7.8 Human-computer interaction task of the intelligent voice question answering system

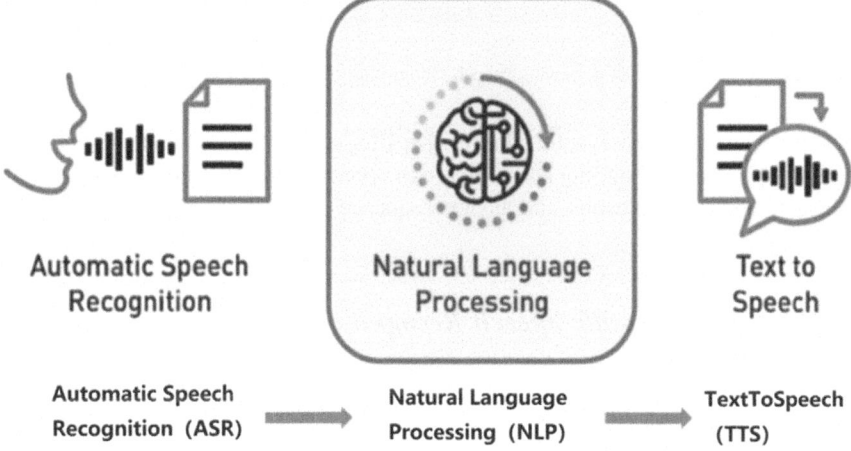

Fig. 7.9 Three sub-tasks of the intelligent voice question answering system

into synthetic speech, allowing the computer to vocalize responses, thus acting as the computer's oral interface. In this particular case, TTS generates speech in standard Mandarin.

The typical operation of an intelligent voice question answering system is depicted in Fig. 7.10. Users initiate interaction by posing questions vocally. The ASR stage involves speech feature extraction, the application of acoustic models and decoders, and the utilization of language models to transcribe speech into text. During the NLP stage, the textual data undergoes corpus text processing, query searches, automatic corrections, search ranking, and question answering to infer the

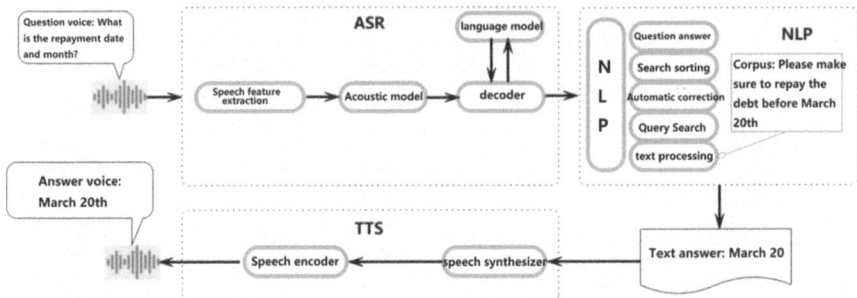

Fig. 7.10 Workflow of the intelligent voice question answering system

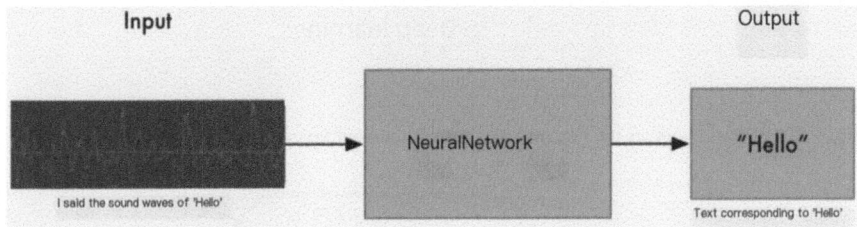

Fig. 7.11 Simple process of automatic speech recognition

most relevant answers. Finally, during the TTS stage, these textual answers are converted into corresponding speech through speech synthesis and encoding, completing the auditory question-and-answer sequence.

7.2.2 ASR Automatic Speech Recognition Technology

Figure 7.11 reflects the simple process of automatic speech recognition, which sends the input speech sound wave into the neural network and outputs the corresponding language text. It usually needs to go through the steps of sound wave digital sampling, framing, Fourier transform, feature vector extraction, acoustic network reasoning, text alignment, etc., which are briefly introduced below.

Figure 7.12 illustrates a schematic chart elaborating the principles of digital sampling and the application of Fourier transform. Given that sound exists as an analog waveform, its time domain representation merely expresses variations in sound pressure over time. This format fails to encapsulate the comprehensive characteristics of the auditory signal. Consequently, an essential step involves converting the analog waveform into a digital format through sampling operations. This conversion process enables effective feature extraction in subsequent stages. The conversion process incorporates two primary stages: sampling and quantization. Sampling requires setting a specific sampling rate while quantization involves the

7.2 Intelligent Q&A Technology

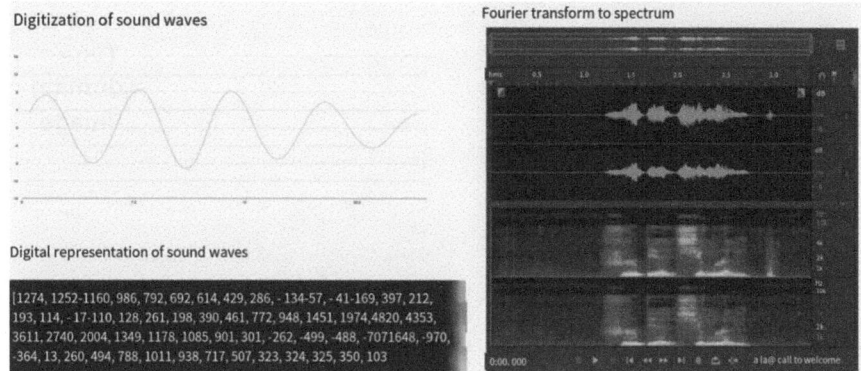

Fig. 7.12 Digital sampling and Fourier transform

determination of sampling bits. This procedure transitions the continuous analog signal into a series of discrete data points. For instance, a sampling frequency of 16 kHz signifies that 16,000 samples are captured per second. These samples are recorded as amplitude values within the range of integers from -32768 to 32767.

Variations in mouth shape introduce inconsistencies in the sound signal, necessitating the application of frame processing for effective analysis. Framing smooths the signal within each discrete segment. Commonly, a frame duration of 20 ms is used. Following framing, we perform a Fourier transform on the amplitude data, converting it from the time domain to the frequency domain. Time domain data represents observations over time, whereas frequency domain data represents observations over frequency. The Fourier transform decomposes the frame's time-sampled amplitude function into its constituent frequencies' amplitudes. In essence, it changes the observational dimension from time to frequency. Since any continuous function can be expressed as a superposition of sinusoidal functions at varying frequencies, aligning these sinusoidal functions yields a frequency domain representation. This frequency domain image uses frequency as the primary observational coordinate. While the time domain image details signal changes over time, the frequency domain image illustrates signal variations across different frequencies, as exemplified in Fig. 7.13.

A frequency domain representation of an image exhibits the frequency attributes intrinsic to the signal. Converting a sound signal from the time domain to the frequency domain is crucial because it unveils sound characteristics that are directly correlated with the signal's inherent properties. For instance, in the illustration on the right side of Fig. 7.12, the densely concentrated light areas in the low-frequency regions of the spectrogram signify a pronounced low-frequency sound signal, typically indicating a male voice.

By utilizing a spectrogram, the data structure of the sound undergoes a transformation; essentially, the amplitude data from a time series is converted into a two-dimensional visual representation. Recognizing such a two-dimensional image as text essentially becomes an image classification challenge. The most sophisticated

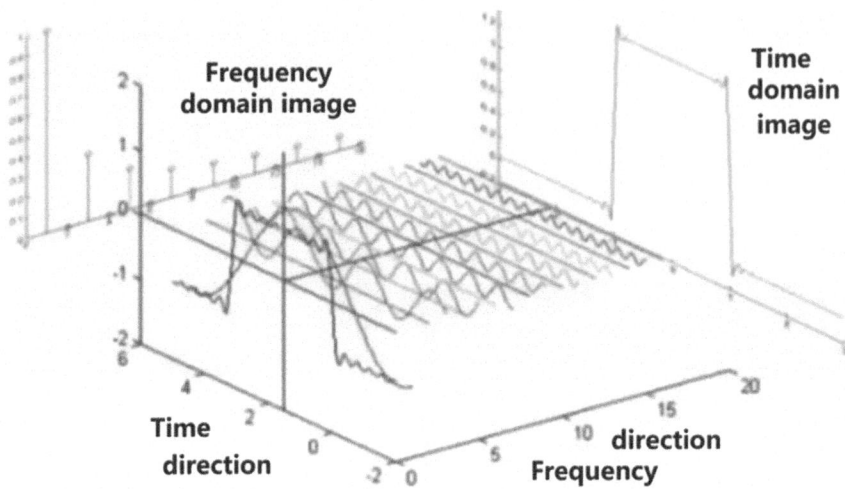

Fig. 7.13 Fourier transform frequency domain image and time domain image

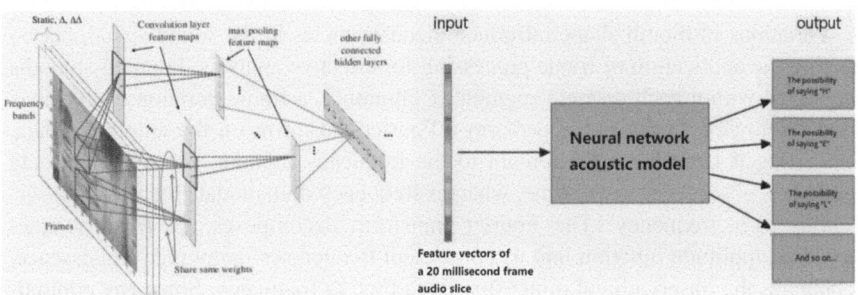

Fig. 7.14 Schematic diagram of sound feature vector extraction and acoustic network inference

solution to this particular problem is the application of convolutional neural networks (CNNs). Figure 7.14 illustrates the process of sound feature vector extraction and acoustic network inference. Inputting a frame of the spectrogram image, the convolutional layer extracts local features of the image. Subsequently, the maximum pooling layer captures predominant features over a broader scope, discarding insignificant details to produce a more compact feature map. The fully connected layer then processes this map to derive the speech feature vector. This feature vector encapsulates the core attributes of the spectrogram and is fed into an acoustic neural network classifier. Once trained, the acoustic neural network can categorize the feature vector, generating probability scores for each word. The word with the highest probability score is then rendered as the text output for the frame.

Figure 7.15 shows that when we run the complete audio and connect the corresponding text output of each frame, we get a text sequence, where the horizontal axis represents the time and the vertical axis represents the corresponding text output.

7.2 Intelligent Q&A Technology

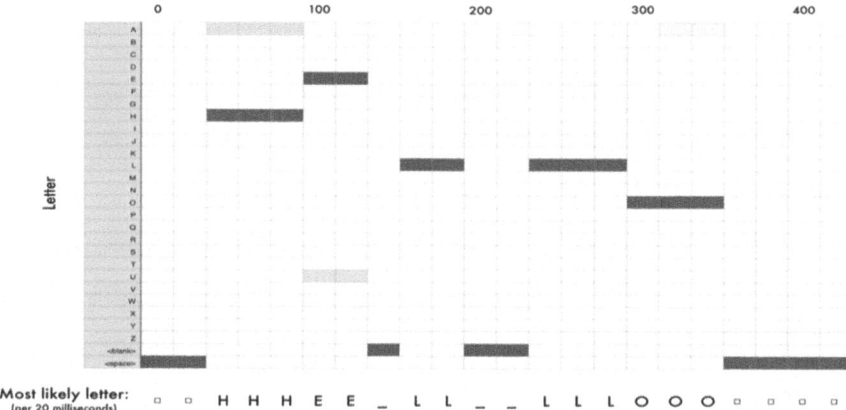

Fig. 7.15 Schematic diagram of sound feature vector extraction and acoustic network inference

Fig. 7.16 Schematic diagram of automatic text alignment in the CTC algorithm

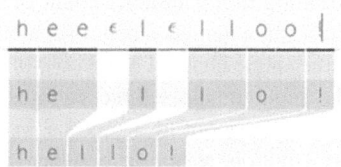

It is worth noting that due to the different pronunciation duration of each phoneme, the text sequence obtained at this time has repeats and intervals. At this time, we need to use Connectionist Temporal Classification (CTC) algorithm for end-to-end automatic alignment, so as to merge the repeats and intervals. Finally output HEE_L__LLOO as HELLO, as shown in Fig. 7.16.

The frame-by-frame recognition of spliced sentences may not be smooth, we then use a Decoder for decoding correction, such as greedy algorithm, bunching search and other algorithms, and finally complete the optimization of the output text. At this point, we have completed the conversion from speech to text, and the automatic speech recognition is complete.

7.2.3 QuartzNet Model

To enhance operational speed and precision, the QuartzNet model is incorporated within the Nemo framework. Before delving into this model, it's essential to understand the deep separable point convolution operation.

In the conventional convolution operation illustrated on the left side of Fig. 7.17, four convolutional kernels perform the convolution on a three-channel image. Each 3×3 convolutional kernel aligns with the three channels of the input image,

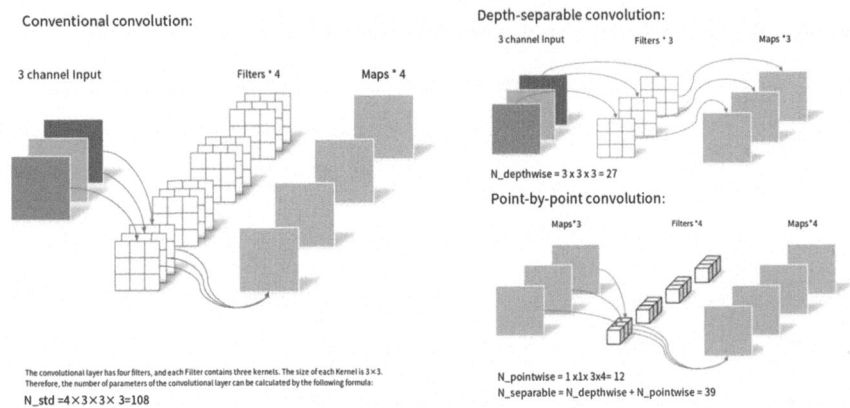

Fig. 7.17 Schematic diagram of deep separable point convolution operation

resulting in a parameter count of $4 \times 3 \times 3 \times 3 = 108$. Contrastingly, the top right illustration depicts a depth-separable convolution. Its key feature involves the use of three single channel 3×3 convolutional kernels that convolve independently across the input image's three channels. This yields three feature maps, reducing the parameter count to $3 \times 3 \times 3 = 27$. The lower right figure shows the subsequent convolution operation between these feature maps and four 1×1 convolutional kernels, each encompassing three channels. This step effectively merges the channel information from the feature maps, with a parameter capacity of $1 \times 1 \times 3 \times 4 = 12$. The combined process—illustrated in the top right and lower right figures—of channel separation followed by channel merging results in a total parameter count of $27 + 12 = 39$. This technique, which ultimately produces a feature map comparable to that of conventional convolution operations but with significantly fewer parameters, transitions from a parameter capacity of 108 to just 39. Thanks to the reduced parameter count, the model trains faster, allows for rapid inference, and can be made deeper. Consequently, it is well-suited for deployment in edge computing scenarios characterized by low power consumption and limited computational resources.

As previously described, Nemo constructs deep learning architectures utilizing neural modules as its foundational principle. This modular design philosophy is exemplified in the structure of QuartzNet, particularly illustrated in Fig. 7.18. The TCSConv-BN-Relu xR module, also known as the Time Channel Separable Convolution module, comprises R convolutional blocks. Each block consists of a 1-dimensional depth-separable convolution layer, followed by a point convolution layer, a batch normalization layer, and a rectified linear unit (ReLU) activation function layer, sequentially arranged. These convolutional blocks are interconnected through a residual network. The residual network's reliance on identity transformation as a foundational premise ensures that the learning efficiency of each convolutional block is progressively optimized, addressing issues such as vanishing gradients, exploding gradients, and network degradation. The comprehensive architecture of QuartzNet is outlined as follows: Initially, a TCSConv-BN-Relu time-

7.2 Intelligent Q&A Technology

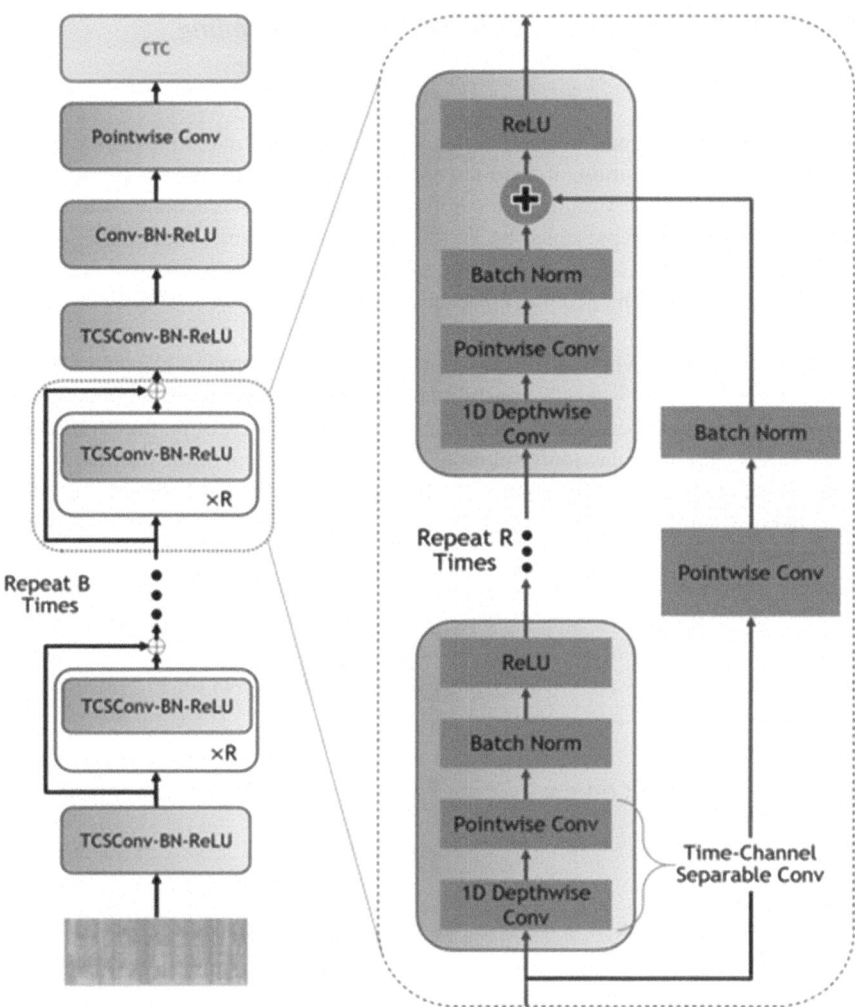

Fig. 7.18 Schematic diagram of the QuartzNet model structure

separated convolution block devoid of residual connections is employed. Subsequently, a series of B time-separated convolution blocks—incorporating residual TCSConv-BN-Relu xR structures—are connected in a residual network configuration. This design enhances network stability and facilitates effective deep learning model training. Then through a TCSConv-BN-Relu time-separated convolution block without residuals, then through an ordinary convolution block CONV-BN-relu, and then PointWise Conv via a point convolution block pointwise Conv output to CTC loss function. The model of QuartzNet structure can be used in Nvidia's NGC container, taking aishell2_quartznet15x5 model as an example, B equals 15, R equals 5, Download address to https://ngc.nvidia.com/catalog/models/nvidia: aishell2_quartznet15x5.

7.2.4 Q&A Technology Based on Free Text Reading Comprehension

After the question text and free text are cleaned and divided into words, we need to express the text as a string of numbers, so that downstream tasks can carry out semantic computation. Common methods include one-hot encoding and word embedding technique represented by Word2vector. As shown in Fig. 7.19.

One-hot encoding assigns unique categorical identifiers to words within a text, where the length of each encoding correlates with the total number of distinct words present in the paragraph. In this encoding scheme, each word vector comprises only one element set to 1, while all remaining elements are 0, forming a sparse matrix. This method, however, lacks the capability to represent the semantic relationships or degrees of correlation between synonyms and antonyms. Conversely, the Word2Vec model offers a compact encoding solution, where each element of the resultant word vector is denoted as a floating-point number. When the word vector dimension is N, the generated word vectors are mapped into an N-dimensional space. In this space, the proximity between word vectors of different words is reflective of their semantic similarity; similar words, such as synonyms, have vectors that lie close together, whereas antonyms are spatially distant. For example, the spatial distance between the word vectors for "run" and "jump" is minimal, while the distance between "dog" and "flower" vectors is significantly larger. Word2Vec derives these word vectors

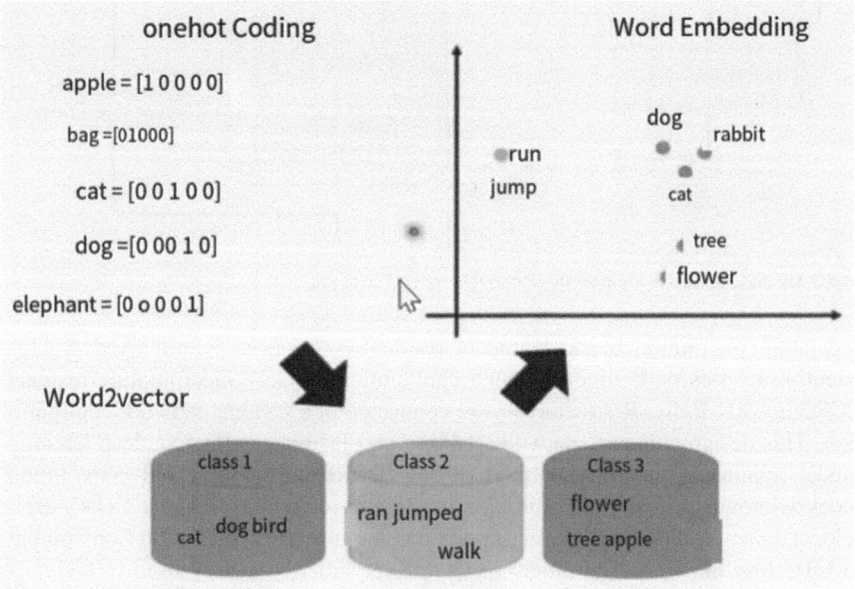

Fig. 7.19 How one-hot encoding compares to Word2Vector in terms of semantic correlation

7.2 Intelligent Q&A Technology

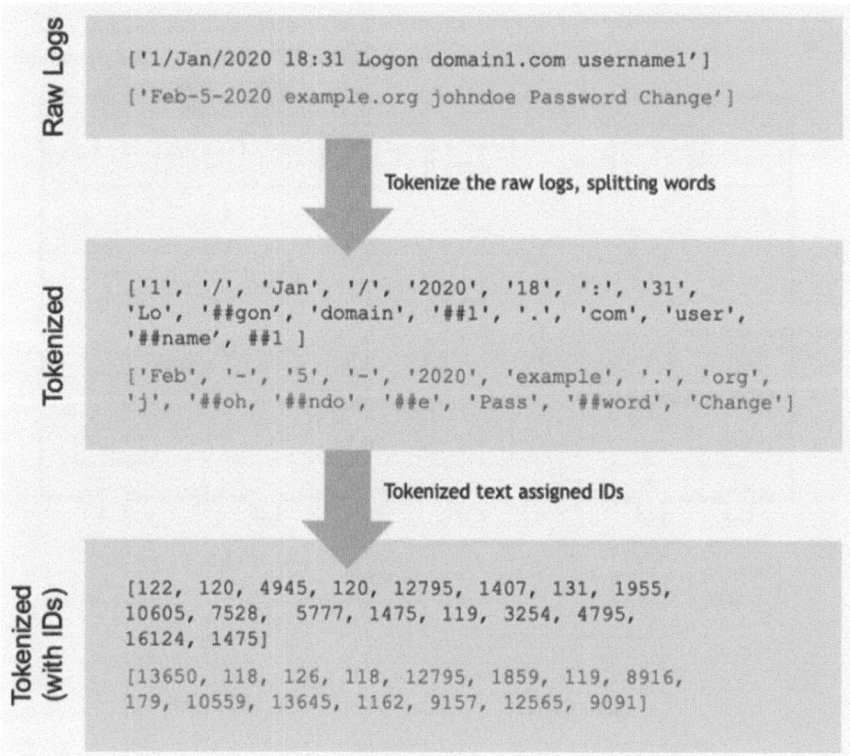

Fig. 7.20 Schematic diagram of text vectorization

through training on extensive corpora, considering the context positions of words within the text. As a result, these vectors encapsulate not only the positions of words in the corpus but also their semantic meanings to a discernable extent. In comparison with one-hot encoding, Word2Vec yields denser and lower-dimensional word vectors, thus enhancing computational efficiency and providing superior capability in evaluating word meaning similarities.

Once you understand word vectors, move on to text vectorization. Firstly, the input text is divided into words, and then the ID number of each word is found in the word list and converted into a text embedding vector of type Tensor data, as shown in Fig. 7.20.

In the Bidirectional Encoder Representations from Transformers (BERT) model, the embedding vector assigned to a paragraph of text encapsulates the semantic representation of both the problem text and the corresponding free text. Additionally, a position embedding vector signifies the positional information of each individual word within the text. These embedding vectors—token embeddings, segment embeddings, and position embeddings—are combined and subsequently input into the Transformer's encoder stack. Within this structure, the model processes and

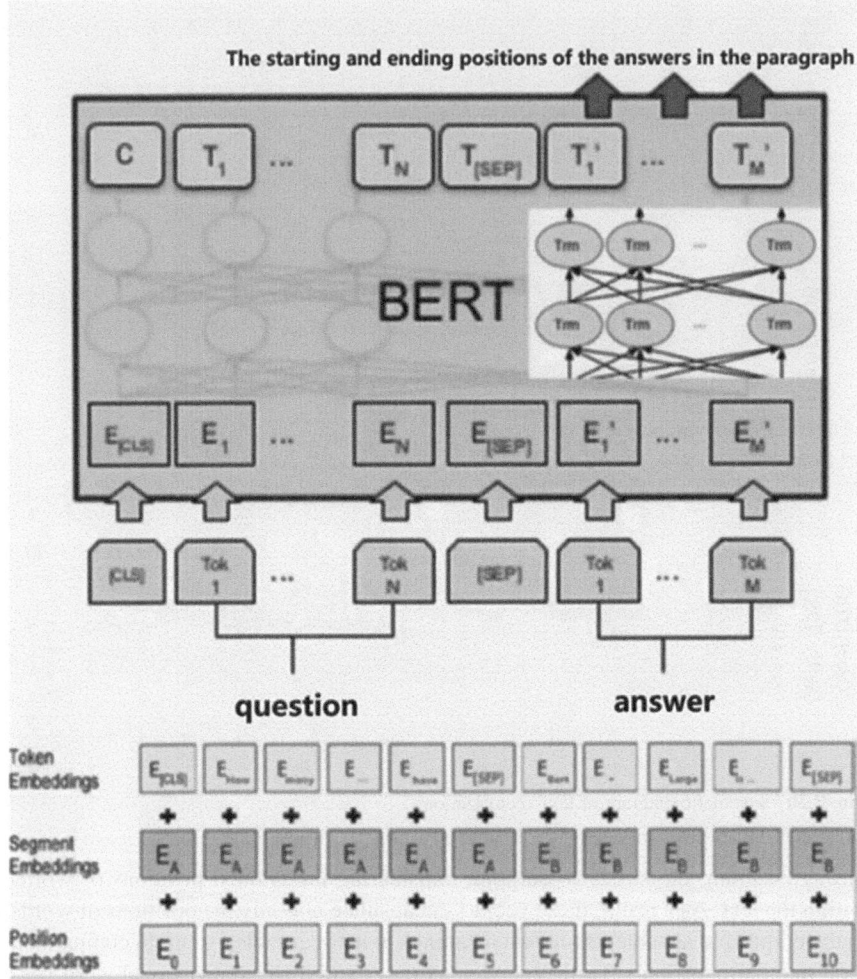

Fig. 7.21 Schematic diagram of reading comprehension question and answer

analyzes the contextual relationships, computing the similarity scores between words in the problem text and those in the free text. This comparative analysis allows the model to identify which words in the problem text exhibit the highest semantic correlation with words in the free text. Through this understanding, it can discern the relational dynamics between the question text and the free text. To pinpoint the optimal answer, the softmax activation function is employed. This function determines the most probable starting and ending positions of the answer within the free text, effectively extracting the precise answer to the posed question. The implementation of this methodology is illustrated in Fig. 7.21.

7.2.5 Text-to-Speech Synthesis Techniques

In Sect. 7.3.4, the answer obtained is in text form, and we need to convert this text into speech to play it to the customer, which requires the use of TTS technology. Typically, the first step of TTS involves preprocessing the text, which includes pauses for punctuation (commas), intonation (question marks), emphasis (exclamation points), etc., and generating language features represented in text vector form. The second step of TTS involves using the encoder and decoder of an acoustic model to generate corresponding spectrograms (i.e., speech features) from the language features. The third step of TTS involves using a vocoder to convert the spectrograms into waveforms for speech playback. The process principle of TTS is illustrated in Fig. 7.22.

7.2.6 Transfer Learning

Transfer learning can be illustrated through a vivid metaphor involving historical royalty. Imagine traveling back in time to become a prince. To govern effectively, you need extensive knowledge. Learning everything from scratch would be impossible due to time constraints. Instead, you would seek out your emperor father, absorbing his accumulated wisdom directly into your mind. This transference enables you to solve problems efficiently by leveraging existing knowledge and forming new insights suited to your reign. Applied to artificial intelligence, this concept reveals that new knowledge systems need not be constructed from the ground up. A foundation of pre-existing, mature models can undergo secondary training or fine-tuning to adapt to specialized needs. In technical terms, to expedite the development of an AI model tailored to a specific domain, we create a specialized dataset. We then use an established AI model as the base, refining its high-level structure and network weights through secondary training. This new model retains the original model's capabilities while gaining proficiency in the specialized field. Transfer learning excels particularly in scenarios involving limited data. Specialized fields often lack large-scale datasets, offering only small-scale data samples. Training a model from scratch under these constraints typically yields subpar

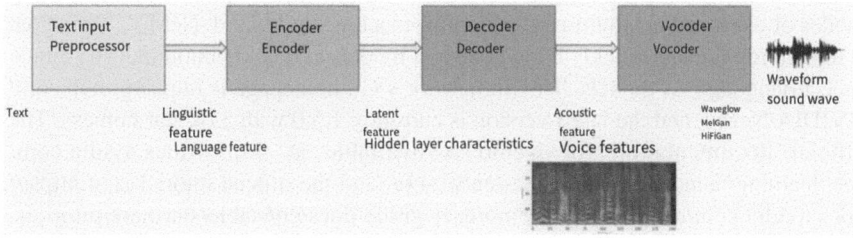

Fig. 7.22 Schematic diagram of TTS speech synthesis process

performance. However, leveraging a pre-trained model and undertaking secondary training with a specialized dataset significantly enhances the model's effectiveness. Thus, transfer learning epitomizes the concept of "standing on the shoulders of giants, seeing farther."

Given that dialect-specific data sets are not readily accessible in initial data collection phases, it becomes necessary to generate them independently. Constraints such as budget, manpower, and time limit the feasibility of producing extensive voice text data sets. Consequently, transfer learning emerges as the optimal strategy to enhance performance under these limitations. A crucial criterion for selecting a base model in transfer learning is its relevance to the task and foundational reliance on expansive data sets. In this specific scenario of identifying Chinese dialects (automatic speech recognition tasks), an English ASR model cannot serve as the base. Instead, a Chinese ASR model must be chosen. We have opted for QuartzNet15x5Base-Zh by Nvidia, which is a sophisticated ASR model comprising 79 layers and 18.9 million parameters. This model is trained extensively, through hundreds of epochs on a multi-GPU setup, using a corpus of approximately 1000 h of Mandarin Chinese transcriptions. Given its considerable capacity and the extensive dataset it leverages, QuartzNet15x5Base-Zh serves as a robust base model. For this application, transfer learning is implemented using this model, leveraging its pre-trained capabilities directly for Mandarin speech recognition. The detailed methodology for transfer learning can be found in Sect. 7.4.2.

7.3 Development Framework

7.3.1 Nvidia NEMO Conversational AI Framework

Nemo is an open-source conversational AI framework produced by—the world's leading artificial intelligence computing company. It uses Pytorch as the deep learning backend, follows the Apache-2.0 license agreement, and builds the entire AI network with neural module as the core concept. It provides three functional modules: automatic speech recognition (ASR), natural language processing (NLP), and text-to-speech synthesis (TTS). As a software framework for GPU chip design companies, Nemo innately supports the Tensor Core of NVIDIA Gpus for maximum performance through mixed precision calculation, supporting distributed training modes of one-machine multi-card and multi-machine multi-card. NeMo's main goal is to help researchers quickly build code and models to quickly build industry-grade speech-language AI models. The framework's Git homepage is https://github.com/NVIDIA/NeMo, and the latest version is currently 1.5.0 with 3.6k star hotness. The official documentation for Nemo is available at https://docs.nvidia.com/deeplearning/nemo/user-guide/docs/en/v1.4.0/, and the official tutorial is at https://docs.nvidia.com/deeplearning/nemo/user-guide/docs/en/stable/starthere/tutorials.html.

7.3 Development Framework

Nemo's main modules include: speech-to-text, speech classification, speech activity detection, speaker recognition, multilingual processing, machine translation, text classification, question answering dialogues, dialogue state tracking, information retrieval, intent recognition, spectrogram generation, and end-to-end speech synthesis. The strength of Nemo lies not only in its own functionality and performance but also in its provision of a range of large pre-trained models. Users can build their own custom datasets and easily perform transfer learning and model fine-tuning on these pre-trained models, thereby swiftly constructing language models tailored to a specific domain. The website for pre-trained models is https://catalog.ngc.nvidia.com/models, and Nemo's models are all marked with the icon .

7.3.2 ESPnet End-to-End Voice Processing Framework

ESPnet is an end-to-end speech processing deep learning framework, providing functionalities such as end-to-end speech recognition (ASR), end-to-end text-to-speech conversion (TTS), end-to-end speech semantic understanding, speech enhancement, speech separation, speech translation, and speech conversion. Its commercial logo can be seen in Fig. 7.23. ESPnet utilizes Chainer and PyTorch as deep learning engines, implements Kaldi-style data processing, follows the Apache-2.0 open-source license agreement, currently has a popularity on Git with 4.5K stars, and its code repository is located at https://github.com/espnet/espnet, with the highest version being 0.10.4. This case uses ESPnet for text-to-speech functionality.

ESPnet offers a range of pre-trained models that can be conveniently downloaded and used. You can find the download link at https://github.com/espnet/espnet_model_zoo/blob/master/espnet_model_zoo/table.csv. These models cover multilingual support for two major tasks: ASR and TTS, including English, Chinese, and Japanese, and support both male and female voices. ESPnet also provides sample code for model usage for quick development, making it very convenient. For more details on ESPnet's model_zoo page, refer to Fig. 7.24.

7.3.3 Transformers Model Library

Hugging Face, headquartered in New York, is a chatbot startup service provider. The name "Hugging Face" in Chinese means "embrace smiling face." In comparison to

Fig. 7.23 Commercial LOGO of ESPnet

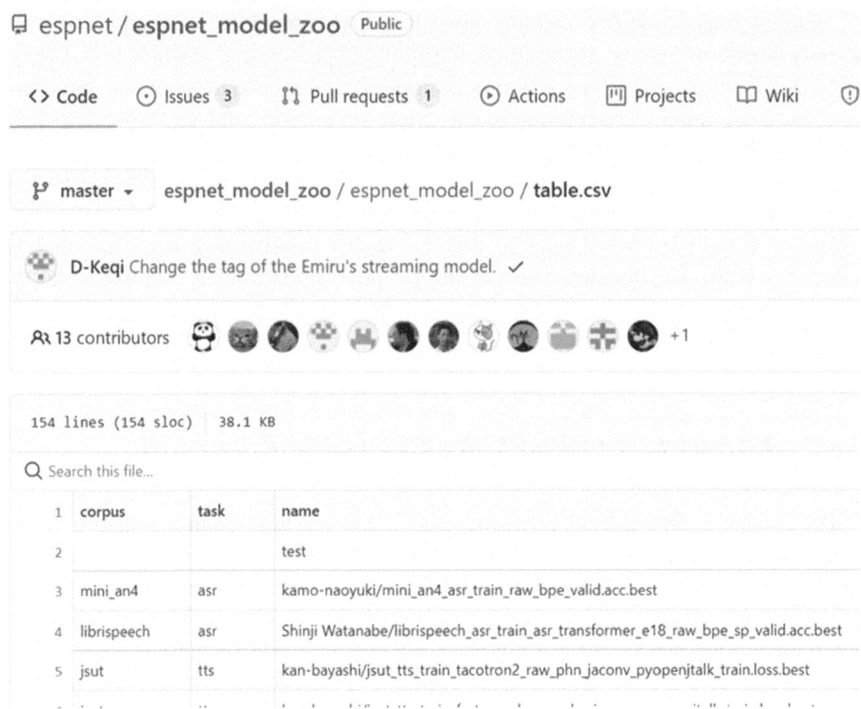

Fig. 7.24 ESPnet's model_zoo page

other companies, Hugging Face places greater emphasis on the emotional and environmental impact of its products. Specializing in NLP technology, Hugging Face boasts a large open-source community. What sets Hugging Face apart is its Natural Language Processing pre-training model library called Transformers, which has gained immense popularity on GitHub with over 55k stars, 13k forks, 86 releases, and 298 branches. The Transformers library offers numerous state-of-the-art pre-trained language models and frameworks, following the Apache-2.0 open-source license. The latest version is 4.12.5, and the code repository can be found at https://github.com/huggingface/transformers. Continuously evolving, the Transformers library now provides thousands of pre-trained models that can perform tasks across different modalities such as text, vision, and audio. Transformers are supported by the three most popular deep learning libraries—Jax, PyTorch, and TensorFlow—ensuring seamless integration across them. For more information, you can visit the official website of Hugging Face at https://huggingface.co/ and access the model downloads at https://huggingface.co/models. Refer to Fig. 7.25 for Hugging Face's commercial logo and official homepage.

In this case, the uer/roberta-base-chinese-extractive qa model was used for Hugging Face inference, while the uer/gpt2-chinese-cluecorpussmall model was used to generate the Chinese text. Hugging Face's model is simple and convenient to use. On its model download page, search for the model, click on it, and you can

7.3 Development Framework

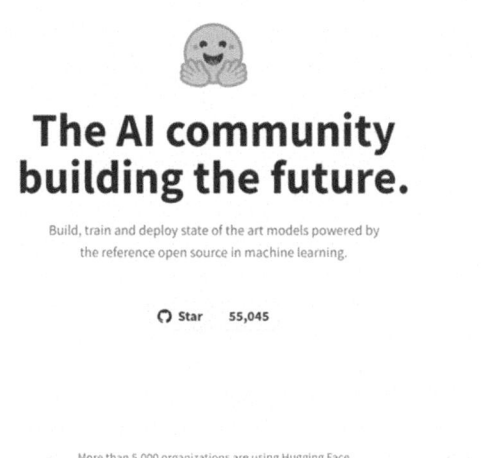

Fig. 7.25 Hugging face's business LOGO and official homepage

```
>>> from transformers import BertTokenizer, GPT2LMHeadModel, TextGenerationPipeline
>>> tokenizer = BertTokenizer.from_pretrained("uer/gpt2-chinese-cluecorpussmall")
>>> model = GPT2LMHeadModel.from_pretrained("uer/gpt2-chinese-cluecorpussmall")
>>> text_generator = TextGenerationPipeline(model, tokenizer)
>>>Text_ Generator ("This was a long time ago," max_length=100, do_sample=True)
    [{'generated_text ':' This was a long time ago, and I used to see it as a transmission of ideas
```

Fig. 7.26 An example of Hugging face's model usage

see the sample code of the model. Figure 7.26 shows a usage example of invoking the above model using the transformers library, simply loading the pre-trained model and then inputting the text to output the prediction results.

7.3.4 PyQt5 Cross-Platform GUI Framework

Qt is a cross-platform GUI framework developed by Trolltech of Norway using C++ language, supports Windows, MacOS, and Linux, and provides cross-platform class libraries, inheritance development tools and cross-platform ides. Its GUI program interface style is exactly the same as the current operating system, and the operation efficiency is high. Code developed using Qt only needs to be recompiled on different operating systems and can be run directly. Trolltech was acquired by Nokia (NOK) in June 2008, and Nokia's Qt business was acquired by Digia, a Finnish company, in August 2012.

PyQt integrates Python with Qt, allowing you to utilize Python language to access the Qt library's API. This not only maintains the efficiency of Qt programs but also enhances development productivity. The official website for PyQt can be found at https://www.riverbankcomputing.com/software/pyqt/. PyQt5 is a version of PyQt released under the GPL v3 license, and its development documentation is available at https://www.riverbankcomputing.com/static/Docs/PyQt5/. The current highest version of PyQt is PyQt6 6.2.2. This case employs PyQt5 to create a desktop GUI for demonstrating the interface operations and results of a voice-activated debt collection system.

7.3.5 SIP Protocol and PJSIP Framework

With the progress of computer science and technology, IP data network based on packet switching technology has replaced the traditional telephone network based on circuit switching in the communication field with its convenience and low cost. SIP is a multimedia communication protocol developed by the Internet Engineering Task Force (IETF). It is an application layer signaling control protocol that provides complete session creation and session change services for a variety of instant messaging services. It is used to create, modify, and release sessions of one or more participants, such as Internet multimedia conferencing, IP telephony, or multimedia distribution. Participants in a session can communicate via Multicast, Unicast, or a mixture of the two. SIP is flexible, easy to implement, and easy to scale.

Session initiation protocol (SIP) originated in the mid-1990s, stemming from the pioneering efforts of Henning Schulzrinne, an associate professor in the Department of Computer Science at Columbia University, and his dedicated research team. A significant attribute of the SIP protocol is its agnosticism toward the type of session being established, focusing solely on the mechanisms of session management. This inherent adaptability enables SIP to be utilized across a broad spectrum of applications and services. These range from interactive gaming and multimedia on demand, like music and video, to various forms of conferencing—voice, video, and web alike. SIP messages employ a text-based format, which facilitates readability and debugging processes, enhancing accessibility for programmers. This new service paradigm simplifies programming efforts, offering an intuitive interface for developers. By leveraging MIME type descriptions akin to email clients, SIP ensures session-related applications can auto-initiate seamlessly. Additionally, SIP capitalizes on numerous pre-existing and well-established Internet services and protocols, such as the domain name system (DNS), real-time transport protocol (RTP), and resource reservation protocol (RSVP). Consequently, the SIP architecture does not necessitate the introduction of novel services, given that many infrastructural components are either pre-existing or readily obtainable. Furthermore, SIP's operational independence from the transport layer is noteworthy. This independence means that SIP can function over various transport networks, including IP and ATM. It employs both the user datagram protocol (UDP) and the transmission control protocol (TCP),

7.3 Development Framework

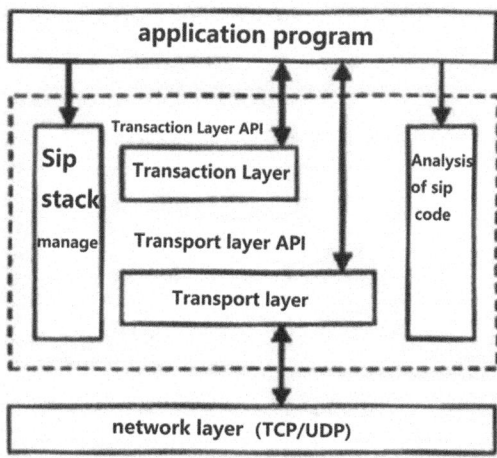

Fig. 7.27 SIP stack architecture

providing flexible connectivity independent of the underlying transport infrastructure. The construction of SIP messages mirrors that of the hypertext transfer protocol (HTTP), thereby enabling developers to craft applications using familiar programming languages. The structural details of the SIP stack can be visualized in Fig. 7.27, illustrating its layered architecture.

PJSIP stands out as the premier open-source SIP framework within the developer community, boasting support for a wide range of SIP extension features. Designed as a comprehensive multimedia communication framework grounded in the SIP protocol, PJSIP offers an exceptionally clear API alongside robust NAT traversal capabilities. The framework's remarkable portability ensures compatibility across nearly all modern systems, including desktop environments, embedded systems, and smartphones. Furthermore, PJSIP excels in multimedia support, encompassing voice, video, presence, and instant messaging functionalities. Commendably documented, PJSIP provides an environment that significantly enhances developer productivity. The framework's inception traces back to 2005, credited to the pioneering efforts of Benny Prijono and Perry Ismangil, followed by the invaluable contributions of Nanang Izzuddin and Sauw Ming. In the subsequent year, Teluu Ltd. emerged as the entity responsible for the ongoing development and maintenance of PJSIP. Notably, PJSIP operates under a dual-licensing model, featuring both the GPLv2 and a commercial license, allowing developers the flexibility to select a licensing option that best suits their project requirements.

Structurally, PJSIP comprises two primary components: the SIP stack, responsible for SIP message processing, and the media stream processing module, which handles RTP packet processing. The SIP stack is intricately layered, offering granularity and modularity. At the foundation lies PJLIB, a fundamental library that manages low-level operations such as data structures, memory allocation, file I/O, threading, and thread synchronization, forming the bedrock of the SIP stack. Building upon this foundation, PJLIB-UTIL enhances functionality with commonly used algorithms including MD5 and CRC32, as well as APIs for string manipulation and

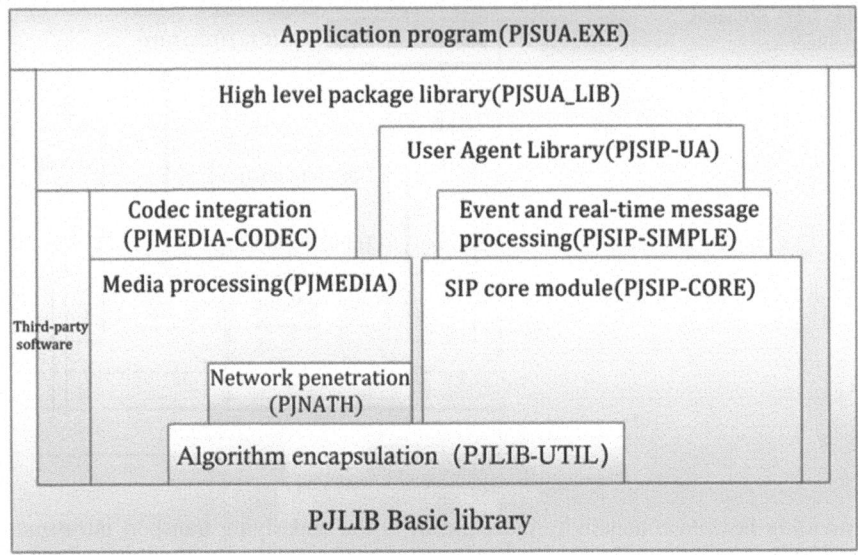

Fig. 7.28 PJSIP module architecture

file format parsing. Central to the framework is PJSIP-CORE, the core module of the SIP protocol stack, which ensures reliable SIP message handling. Complementing this, PJSIP-SIMPLE provides SIP event and real-time message processing capabilities. For session and registration management, PJSIP-UA serves as the dedicated SIP user agent library. Enhancing usability, PJSUA-LIB offers a high-level abstraction, integrating seamlessly with PJSIP-UA and underlying libraries. Addressing network-related challenges, PJNATH specializes in network layer conversion, while PJMEDIA-CODEC brings together various encoding and decoding algorithms to handle multimedia data efficiently. The architectural overview, as depicted in Fig. 7.28, illustrates the intricate yet cohesive structure of the PJSIP modules.

In this case, PJSIP is used to build a network telephone system based on the internal LAN, which realizes the conversion of analog signal and digital signal of the telephone communication network and the management of call users.

7.4 Case Practice

This section introduces the development and operation environment construction, key code writing, and operation results of this case. After reading this section, you will understand the specific implementation process of this case in detail.

7.4.1 Hardware and Software Environment Setup and Case Runs

Dialect Speech Phone Robot

Hardware configuration: I9-10900K processor, RTX2080 Super 8GB graphics card, 32GB RAM, 1TB SSD.

Software environment: Windows 10 64-bit, Anaconda3, Cuda 10.1, Cudnn7.6.4, Pytorch 1.7.0+cu101, Nemo 1.0.0b1.

Since Nemo framework requires an NVIDIA GPU to run, NVIDIA graphics card is essential in this scenario. Please refer to Sect. 1.3.1 for Anaconda installation instructions. To install Nemo, download the source code from https://github.com/NVIDIA/NeMo/tags?after=v1.0.0rc1, and run "python setup.py install" for installation. Use "pip install espnet" to install ESPnet. Various models need to be downloaded from the internet and placed in the corresponding directories.

Double-click on the "asr transfer learning" directory in the attachment of this book and install the Audacity 2.3.1.exe recording software, setting the recording to mono, 44,100 Hz as shown in Fig. 7.29.

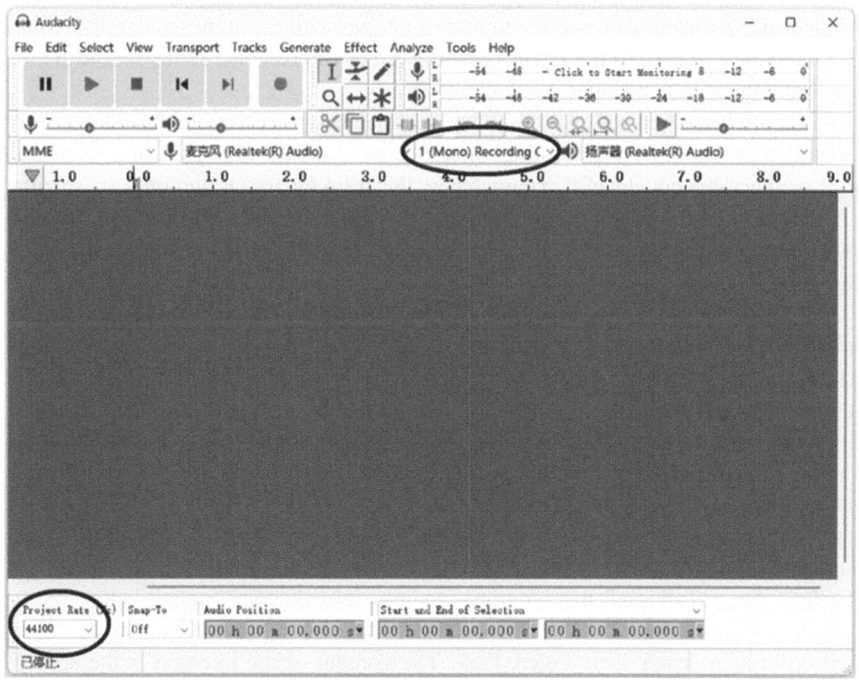

Fig. 7.29 Schematic diagram of recording software

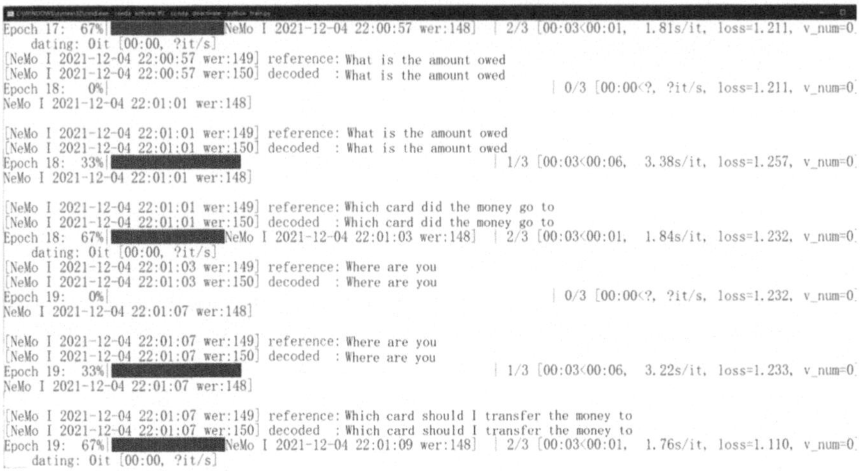

```
Epoch 17:  67%|████████████|NeMo I 2021-12-04 22:00:57 wer:148]  | 2/3 [00:03<00:01,  1.81s/it, loss=1.211, v_num=0
    dating: 0it [00:00, ?it/s]
[NeMo I 2021-12-04 22:00:57 wer:149] reference: What is the amount owed
[NeMo I 2021-12-04 22:00:57 wer:150] decoded  : What is the amount owed
Epoch 18:   0%|                                                  | 0/3 [00:00<?, ?it/s, loss=1.211, v_num=0
NeMo I 2021-12-04 22:01:01 wer:148]

[NeMo I 2021-12-04 22:01:01 wer:149] reference: What is the amount owed
[NeMo I 2021-12-04 22:01:01 wer:150] decoded  : What is the amount owed
Epoch 18:  33%|██████|                                           | 1/3 [00:03<00:06,  3.38s/it, loss=1.257, v_num=0
NeMo I 2021-12-04 22:01:01 wer:148]

[NeMo I 2021-12-04 22:01:01 wer:149] reference: Which card did the money go to
[NeMo I 2021-12-04 22:01:01 wer:150] decoded  : Which card did the money go to
Epoch 18:  67%|████████████|NeMo I 2021-12-04 22:01:03 wer:148]  | 2/3 [00:03<00:01,  1.84s/it, loss=1.232, v_num=0
    dating: 0it [00:00, ?it/s]
[NeMo I 2021-12-04 22:01:03 wer:149] reference: Where are you
[NeMo I 2021-12-04 22:01:03 wer:150] decoded  : Where are you
Epoch 19:   0%|                                                  | 0/3 [00:00<?, ?it/s, loss=1.232, v_num=0
NeMo I 2021-12-04 22:01:07 wer:148]

[NeMo I 2021-12-04 22:01:07 wer:149] reference: Where are you
[NeMo I 2021-12-04 22:01:07 wer:150] decoded  : Where are you
Epoch 19:  33%|██████|                                           | 1/3 [00:03<00:06,  3.22s/it, loss=1.233, v_num=0
NeMo I 2021-12-04 22:01:07 wer:148]

[NeMo I 2021-12-04 22:01:07 wer:149] reference: Which card should I transfer the money to
[NeMo I 2021-12-04 22:01:07 wer:150] decoded  : Which card should I transfer the money to
Epoch 19:  67%|████████████|NeMo I 2021-12-04 22:01:09 wer:148]  | 2/3 [00:03<00:01,  1.76s/it, loss=1.110, v_num=0
    dating: 0it [00:00, ?it/s]
```

Fig. 7.30 Asr transfer learning diagram

Click the Record button to record the dialect. In the asr transfer Learning/voice_data directory, create multiple subdirectories named after the dialect content. The recording files of the corresponding content are saved in the subdirectories. When making dialect data set, try to record people with different gender and voice characteristics, and keep the environment as quiet as possible. There should be no less than 100 corpus files per subdirectory.

After you have finished making the dialect dataset, run generate dialect datasets.bat under the asr transfer learning directory to generate train.json and test.json. Modify the max_epochs=20 parameter in the train.py file as required to specify the epochs to be trained. Then run python train.py on the command line to start the transfer learning to train the dialect model, and you will see the screen as shown in Fig. 7.30:

Note that the loss value should converge to a smaller value. The trained dialect model, dialect.nemo, will be generated in the current directory. The model file is actually a compressed package, opened with winrar and found that it contains a model configuration file model_config.yaml and a model weights file model_weights.ckpt. After the training is over, you can delete the pytorch_lightning framework's training log directory lightning_logs. Then copy the migration learning model dialect.nemo to the "Question Answering Robot" directory. Run python question answering robot.py on the command line and you will see the interface shown in Fig. 7.31:

Click "Start Talking," the program receives human voice signals, click "Stop Talking," the program will search for the best answer to the question in the "Free Text.txt" file and play it by voice. Play. The content of the free text is the speech related to the customer's collection task, for example:

7.4 Case Practice

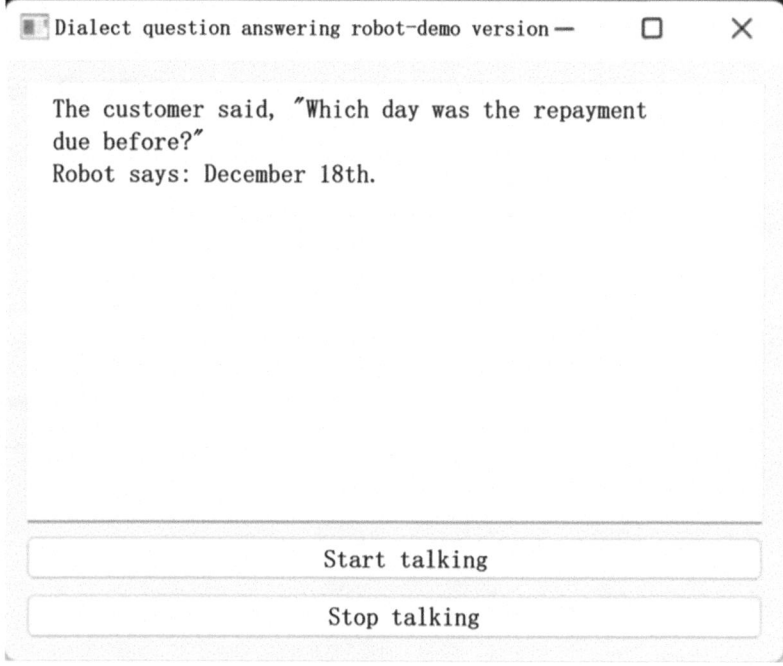

Fig. 7.31 Demo program interface of this case

```
Listen for...
[NeMo I 2021-12-05 09:51:58 collections:173] Dataset loaded with 1 files totalling 27.78 hours
[NeMo I 2021-12-05 09:51:58 collections:174] 0 files were filtered totalling 0.00 hours
[NeMo I 2021-12-05 09:51:59 collections:173] Dataset loaded with 1 files totalling 27.78 hours
[NeMo I 2021-12-05 09:51:59 collections:174] 0 files were filtered totalling 0.00 hours
Mandarin recognition: On which day will the repayment be made from scratch. ppl1=810.153 Dialect recognition:
repayment before which day ppl2=244.197 Dialect recognition is better, using dialect models.
{'score': 0.16354086995124817, 'start': 27, 'end': 33, 'answer': 'December 18th'}
The customer said, "Which day was the repayment due before?"
Robot says: December 18th
```

Fig. 7.32 Shows the program console interface in this example

Mr. X, I am from XX Bank. You have seriously breached the contract now. Please make sure to repay the outstanding amount of 8000 yuan before December 18th. The money will be transferred to your debit card with the ending number 78223, otherwise it will affect your credit record. If there is a problem with your bank card, you can bring your original ID card during working hours and go to the nearby XX bank branch for card replacement or other processing.

Meanwhile, in the console interface, you will see the respective speech recognition results of the Mandarin model and the dialect model, and the program will take the more fluent sentence as the input to the Q&A system and display the questions and answers, as shown in Fig. 7.32.

When the program is running, the GPU-Z software shows that the GPU computing core is used to speed up the model inference, as shown in Fig. 7.33.

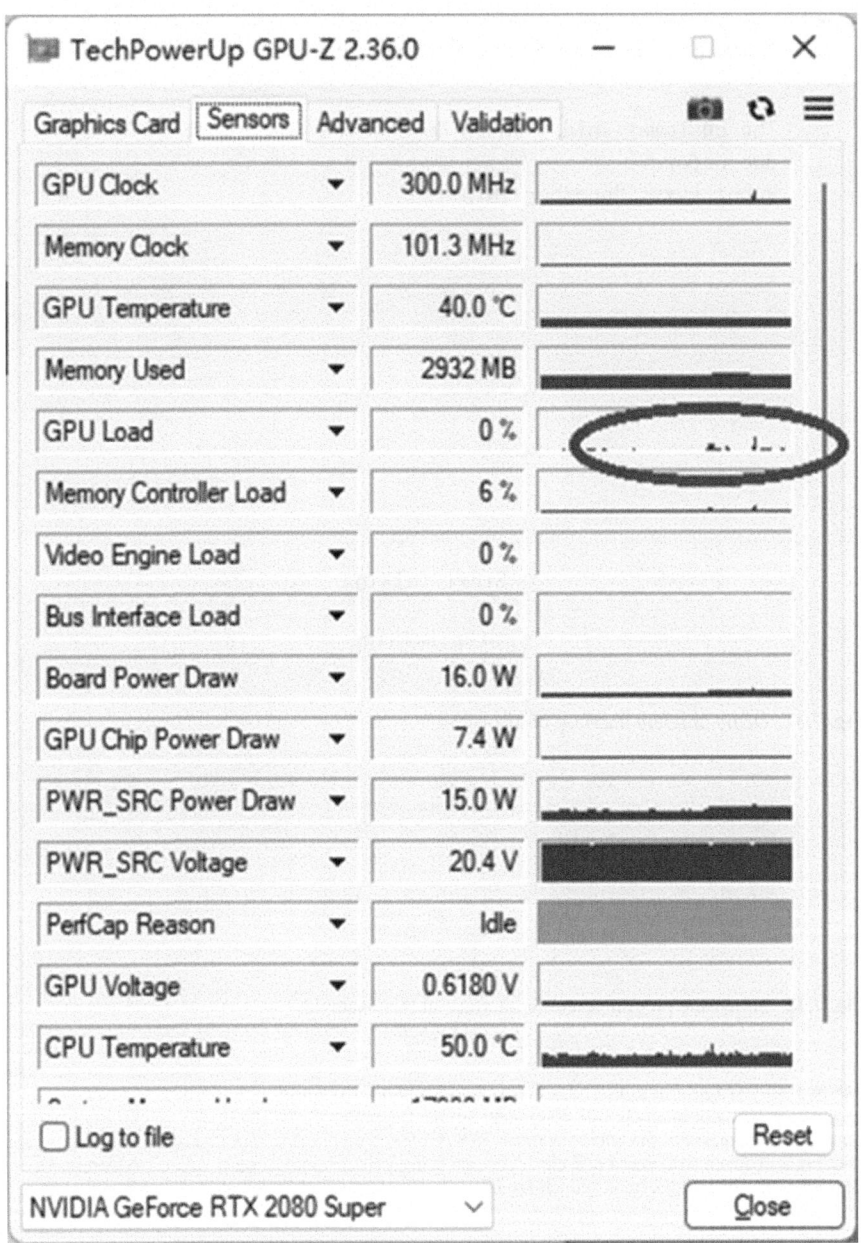

Fig. 7.33 This example demonstrates the use of the program GPU

7.4 Case Practice

At this point, the dual-mode voice robot program has been successfully run, readers can carry out secondary development on this basis, and quickly build their own exclusive voice question answering system.

SIP Network Phone System

There are a variety of options for voice gateway hardware and SIP server software. The voice gateway hardware model chosen by the author in the actual work is Maipu VG2000-32, which is a device that supports multiple concurrent calls. The computer on which the SIP client program is installed is connected to the network entrance (FE0) of the device through a network cable, and the telephone interface (FX0) of the device is inserted into the telephone line to access the telephone communication network of the carrier, as shown in Fig. 7.34.

It is important to note that the FX0 interface is converted to a telephone line via a network cable. Peel the network cable and phone line, connect two of the branch wires, and plug the RJ45 connector into the device. Figure 7.35 shows how to connect the cables.

In actual work, the author uses MyPowerVC8100, the SIP server software provided by Maipu, and does not need to be developed. Obtain the SIP server VM (CentOS system) from Maipu, log in to the IP address of the SIP server using a browser, and configure the IP address of the gateway hardware in resource management—common gateway management. Figure 7.36 shows the IP address.

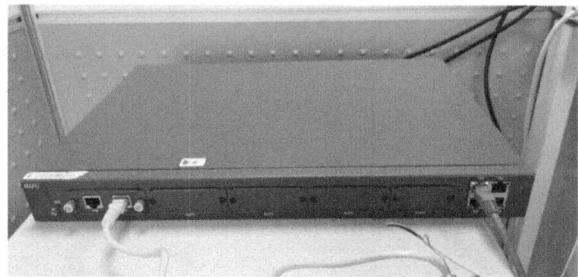

Fig. 7.34 Maipu VG2000-32 Voice gateway device

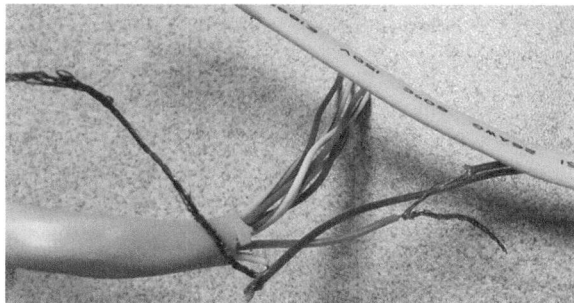

Fig. 7.35 Connecting the two network cables to the telephone cables

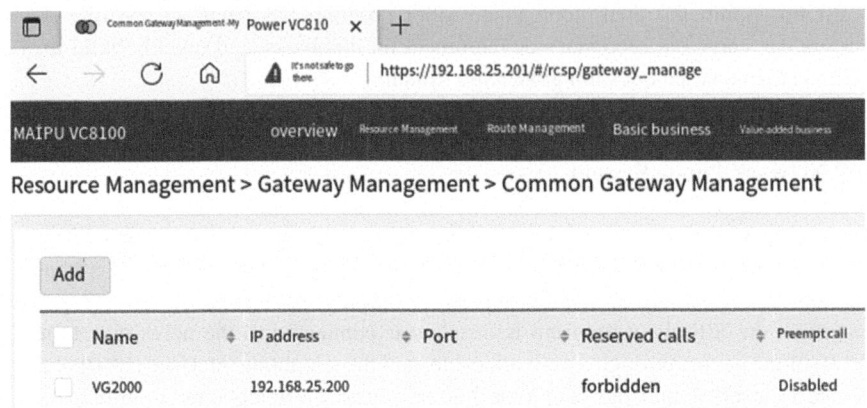

Fig. 7.36 Configuring the gateway address

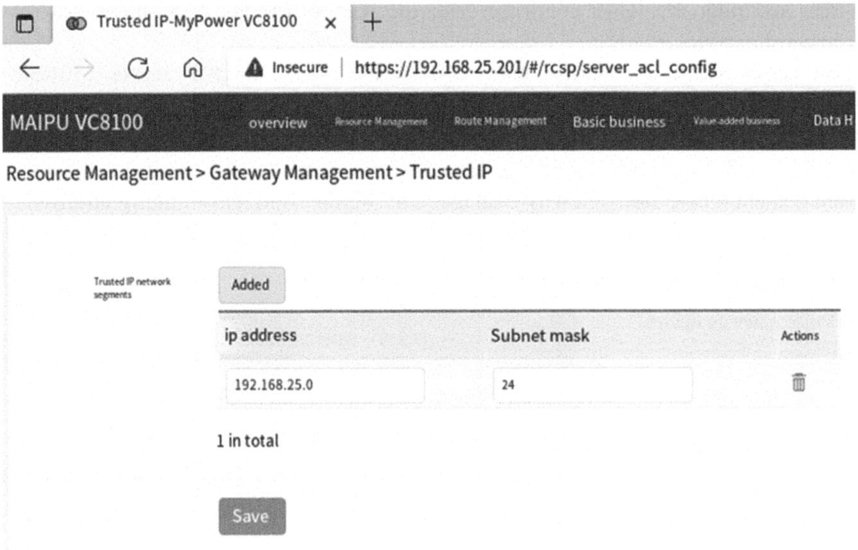

Fig. 7.37 Configuring trusted IP addresses

Under resource management—trusted IP address, configure the allowed IP address segment, as shown in Fig. 7.37.

In route management—outgoing route, configure an outgoing policy. An outgoing call means that the call is forwarded to the telephone network provided by the carrier, as shown in Fig. 7.38.

Configure user information in resource management—number management. The number corresponds to the user name, and the registration password corresponds to the user password, as shown in Fig. 7.39.

7.4 Case Practice

Fig. 7.38 Configuring an outgoing route

Next, we test the availability of the SIP protocol by using the pjsua client compiled by the authors (pjsua.exe in the SIP directory attached to this book) to submit a communication request to the SIP server. Replace the SIP server address, user name, and password in the pjsua.cfg file according to the actual situation. Then go to the SIP directory on the command line and run the following command:

```
pjsua --config-file pjsua.cfg
```

If the message underlined in Fig. 7.40 is displayed, the SIP server is successfully connected.

After entering the Pjsua home screen, type m and press enter, as shown in Fig. 7.41.

Enter the URI of the called phone after the prompt in the format sip: Called number @sip server address, as shown in Fig. 7.42.

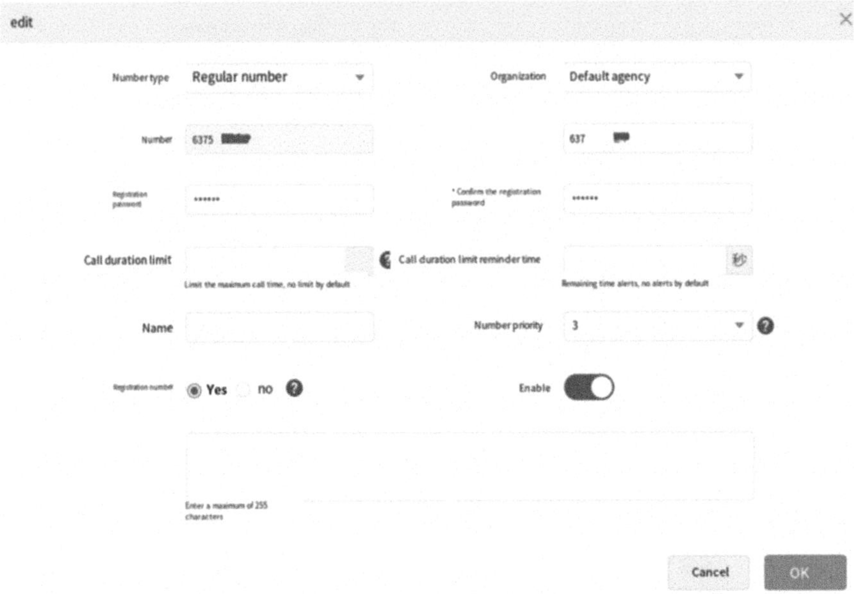

Fig. 7.39 Configuring user information

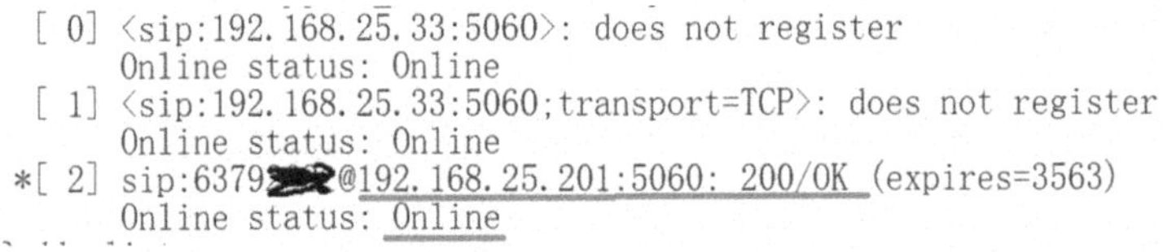

```
[ 0] <sip:192.168.25.33:5060>: does not register
     Online status: Online
[ 1] <sip:192.168.25.33:5060;transport=TCP>: does not register
     Online status: Online
*[ 2] sip:6379███@192.168.25.201:5060: 200/OK (expires=3563)
     Online status: Online
```

Fig. 7.40 Pjsua successfully connected message

At this time, the called number rings. After the call is connected, the call can be made normally, and the Pjsua client is displayed outside the sound card of the computer. Type quit to exit pjsua. Next, let's test if the code provided in this book will run. Unzip pjsua.rar into the anaconda installation path C:\Anaconda\envs directory, and in the SIP directory execute the following command to activate the environment:

```
conda activate pjsua
```

Change line 59 of test.py to the actual sip server address, username, and password. Execute the following command:

```
python test.py sip: called number @sip server address
```

7.4 Case Practice   233

```
+==============================================================================+
|          Call Commands:        |    Buddy, IM & Presence:    |      Account:          |
|                                |                             |                        |
|   m   Make new call            | +b  Add new buddy       .   | +a  Add new accnt.     |
|   M   Make multiple calls      | -b  Delete buddy            | -a  Delete accnt.      |
|   a   Answer call              |  i  Send IM                 | !a  Modify accnt.      |
|   h   Hangup call   (ha=all)   |  s  Subscribe presence      | rr  (Re-)register      |
|   H   Hold call                |  u  Unsubscribe presence    | ru  Unregister         |
|   v   re-inVite (release hold) |  t  ToGgle Online status    |  >  Cycle next ac.     |
|   U   send UPDATE              |  T  Set online status       |  <  Cycle prev ac.     |
|  ],[  Select next/prev call    +-----------------------------+------------------------+
|   x   Xfer call                |      Media Commands:        |   Status & Config:     |
|   X   Xfer with Replaces       |                             |                        |
|   #   Send RFC 2833 DTMF       | cl  List ports              |  d   Dump status       |
|   *   Send DTMF with INFO      | cc  Connect port            | dd   Dump detailed     |
|  dq   Dump curr. call quality  | cd  Disconnect port         | dc   Dump config       |
|                                |  V  Adjust audio Volume     |  f   Save config       |
|   S   Send arbitrary REQUEST   | Cp  Codec priorities        |                        |
+--------------------------------+-----------------------------+------------------------+
|   q   QUIT    L   ReLoad    sleep MS    echo [0|1|txt]     n: detect NAT type        |
+==============================================================================+
You have 0 active call
>>> m
```

Fig. 7.41 Pjsua home screen

Fig. 7.42 Pjsua call command

```
Choices:
   0              For current dialog.
  -1              All 0 buddies in buddy list
  [1 - 0]         Select from buddy list
  URL             An URL
  <Enter>         Empty input (or 'q') to cancel
Make call: [sip:9637𝟛𝟚𝟘𝟛𝟘@192.168.25.201_]
```

```
(pjsua) E:\SIP>d:\python37\python test.py sip:913▆▆▆▆▆▆@192.168.25.201
20:39:16.011 os_core_win32.   !pjlib 2.11.1 for win32 initialized
20:39:16.031 sip_endpoint.c   .Creating endpoint instance...
20:39:16.031        pjlib    .select() I/O Queue created (00DD7E58)
20:39:16.031 sip_endpoint.c   .Module "mod-msg-print" registered
20:39:16.041 sip_transport.   .Transport manager created.
20:39:16.041 pjsua_core.c     .PJSUA state changed: NULL --> CREATED
b'20:39:16.041 sip_endpoint.c   .Module "mod-pjsua-log" registered\n' b'20:39:
er" registered\n' b'20:39:16.041 sip_endpoint.c   .Module "mod-stateful-util"
.Module "mod-ua" registered\n' b'20:39:16.041 sip_endpoint.c   .Module "mod-
dpoint.c   .Module "mod-pjsua" registered\n' b'20:39:16.041 sip_endpoint.c   .M
```

Fig. 7.43 Prompt message for a code call

At this time, the protocol work log screen shown in Fig. 7.43 is displayed. The called number rings, and the call is connected normally. The program plays a TTS prompt and records the call to the record-wav file in the current directory.

At this point, the test for this item is successful.

7.4.2 Code Practice

Nemo_asr Transfer Learning Practice

The task of this section is to use custom dialect data set to train dialect asr model on the basis of Mandarin asr model. Firstly, create a dialect data set in voice_data according to the preceding method. Figure 7.44 shows the directory structure.

The custom dialect dataset uses people of different genders, ages, and voice characteristics to read aloud, and the total duration is recommended to be more than 500 h. The more corpus data, the better. This part of the work can be customized and outsourced or purchased. Establish subdirectories with the dialect text as the name, and store multiple dialect voice files corresponding to the text in each subdirectory, in the format of single channel, 16,000 Hz (or 44,100 Hz). In the process of transfer learning of Nemo framework, it is necessary to specify the basic properties of the voice data file and make a list of the voice data file. The list is a plain text file with json file suffix. Each line describes a voice file. The format is as follows:

```
{"audio_filepath": voice filepath, "duration": voice length, "text": voice text}
```

To quickly generate directory listings, write the following code. First entry package:

```
import os
import librosa
```

Then go through all the files in the data set directory:

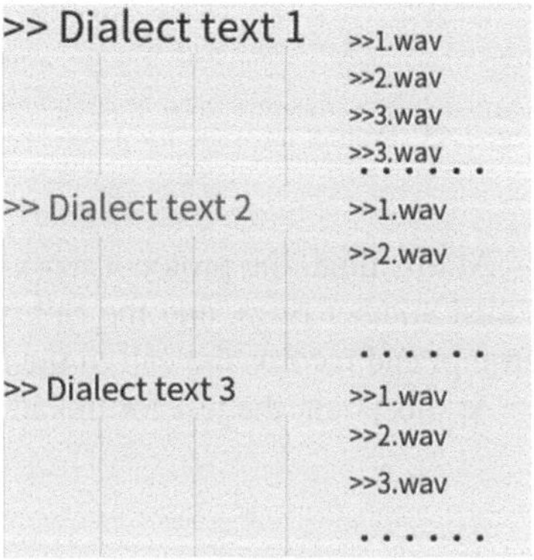

Fig. 7.44 Directory structure of dialect data set

>> Dialect text 1	>>1.wav
	>>2.wav
	>>3.wav
	>>3.wav
	......
>> Dialect text 2	>>1.wav
	>>2.wav
	......
>> Dialect text 3	>>1.wav
	>>2.wav
	>>3.wav
	......

7.4 Case Practice

```
for root, dirs, files in os.walk('./voice_data'):
  for d in dirs:
    rootdir = os.path.join(root, d)
    dir_path, dir_name = os.path.split(d)
    list_files = os.listdir(rootdir) # Lists all files in the folder
```

70% of the voice files in each subdirectory are added to the training set list, and the remaining 30% is added to the validation set list:

```
for i in range(int(len(list_files)*0.7)): # Compile the training set
for i in range(int(len(list_files)*0.7), len(list_files)): # Compile
validation set
```

For each speech file, use the librosa library's get_duration() function to get the length of the speech and generate the data needed for the listing.

```
path = os.path.join(rootdir,list_files[i])
      if os.path.isfile(path):
        time = librosa.get_duration(filename = path)
        print('{"audio_filepath": "' + path.replace('\cr','/') + '",
"duration": ' + str(time) + ', "text": "' + dir_name + '"}')
```

The script file is: generate dialect datasets.bat, call the following command to generate the training set manifest train.json and validation set manifest test.json, respectively:

```
python generates the trainset.py>train.json
python generates the verification set.py>test.json
```

After the data set is built, let's write the transfer learning code. First we need to import the library package, we use the Nvidia Nemo framework for transfer learning:

```
import nemo.collections.asr as nemo_asr
```

Then load the pre-trained model as the base model, which will automatically download the QuartzNet15x5Base-Zh model from Nvidia Corporation's model warehouse to the local directory, The path is C:\Users\ %USERNAME% \.cache \torch\NeMo\NeMo_1.0.0b1\ Quartznet15x5base-en:

```
quartznet = nemo_asr.models.EncDecCTCModel.from_pretrained
(model_name="QuartzNet15x5Base-Zh")
```

In order to demonstrate the effect of transfer learning, let's first take a look at the effect of Mandarin basic model on dialect recognition:

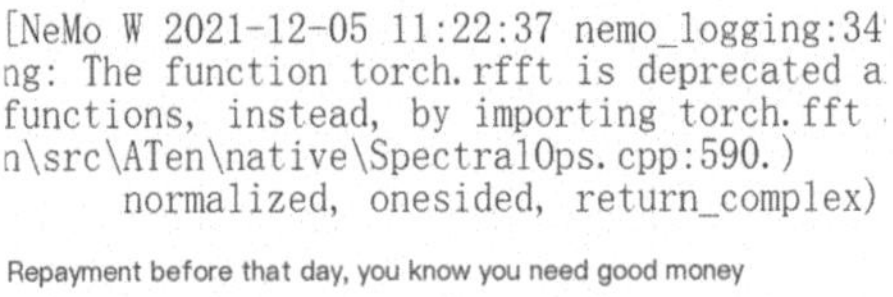

Fig. 7.45 Effect of the basic model on dialect recognition

```
[NeMo W 2021-12-05 11:22:37 nemo_logging:34
ng: The function torch.rfft is deprecated a
functions, instead, by importing torch.fft
n\src\ATen\native\SpectralOps.cpp:590.)
            normalized, onesided, return_complex)
```
Repayment before that day, you know you need good money

```
print(quartznet.transcribe(paths2audio_files=["./voice_data/ Which
day to repay na /48.wav","./voice_data/ A lot of money to repay na /28.
wav"]))
```

The output of the preceding code is shown in Fig. 7.45.

The corpus of dialects is "repay by which date" and "repay a lot of money this time." As can be seen, the recognition effect of the basic model is relatively poor, after all, this model is not trained using the dialect.

Now, here's the kicker. Let's first specify the list of speech datasets:

```
train_file = "./train.json" # specifies the list of training sets for the
custom corpus
 test_file = "./test.json" # specifies the list of validation sets for the
custom corpus
```

During transfer learning, you need to modify the parameters of QuartzNet, such as specifying the training set and the validation set. This requires getting QuartzNet's parameters in advance. This parameter is in the configuration file quartznet_15x5_zh.yaml, which is in the examples\asr\conf\quartznet directory of the Nemo framework source. In the following address download Nemo source: https://github.com/NVIDIA/NeMo/releases/tag/v1.0.0b1, get quartznet_15x5_zh. Yaml files. Here is a brief introduction to the content of the configuration file, as shown in Fig. 7.46, to help us understand more.

First define an audio sample rate of 16,000 (it turns out that 44,100 is also OK); repeat indicates that the number of convolutional blocks is 5; The dropout and separable parameters specify random deletion and separable convolution of neurons during training, respectively; labels specify the output text range of the model, which is 5206 letters and Chinese characters. The next step is to configure the training set and verification set parameters, as shown in Fig. 7.47, where "???" indicates that this parameter can be passed in through code. We need to pass this parameter in the code that follows.

Look down there are preprocessor parameters, encoder parameters, trainer parameters, limited space will not be introduced one by one. It is worth noting that the configuration file uses cpu for training by default. We need to modify the value of gpus parameter in trainer according to the configuration of gpu, which means that several Gpus are used for training. In addition, if there is insufficient video memory in the training process, it is necessary to reduce the batch_size parameter value of training set and verification set in the QuartzNet configuration file. Then, we use the

7.4 Case Practice

```
name: &name "QuartzNet15x5"

model:
  sample_rate: &sample_rate 16000
  repeat: &repeat 5
  Dropout:&dropout 0.0
  separable: &separable true
  labels: &labels [' ', '''', A, B, C, D, E, F, G, H, I, J,
                  S, T, U, V, W, X, Y, Z, Bing, Yi, Ding, Shi, Qiu, C, Ye,
                  Cong, Dong, Si, Cheng, Di, Liang, Jiu, Mo, Yi, Zhi,
                  Wu, Zha, Hu, Fa, Le, Ping, Li, Yu, Controversy, Shi,
                  Er, Yu, Loss, Yun, Mutual, Five, Ting, Liang, Qin, Hao,
                  Di, Ren, Ren, Ren, Qian, Ge, Dai, Ling, Yi, San, Yi, We,
                  Yang, You, Hu, Hui, Umbrella, Wei, Chuan, Ya, Injury,
                  Lun, pseudo,
```

Fig. 7.46 QuartzNet configuration file (1)

Fig. 7.47 QuartzNet configuration file (2)

```
train_ds:
  manifest_filepath: ???
  sample_rate: 16000
  labels: *labels
  batch_size: 32
  trim_silence: True
  normalize: False
  max_duration: 16.7
  shuffle: True
  is_tarred: False
  tarred_audio_filepaths: null
  tarred_shard_strategy: "scatter"

validation_ds:
  manifest_filepath: ???
  sample_rate: 16000
  normalize: False
  labels: *labels
  batch_size: 32
  shuffle: False
```

YAML module to put the model parameters into the patams dictionary from the QuartzNet configuration file:

```
from ruamel.yaml import YAML
config_path = './quartznet_15x5_zh.yaml' # Specify the configuration file for the QuartzNet model
yaml = YAML(typ='safe')
with open(config_path, encoding='utf-8') as f:
    params = yaml.load(f) # Get parameters from the configuration file of the QuartzNet model
```

For easy training, we import the pytorch framework's advanced wrapper pytorch_lightning module for efficient execution:

```
import pytorch_lightning as pl
```

We create a trainer instance, using a GPU, and train 256 epochs:

```
trainer = pl.Trainer(gpus=1, max_epochs=256)
```

Next, set the custom training set and validation set paths for the params dictionary:

```
params['model']['train_ds']['manifest_filepath'] = train_file
params['model']['validation_ds']['manifest_filepath'] = test_file
```

Pass the updated parameters to QuartzNet:

```
Quartznet.setup_training_data(train_data_config=params['model']['train_ds']) # Updates QuartzNet's training set
Quartznet.setup_validation_data(val_data_config=params['model']['validation_ds']) # Updates QuartzNet's validation set
```

Because the data set list specifies the voice file (input data) and its corresponding text (supervisory data), the network can be propagated forward and back. Once these Settings are complete, transfer learning is initiated:

```
trainer.fit(quartznet)
```

The trainer provided by pytorch_lightning is very simple to use. You only need to pass QuartzNet after updating parameters. After the training starts, you will see the training interface as shown in Fig. 7.48:

Pay attention to the decline of loss. Normally, the network will continue to converge and the loss will tend to a small value. In the case of a large number of samples, the data has better semantic representation, and the decline of loss will be more ideal. This book shows the process with only a small number of sample

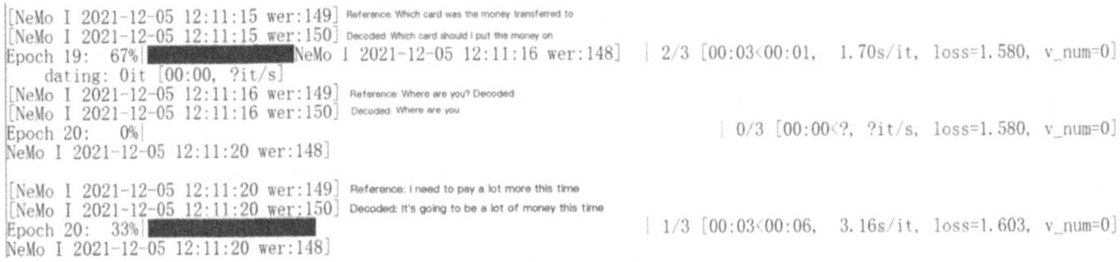

Fig. 7.48 Transfer learning training process

7.4 Case Practice

```
[NeMo I 2021-12-05 14:31:12 collections:173] Dataset
[NeMo I 2021-12-05 14:31:12 collections:174] 0 files
['Which day is the repayment due before','How much money should be returned']
```

Fig. 7.49 The effect of transfer learning model on dialect recognition

samples, and readers may get better convergence in real project development than the figure above. After the training is complete, we use the migrated model to identify the previous dialects:

```
print(quartznet.transcribe(paths2audio_files=["./voice_data/ Which day to pay back na /48.wav","./voice_data/ A lot of money to pay back na / 28.wav"])) # Use a trained transfer learning model to identify dialects
```

The recognition effect is shown in Fig. 7.49. As can be seen, dialects that were previously identified poorly using the Mandarin model have been identified completely correctly after transfer learning.

We save the trained migration model to the dialect.nemo file, which is a compressed file opened with winrar that contains a model configuration file model_config.yaml and a model weights file model_weights.ckpt.

```
quartznet.save_to('./dialect.nemo')
```

In the follow-up task, if the transfer learning model needs to be used to identify dialects, just call restore_from() function of Nemo framework to load the model from the model file, and call transcribe() function for inference prediction, which is very convenient:

```
quartznet = nemo_asr.models.EncDecCTCModel.restore_from('./dialect.nemo')
print(quartznet.transcribe(paths2audio_files=["./voice_data/ Which day before repayment Na /48.wav","./voice_data/ This time to pay a lot of money Na /28.wav"]))
```

Since it is the same model, the inference result of loading transfer learning model is exactly the same as Sect. 6.4.2.1.5. See "asr Transfer Learning /train.py" for the code.

Voice Q&A Module Practice

Below, we write the main code of the voice question answering system. First import the relevant library package:

```
from pyaudio import paInt16 as paInt16
from pyaudio import PyAudio as PyAudio
from wave import open as wopen
import nemo.collections.asr as nemo_asr
from PyQt5.QtWidgets import QApplication,QWidget,QTextEdit,
QVBoxLayout,QPushButton
from transformers import AutoModelForQuestionAnswering,
AutoTokenizer,pipeline, BertTokenizer, GPT2LMHeadModel
import soundfile
from espnet2.bin.tts_inference import Text2Speech
from playsound import playsound
```

Among them, pyaudio is a cross-platform audio I/O library that allows real-time recording; wave and soundfile libraries can process wave waveform files; Use PyQt5 to define forms, buttons, text controls, etc. for sample programs; Use Nemo for ASR tasks; Import the NLP Q&A module of transformers library to build the Q&A system; Import espnet2 module to build TTS task; Import the playsound module for playing wav waveforms.

Next, add PyQt's interface form to the main code, bind the TextEditDemo class.

```
if __name__ == '__main__':
    app=QApplication(argv)
    win=TextEditDemo()
    win.show()
    Exit(app.exec_())
```

The TextEditDemo class is the main interface of the sample program. We set some interface elements in the __init__() function, including the form title, form size, buttons, and text box layout:

```
self.setWindowTitle(' Demo version of Chapter 7 Dialect Collection
Voice Answering Robot ---- ')
# Define the initial size of the window
The self. The resize (400300).
# Create multi-line text box
self.textEdit=QTextEdit()
# Create two buttons
self.btnPress1=QPushButton(' Start talking ')
# mouse-over button prompt
Self. BtnPress1. SetToolTip (" quiet environment, read aloud mandarin
")
self.btnPress2=QPushButton(' Stop talking ')
# mouse-over button prompt
Self. BtnPress2. SetToolTip (" start transcription text)"
```

7.4 Case Practice

```
# Instantiate the vertical layout
layout=QVBoxLayout()
# Related controls added to vertical layout
layout.addWidget(self.textEdit)
layout.addWidget(self.btnPress1)
layout.addWidget(self.btnPress2)
# Set layout
self.setLayout(layout)
```

We set up a "Start talking" button and a "Stop talking" button. After clicking, the recording operation and question and answer operation are performed respectively, and the slot function of the response needs to be set, respectively:

```
# Bind the click signal of the button to the related slot function, and
click to trigger
    self.btnPress1.clicked.connect(self.btnPress1_clicked)
    self.btnPress2.clicked.connect(self.btnPress2_clicked)
```

Since the __init__() function of the TextEditDemo class is automatically executed when the program starts, we need to load the various models used in the next steps here:

```
model = AutoModelForQuestionAnswering.from_pretrained('./roberta-base-chinese-extractive-qa')
tokenizer = AutoTokenizer.from_pretrained('./roberta-base-chinese-extractive-qa')
```

The above two lines of code load Hugging Face's Chinese bert question-and-answer model and marker. For the private deployment, we download the model locally, and the code directly specifies the local path to use it. Next, we associate the model with the tagger using the pipeline() function:

```
self.QA = pipeline('question-answering', model=model,
tokenizer=tokenizer)
```

Get the free text, which is the text of all the corpus relevant to that customer's collection task:

```
f = open(' free text.txt', 'r', encoding='utf-8')
self.txt_data = f.read()
f.close()
```

Construct a question and answer structure, filling in the question text and the free text, respectively:

```
self.qa_input = {'question':'','context': self.txt_data}
```

Next, we load an Nvidia Mandarin ASR model Quartznet15x5base-en and the migration model we trained earlier.nemo:

```
Self.Asr_model1 = nemo_asr.Models.EncDecCTCModel.From_pretrained
(model_name = "QuartzNet15x5Base - useful") # load model of mandarin
Self.Asr_model2 = nemo_asr.Models.EncDecCTCModel.Restore_from ('.
/ the dialect. Nemo ') # load transfer model of learning
```

Find the high-quality (44,100 Hz) Chinese female voice TTS model kan-bayashi/csmsc_full_band_vits in the model list of ESPnet, download it to your local and load the model. We put it on video memory and take advantage of the GPU to speed it up:

```
self.text2speech = Text2Speech("config.yaml", "train.total_count.
ave_10best.pth", device="cuda")
```

In order to evaluate the quality of the text sentences obtained by the Mandarin ASR model and the dialect ASR model, respectively, we used the Chinese gpt2 model of Hugging Face to calculate, loading the markers and the model, respectively:

```
self.b_tokenizer = BertTokenizer.from_pretrained("./gpt2-chinese-
cluecorpussmall")
self.b_model = GPT2LMHeadModel.from_pretrained("./gpt2-chinese-
cluecorpussmall")
```

After the call is connected, the robot plays a voice prompt (in practice, it first asks if it is me, but this book is just a demonstration):

```
speech = self.text2speech(self.txt_data)["wav"]
soundfile.write("./answer.wav", speech.cpu().numpy(), self.
text2speech.fs, "PCM_16")
playsound('./answer.wav')
```

Don't forget to record the text information of the entire conversation and create a file (in practice, a database):

```
Self.Fw = open ('/call records/liu '+ time. The strftime (" % % % m % d_Y
H % m % S ", the time the localtime ()) +'. TXT ', 'w', encoding = "utf-8")
```

At this point, the initialization is complete. Next, we need to define the button's response slot function. For the "Start Talking" button, start a child thread to record:

7.4 Case Practice

```python
def btnPress1_clicked(self):
    print(' Listen to... ')
    self.c = ChildThread()
    self.t = Thread(target=self.c.run)
    self.t.start()
```

Receive audio device input in ChildThread() function, make real-time recording and save to question.wav file, set isend flag to judge whether "Stop talking" has been clicked, once clicked, the recording will be interrupted and enter the question and answer stage:

```python
# Define the child thread class
class ChildThread:
    def __init__(self):
        self.isend = False
    def terminate(self):
        self.isend = True
    def run(self):
        CHUNK = 1024 #wav files are composed of several chunks, CHUNK we understand as packets or data fragments.
        FORMAT = paInt16 # pyaudio.paInt16 indicates that we use 16 quantized bits for recording.
        CHANNELS = 1 # represents the sound channel, the mono channel used here.
        RATE = 16000 # Sample rate 16k
        frames = []
        p = PyAudio()
        stream = p.open(format=FORMAT,
                channels=CHANNELS,
                rate=RATE,
                input=True,
                frames_per_buffer=CHUNK)
        while True:
            data = stream.read(CHUNK)
            frames.append(data)
            if self.isend == True:
                break
        stream.stop_stream()
        stream.close()
        p.terminate()
        wf = wopen('question.wav', 'wb')
        wf.setnchannels(CHANNELS)
        wf.setsampwidth(p.get_sample_size(FORMAT))
        wf.setframerate(RATE)
```

```
wf.writeframes(b''.join(frames))
wf.close()
```

Next, we write the response slot function for the "Stop recording" button. First terminate the recording and wait for the subthread to end:

```
def btnPress2_clicked(self):
    self.c.terminate() # terminate the recording
    self.t.join()
```

Then, we use the Mandarin ASR model and the dialect ASR model respectively to translate speech into text:

```
result1 = self.asr_model1.transcribe(paths2audio_files=
["question.wav"]) # Identify Mandarin with the Mandarin model
    result2 = self.asr_model2.transcribe(paths2audio_files=
["question.wav"]) # Identify dialects with a dialect model
```

Note that due to non-standard pronunciation, environmental noise and other problems, the text obtained at this time is likely to be not smooth, we need to evaluate the text obtained by these two models, who is better in quality. We use the confusion index (PPL) to evaluate sentence smoothness. It basically estimates the probability of a sentence occurring on a per-word basis and normalizes the sentence length. PPL is calculated by the following formula:

$$PP(S) = P(\omega_1\omega_2...\omega_N)^{-\frac{1}{N}}$$
$$= \sqrt[N]{\frac{1}{p(\omega_1\omega_2...\omega_N)}}$$
$$= \sqrt[N]{\prod_{i=1}^{N}\frac{1}{p(\omega_i|\omega_1\omega_2...\omega_{i-1})}}$$

where S represents the current sentence, N represents the sentence length, p-(wi) represents the probability of the i-th word, $p(wi|w1w2w3... wi\text{-}1)$ represents the probability of calculating the i-th word based on the previous $i\text{-}1$ words. A smaller PPL means a smoother sentence. For example, "I am back to school" has a much lower PPL than "I am to back school," indicating a higher quality sentence. The main code of the PPL function is as follows:

```
sens = [sentence]
inputs = self.b_tokenizer(sens, padding='max_length',
max_length=50, truncation=True, return_tensors="pt")
    bs, sl = inputs['input_ids'].size()
```

7.4 Case Practice

```
outputs = self.b_model(**inputs, labels=inputs['input_ids'])
logits = outputs[1]
shift_logits = logits[:, :-1, :].contiguous()
shift_labels = inputs['input_ids'][:, 1:].contiguous()
shift_attentions = inputs['attention_mask'][:, 1:].contiguous()
loss_fct = CrossEntropyLoss(ignore_index=0, reduction="none")
loss = loss_fct(shift_logits.view(-1, shift_logits.size(-1)),
shift_labels.view(-1)).detach().reshape(bs, -1)
meanloss = loss.sum(1) / shift_attentions.sum(1)
ppl = torch.exp(meanloss).numpy().tolist()
return ppl[0]
```

Then, we evaluate the text quality of the two ASRs, take good quality questions as input to the question answering system and record them in a log:

```
ppl1 = self.ppl(asr_result1) # to get PPL of Mandarin ASR
ppl2 = self.ppl(asr_result2) # to get PPL for dialect ASR
print(' Mandarin recognition: ',asr_result1,'ppl1=',ppl1,' dialect recognition: ',asr_result2,'ppl2=',ppl2)
# If the sentence transcribed by the Mandarin model is more fluent than the dialect, the sentence is transcribed in Mandarin, and vice versa
if ppl1 <= ppl2:
    asr_result = asr_result1
    print(' Mandarin recognition is better, use Mandarin model ')
    Self.Fw.Write(time.strftime(" %%Y-m-H:%d%%m:%S ", the time the localtime()) + 'customer (mandarin) : "+ asr_result +' \ n ')
else:
    asr_result = asr_result2
    print(' dialect recognition is better, use dialect model ')
    Self.Fw.Write(time.strftime(" %%Y-m-H:%d%%m:%S ", the time the localtime()) + 'customer (dialect) :' + + '\ n' asr_result)
```

If the customer responds with an affirmative wish, end the call if:

```
if (' OK 'in asr_result or' received 'in asr_result or' um 'in asr_result or' know 'in asr_result or' get 'in 'in asr_result or' in asr_result or 'right' in asr_result) and len(asr_result) <5:
    answer = 'Thank you and goodbye! '
```

If the customer gives a response to some business inquiry, update the question and enter the question and answer system to get the best answer in the free text.

```
self.qa_input.update({'question': asr_result})
answer = self.QA(self.qa_input, device=0)
```

It is worth noting that answer is a dictionary of the answer results of the question and answer system, in which the "score" key indicates confidence. If the confidence is very low, it means that the quality of the answer evaluated by the algorithm is poor. It may also be that the customer's question is no longer in the free text, then it is necessary to jump to the semantic node of the database defined in advance to match the closer answer. Limited by space, this part is omitted. For answers with low confidence, a fixed speech is played:

```
if answer.get('score') > 1e-08:
   answer = answer.get('answer')
else:
   answer = 'To remind you again, please pay off the loan arrears from the bank as soon as possible, otherwise your credit record will be affected.
'
```

Finally, we refined the log and interface display, and then converted the response text into a waveform output via ESPnet's pre-trained model:

```
Self.Fw.Write(time.Strftime("%%Y-m-H:%d%%m:%S", the time the localtime()) + 'robot:' + + '\n' answer)
self.fw.flush()
asr_result += '\n Robot says: '+answer
print(asr_result)
self.textEdit.setPlainText(asr_result)
speech = self.text2speech(answer)["wav"] # Complete TTS using ESPnet's pre-trained model
soundfile.write("./answer.wav", speech.cpu().numpy(), self.text2speech.fs, "PCM_16")
playsound('./answer.wav')
```

See 'Answer Bot/answer bot.py' for the code. At this point, we have completed all the processes from dialect transfer learning, speech recognition, question answering system and text-to-speech conversion, and the dialect voice phone collection robot has been preliminarily written.

SIP Communication Module Actual Combat

According to the operations in Sect. 7.4.1.2, the voice gateway and SIP server are ready. The following focuses on the development of SIP client. For SIP clients, you can choose from Zoiper, X-Lite, EyeBeam, Bria, Eyebeam, Bria, Blink, PC-Telephone, and many other products. Since this case requires code-level integration or interface interaction between the SIP client and the dialect dual-mode robot, it is necessary to choose products that can be developed secondary. In this

7.4 Case Practice

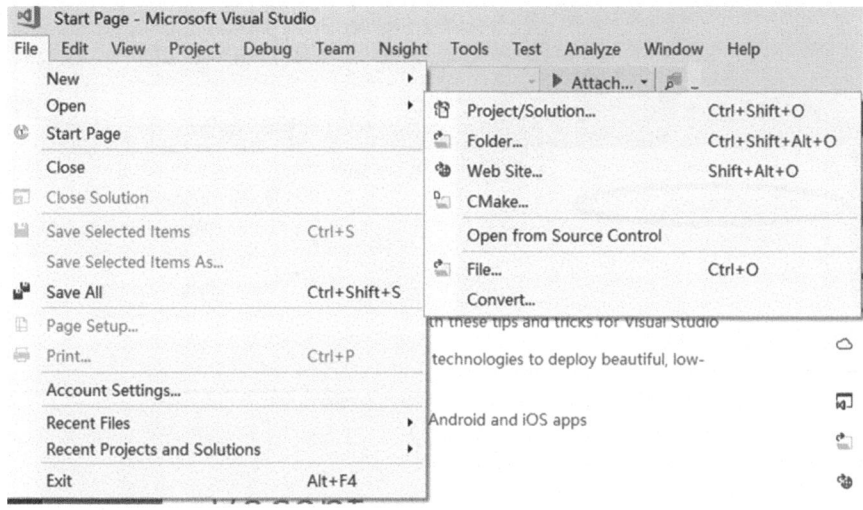

Fig. 7.50 Compiling PJSUA (1)

Fig. 7.51 Compiling PJSUA (2)

case, Blink, known as the best open-source SIP client software, can be selected. The software supports three operating systems, Windows, Linux and MacOS, the underlying protocol uses PJSIP and is developed in Python. The Blink code warehouse for https://github.com/AGProjects/blink-cocoa. Based on this repository, the project development speed can be accelerated. In order to let readers better understand the development of SIP protocol, this chapter uses the lower level of PJSIP to show the actual process. The first step is to compile PJSUA from source code. We take windows platform as an example to explain the compilation process. First, install visual studio2017. Visit PJSIP code warehouse https://github.com/pjsip/pjproject/tags, download the latest 2.11.1 version of the source code. In visual studio2017, click Open/Project/Solution and select the pjproject-vs14.sln project file, as shown in Fig. 7.50.

Set the compile output of your project to Release and Win32. Release represents the release version and Win32 represents the 32-bit platform. Select it at the top of the interface, as shown in Fig. 7.51.

In the pjlib/include/pj directory, create the empty file config_site.h. Select all items in Solution Explorer and right-click Regenerate Solution, as shown in Fig. 7.52.

After the compilation succeeds, a message is displayed in the bottom of the output window, as shown in Fig. 7.53.

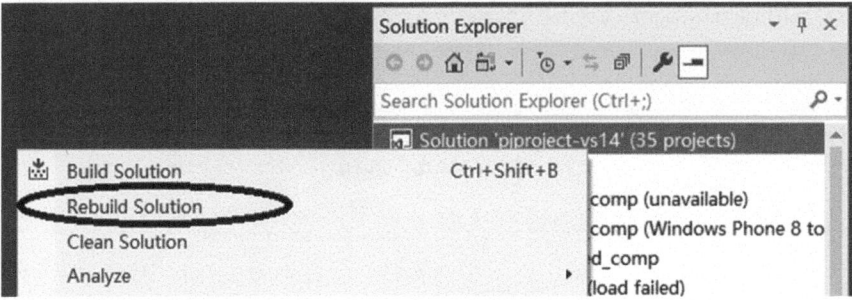

Fig. 7.52 Compiling PJSUA (3)

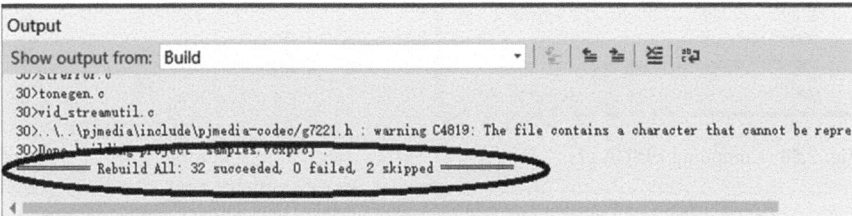

Fig. 7.53 Compiling PJSUA (4)

Next, compile the Python wrapper for PJSUA. It will not compile successfully in 64-bit environments due to version issues. So we need to build a 32Python environment. Command line execution:

```
set CONDA_FORCE_32BIT=1
```

After setting this environment variable, the newly created virtual environment in conda will be built in 32-bit form. Then execute the following command to create the virtual environment named pjsua:

```
conda create -n pjsua python=3.7 -y
```

Clone the code repository locally by executing the following command:

```
git clone https://github.com/mgwilliams/python3-pjsip.git
```

Change the original pjsip-apps/src/python directory to python_bk, and rename the downloaded python3-pjsip directory to python. Edit the _pjsua.def file in the python directory and change init_pjsua to PyInit__pjsua. Reopen visual studio2017, locate the python_pjsua project, click "right"—"Properties," in "c/ C ++"—"-General"—"Additional include directory," add the include directory under the python3 installation path in the virtual environment, as shown in Fig. 7.54.

7.4 Case Practice

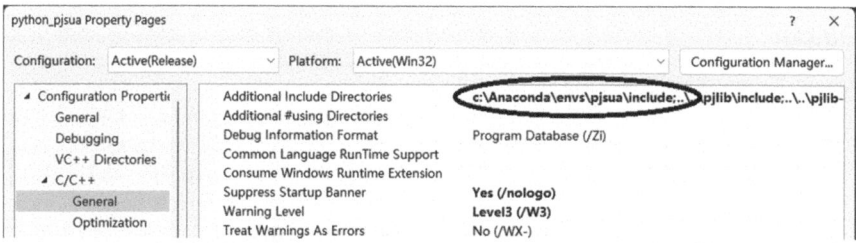

Fig. 7.54 Compiling PJSUA (5)

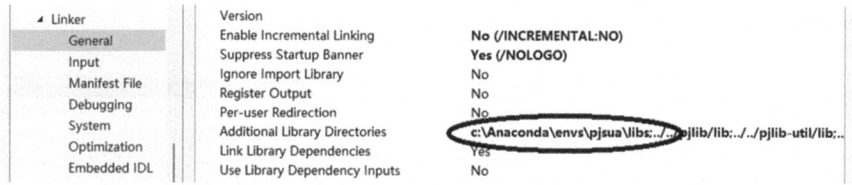

Fig. 7.55 Compiling PJSUA (6)

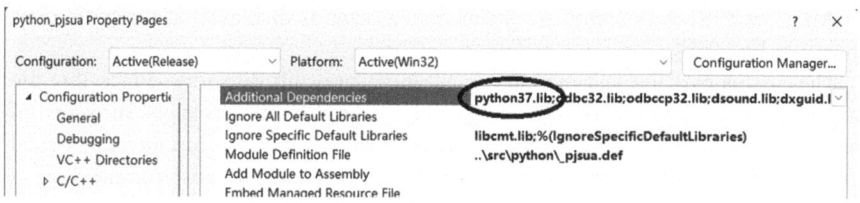

Fig. 7.56 Compiling PJSUA (7)

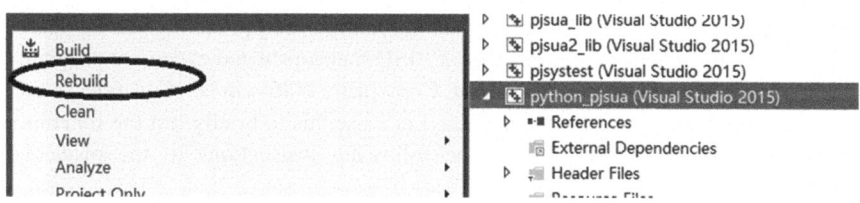

Fig. 7.57 Compiling PJSUA (8)

Select Linker—General—Additional Library Directory to add the libs directory under the python3 installation directory in the virtual environment, as shown in Fig. 7.55.

Select "linker"—"Input"—"Additional Dependencies" and change python24.lib to python37.lib, as shown in Fig. 7.56.

Find the python_pjsua project in Solution Explorer and right-click Rebuild, as shown in Fig. 7.57.

Fig. 7.58 Compiling PJSUA (9)

Fig. 7.59 Compiling PJSUA (10)

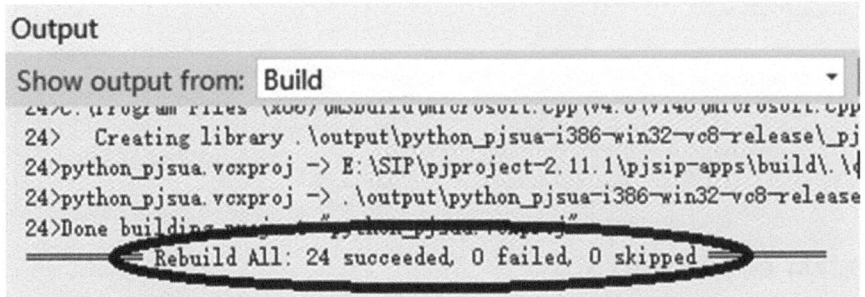

After the PJSUA is compiled, a success message is displayed at the bottom, as shown in Fig. 7.58.

The _pjsua.pyd file is generated in the pjsip-apps/lib directory. Place this file together with pjsip-apps/src/python/pjsua.py in the /lib /site-packages directory of the virtual environment. Run conda activate pjsua on the CLI to enter the virtual environment and import pjsua into the python interpreter. If no error message is displayed, it indicates success, as shown in Fig. 7.59. At this point, the compilation of Pjsua is complete. The compilation project is detailed in the SIP/pjproject-2.11.1 directory attached to this book.

After the above steps, the pjsip-apps\bin directory is compiled to generate pjsua-i386-Win32-vc14-Release.exe, which is a PJSUA-encapsulated executive program, which is a full-featured console sip client. Copy pjsua-i386-win32-vC14-release.exe into the SIP directory and rename it pjsua. Let's use this to briefly test the functionality of the sip client. First, write the following instructions to the pjsua.cfg configuration file:

```
--id sip:63791195@192.168.25.201:5060
--registrar sip:192.168.25.201:5060
--realm *
--username 637xxxxx (Please replace xxxxx with the number of the outgoing call and delete this tip)
--password 666666
```

Pjsua uses a uniform resource identifier (URI) to register each sip resource in the format "sip: Phone number @SIP Server address :SIP server port." The first line of the configuration file indicates the URI identifier of the client on the SIP server, the

7.4 Case Practice

second line indicates the SIP server registration identifier, and the following lines specify the domain, SIP client user name and password on the SIP server respectively. Then execute the command:

```
pjsua --config-file pjsua.cfg
```

Then follow the actions in Sect. 7.4.1.2 and experience the joy of victory without going into details.

Now let's start writing SIP client code. First import package:

```
import sys,time,os
import pjsua as pj
import threading
```

Then, instantiate the pjsua wrapper. This can be understood as creating an access object for pjsua:

```
lib = pj.Lib()
```

Set an end of session identifier:

```
is_end = False
```

Set some parameters of pjsua: log level, log callback function:

```
lib.init(log_cfg = pj.LogConfig(level=4, callback=log_cb))
```

Transfer over UDP, using default port 5060:

```
lib.create_transport(pj.TransportType.UDP, pj.TransportConfig(5060))
```

Start the pjsua run, which by default will create a worker thread:

```
lib.start()
```

The SIP client initiates a login operation to the SIP server, specifying the SIP server address, user account, and password:

```
acc = lib.create_account(pj.AccountConfig("192.168.25.201", "63XXXXXX", "666666"))
```

Pjsua uses an asynchronous mechanism, which means that any operation is returned as soon as it is successfully called, rather than waiting for the operation to finish. Therefore, as soon as the login operation is initiated, the above code returns immediately. If we are going to process the result of the login, we need to introduce callback functions to handle various situations during the login process. Here, we define the MyAccountCallback class to handle the mutual exclusion and change of login status if multiple people log in to an account at the same time:

```
class MyAccountCallback(pj.AccountCallback):
    sem = None
    def __init__(self, account):
        pj.AccountCallback.__init__(self, account)
    # When multiple people log in, use a semaphore mechanism to ensure that only one person logs in
    def wait(self):
        self.sem = threading.Semaphore(0)
        self.sem.acquire()
    # Once the client's login status on the server changes, this function responds
    def on_reg_state(self):
        if self.sem:
            if self.account.info().reg_status >= 200: # Release semaphore after successful login
                self.sem.release()
```

Then, make the callback function effective:

```
acc_cb = MyAccountCallback(acc)
acc.set_callback(acc_cb)
acc_cb.wait()
```

Next, make the call. The make_call() function implements digital-to-analog conversion from the internal LAN to the external telephone communication network, and returns an instance of the call object that will ring on the user's phone:

```
call = acc.make_call(sys.argv[1], MyCallCallback())
```

Note that sys.argv[1] is the URI identifier of the phone being called. MyCallCallback is a callback function that handles asynchronous calls. It is defined as follows:

```
class MyCallCallback(pj.CallCallback):
    def __init__(self, call=None):
        pj.CallCallback.__init__(self, call)
```

7.4 Case Practice

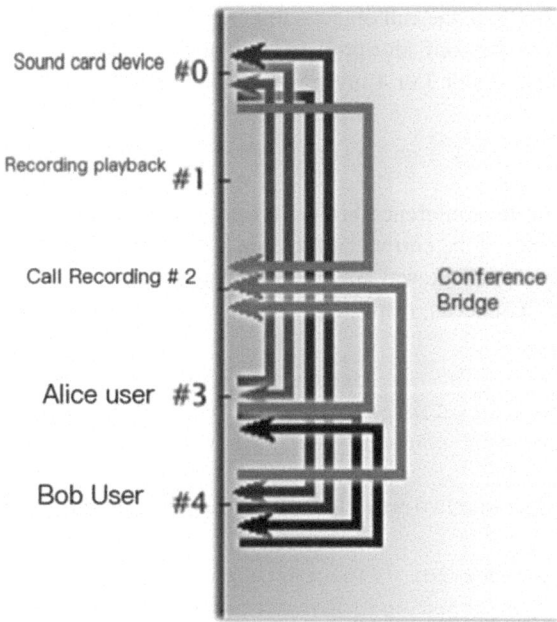

Fig. 7.60 PJSUA's media object, slot number, and conference bridge

```
# This function responds when the call status changes. For example,
the call is connected, in a call, hanging up, and so on
    def on_state(self):
      global is_end
      if self.call.info().state == pj.CallState.DISCONNECTED:
# Set the end of session identifier once a valid hang up occurs
        if time.time()-start_time > 2:
          is_end = True
```

Next, we are going to record the whole process of the call and play the customer's voice out of the sound card of the SIP client computer. First of all, we will clarify a few concepts. In Pjsua, the voice of the customer speaking to the robot, the voice of the robot speaking to the customer, and the recording of the call are all called Media Objects. All media objects are connected to the Conference Bridge via a unique Slot Number. When the corresponding slot number is connected, the media objects can communicate with each other. Figure 7.60 shows that slot 0 is the sound card device of the SIP client, slot 1 is the voice played by the robot to the client, slot 2 is the recording file of the client, slot 3 is the call session of user Alice, and slot 4 is the call session of user Bob. The connection mode in the figure represents: Alice and Bob can talk two-way (see the arrows for connecting #3 and #4), the robot's two-way conversation with Alice and Bob is recorded (see the arrows for connecting #0, #2, #3, #4), and Alice's two-way conversation is played out (see the arrows for connecting #0 and #3). Bob's two-way call is played out (see arrows connecting #0 and #4).

In Pjsua, the call object's info() method returns the call instance information, and access the conf_slot property to obtain the slot number of the current call session (which is slot 3 or 4 in the figure above).

```
call_slot = call.info().conf_slot
```

In the conference bridge, create a slot for a wav file and connect it to the slot number of the current call session. The effect is that the customer will hear an audio playback of the wav file on the phone. Using create_player(), player_get_slot(), and conf_connect(), respectively:

```
player_id = lib.create_player("1.wav")
wave_id = lib.player_get_slot(player_id)
lib.conf_connect(wave_id, call_slot)
```

Record a two-way call between the bot and the customer:

```
recorder_id1 = lib.create_recorder('record.wav') # Create recorder, return recording slot number.
rec_id1 = lib.recorder_get_slot(recorder_id1) # Get the recorder slot number.
lib.conf_connect(call_slot, rec_id1)
lib.conf_connect(wave_id, rec_id1)
```

Connect the current session to the SIP client's sound card slot. This way, the voice for the entire call will be played out of the speaker via the sound card:

```
lib.conf_connect(call_slot, 0)
```

Wait for the call to end and end the program as soon as a hang up signal is found:

```
while True:
  if is_end == True:
    break
  else:
    Time.Sleep(0.5)
```

Finally, free up system resources:

```
lib.conf_disconnect(wave_id, call_slot)
lib.conf_disconnect(call_slot, rec_id1)
lib.conf_disconnect(wave_id, rec_id1)
lib.player_destroy(player_id)
lib.recorder_destroy(recorder_id1) # Releases system resources
```

occupied by the recorder, including deleting all connections associated with the recorder.
```
lib.destroy()
lib = None
```

See SIP/test.py for the code. At this point, the code is written. We execute the command:

```
python test.py sip: called number @sip server address
```

In normal cases, the call log of the sip call is displayed on the computer screen. After a few moments, the call can be generated on the mobile phone. After the call is connected, the customer's voice can be heard from the phone end through the computer speaker. After hanging up, the call recording file is generated in the current directory. In this section, we show the main process of SIP programming to realize network telephony.

7.5 Case Summary

In this particular scenario, commercial banks have effectively utilized acoustic technology, deep learning algorithms, natural language understanding, among other advanced technologies, to enable seamless natural language communication, and customer interaction for loan collection purposes. These technologies facilitate the collection of customer repayment intentions and systematically record the collection processes. This application holds considerable importance in reducing labor costs, enhancing collection efficiency, increasing the operational profits of commercial banks, and preemptively managing potential loan risks. This study is exploratory. Variations in dialect accents, environmental noise, and the completeness of voice data sets pose certain challenges that necessitate further optimization. For those seeking to implement a similar speech robot project, acquiring or generating dialect-specific speech datasets is essential. Currently, specialized companies in China produce these datasets; by submitting semantic text, companies such as Datang and Aisu Wisdom can expand the corpus and generate datasets featuring diverse accents. These companies are recommended for readers, but readers may also undertake this task independently.

This case exemplifies the practical application of artificial intelligence in risk control within commercial banks. It is particularly suitable for scenarios involving a high volume of calls with low individual call value, significantly lowering labor costs and boosting collection efficiency. Beyond collections, voice robots can be effectively deployed in various banking business scenarios, such as marketing financial products, notifying customers of their rights, recommending financial activities, and issuing anti-fraud alerts. Readers can draw on these applications for similar scenarios to continuously enhance customer experience, uphold state-owned asset management practices, contribute to societal harmony, and fulfill the political and social responsibilities incumbent upon financial institutions. The hardware and SIP server software employed in this case from Maipu require formal authorization from the Maipu company. However, readers may opt for other compatible hardware and third-party SIP server software based on their specific requirements. Due to the inherent complexity and extensive nature of a voice outbound robot system—which is highly task-specific—numerous aspects require ongoing iteration. These include dialect speech recognition, semantic understanding, customer semantic node generalization, ASR text correction, knowledge graph matching with semantic nodes, real-time voice streaming media processing over SIP networks, silence detection, noise elimination, and reverse polarity signal setup for call lines. Due to space constraints, this chapter highlights only the primary implementation aspects. Several project integration tasks and peripheral system developments are not detailed here, leaving readers the latitude to explore and apply these technologies independently.

Chapter 8
Chattel Collateral Warehouse Visual Monitoring Project: Image Understanding Technology

In addition to real estate, land, and other immovable properties, the collateral utilized by loan enterprises encompasses a substantial amount of movable property. This includes assets such as automobiles, original wine, machine the browser end, according tory, equipment, and various products. Due to the fixed nature of real estate, the risk of quantity loss remains minimal. Conversely, due to the transferable nature of movable property, the risk of quantity loss is significantly heightened. Under the guidance of national policies, bank loans provide robust support to the real economy, manufacturing enterprises, small and micro enterprises, inclusive finance, and rural enterprises. As a result, commercial banks often acquire more chattel mortgages which include manufacturing products, machine tools, equipment, and similar assets. However, amidst economic downturns, the impact of the epidemic, structural adjustments, and other factors, the management complexity of commercial banks' mortgaged assets has increased. Under the pressures of operational challenges and stringent regulatory supervision, effectively managing chattel collateral to minimize losses from human, moral, and management factors and ensuring that such collateral remains intact or undamaged has long posed a significant challenge for commercial banks. Typically, enterprises' mortgaged movable property is stored in designated warehouses. Commercial banks must assign security personnel for surveillance, and larger warehouses necessitate multiple guards. This arrangement incurs high personnel costs (direct cost), elevated management and communication expenses (indirect cost), and insufficient night-time monitoring (risk cost). Furthermore, in scenarios involving the disposal of non-performing assets, the process demands "asset evaluation, case declaration, court judgment, and execution," leading to protracted timeframes for the realization of these movable collateral assets. The costs of guarding these assets, borne by commercial banks, can become exorbitant and, in extreme cases, may even exceed the asset values themselves. Therefore, there is an urgent need for technological innovations in the management model of movable collateral to address these inefficiencies and risks.

Fortunately, the current environment is immensely favorable for technological advancements. Over recent years, computer vision (CV) technology based on deep learning has seen rapid progression, addressing several pivotal challenges. Terminal devices have resolved computational issues related to data processing. Convolutional neural networks (CNNs) have tackled the complex algorithmic problem of image comprehension. GPU hardware has provided a robust solution to the need for substantial computational power. Today, innovative deep learning techniques have propelled computer vision technology to unprecedented heights. New methodologies such as capsule networks excel at handling image details, while convolutional recurrent neural networks (CRNNs) specialize in processing image text sequence data. Transformers have achieved remarkable success in both language and vision fields. The maturation of deep learning frameworks including TensorFlow, PyTorch, and Caffe has significantly expedited research and development in computer vision applications. Consequently, diverse applications have emerged, such as visual agriculture for crop monitoring, visual transportation for traffic management, and visual monitoring systems for security purposes. Notably, the current capabilities of computer vision often match, if not surpass, human expertise, particularly in specialized domains like medical image recognition. Given this robust technological backdrop, we aspire to leverage computer vision to address the intricacies of monitoring and guarding movable collateral. The maturity of these technologies promises to deliver unprecedented precision and efficiency in this critical area.

8.1 Scheme Design

The objective of this initiative is to develop a comprehensive warehouse video surveillance system aimed at significantly reducing, or even eliminating, the need for personnel-based monitoring. This project will incorporate advanced computer vision technologies within the context of a chattel collateral warehouse environment, thereby facilitating enhanced image analysis capabilities. Key algorithmic components will focus on interpreting video data to: identify individuals entering and exiting the warehouse, monitor their activities, and assess their interactions with warehouse assets. Specifically, the system will be designed to detect and log entries and exits, analyze actions to determine if they impact the condition of secured items, and use this data to perform risk assessments. The objective includes the classification of individuals entering and leaving the premises, detecting contact with collateral, and assigning risk levels accordingly. The system will also implement preemptive alert mechanisms and ensure video documentation of all personnel movements, thereby providing a traceable record of potential risk events. The front-end application will conduct initial risk assessments and facilitate the uploading of video content, while the backend system will handle database storage and manage report generation for visualization purposes. Managers will have the ability to perform real-time video reviews. As illustrated in Table 8.1, risk levels will be determined based on entry actions, the identities of the individuals, and their

8.1 Scheme Design

Table 8.1 Project risk assessment table

Personnel identification Personnel Action	No contact with collateral	Contacted collateral
Registered licensed personnel entering and exiting	Low risk	Medium risk
Unregistered strangers entering and exiting	Medium risk	High risk
No personnel entering or exiting	No risk	No risk

Table 8.2 Overall technical solution design of the project

Scenario subtask	Technical scheme	The algorithm framework adopted
Video capture and recording	Computer vision processing	OpenCV vision processing framework
Where is chattel collateral? Has the quantity been lost?	Target detection	Yolov3 single-stage object detection, ImageAI transfer learning
Does anyone show up? If so, intercept the area the person is in	Object positioning and segmentation	Yolact instance segmentation framework
Who are the people who appear?	Face detection	Face_recognition integration framework
	Face key feature extraction	
	Face recognition	
Did the person touch the collateral?	Person/thing overlap range calculation	Area overlap calculation
Result presentation	Web-based data visualization	Django, Pyecharts

interaction with the collateral. Risk categorization will be adaptable to meet evolving requirements.

Table 8.2 lists the scenario tasks and technical solutions in this case.

In terms of system architecture design, this case is built with three layers of structure: base layer, deep learning layer, and application layer, which are responsible for the underlying hardware and driver, deep learning software framework and application logic processing, respectively, as shown in Fig. 8.1.

In this case, each subtask is processed one by one, and the risk level and disposal actions are finally formed. Figure 8.2 shows the processing logic.

Let's design a simple sqlite database table to record monitoring results, name it monitor_data, and set the self-increasing sequence id as its primary key. The table structure is as follows:

```
CREATE TABLE "monitor_data" (
    "id" INTEGER NOT NULL PRIMARY KEY AUTOINCREMENT,
    "person" varchar(20), -- name of the person entering and leaving
    "enter_datetime" timestamp, -- entry time
    "leave_datetime" timestamp, -- time of departure
    "contact" integer - whether to touch the collateral
);
```

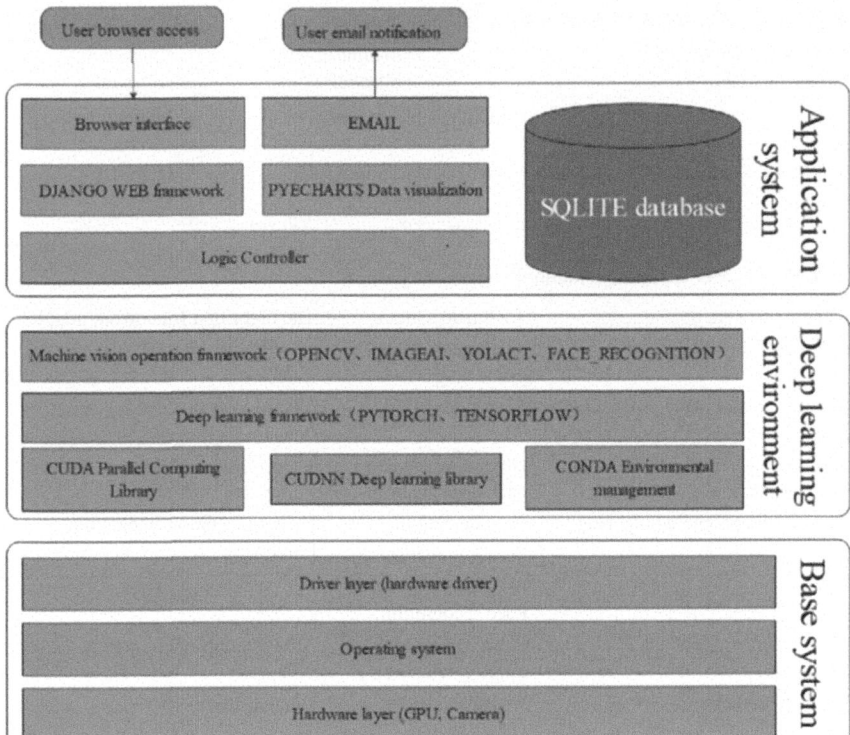

Fig. 8.1 System architecture of this case

And design another table to record when the collateral was lost:

```
CREATE TABLE "goods_loss" (
  "id" INTEGER NOT NULL PRIMARY KEY AUTOINCREMENT,
  "loss_time" timestamp, -- time when the collateral was lost
  "last_count" timestamp, -- The number of detections in the last frame
  "contact" integer -- number of frames detected
);
```

8.2 Development Libraries and Frameworks

This section introduces the relevant technologies used in this case and their open-source frameworks, including OpenCV computer vision, Face_Recognition face recognition, Yolact instance segmentation, ImageAI image processing, Django WEB Services, Pyecharts data visualization.

8.2 Development Libraries and Frameworks

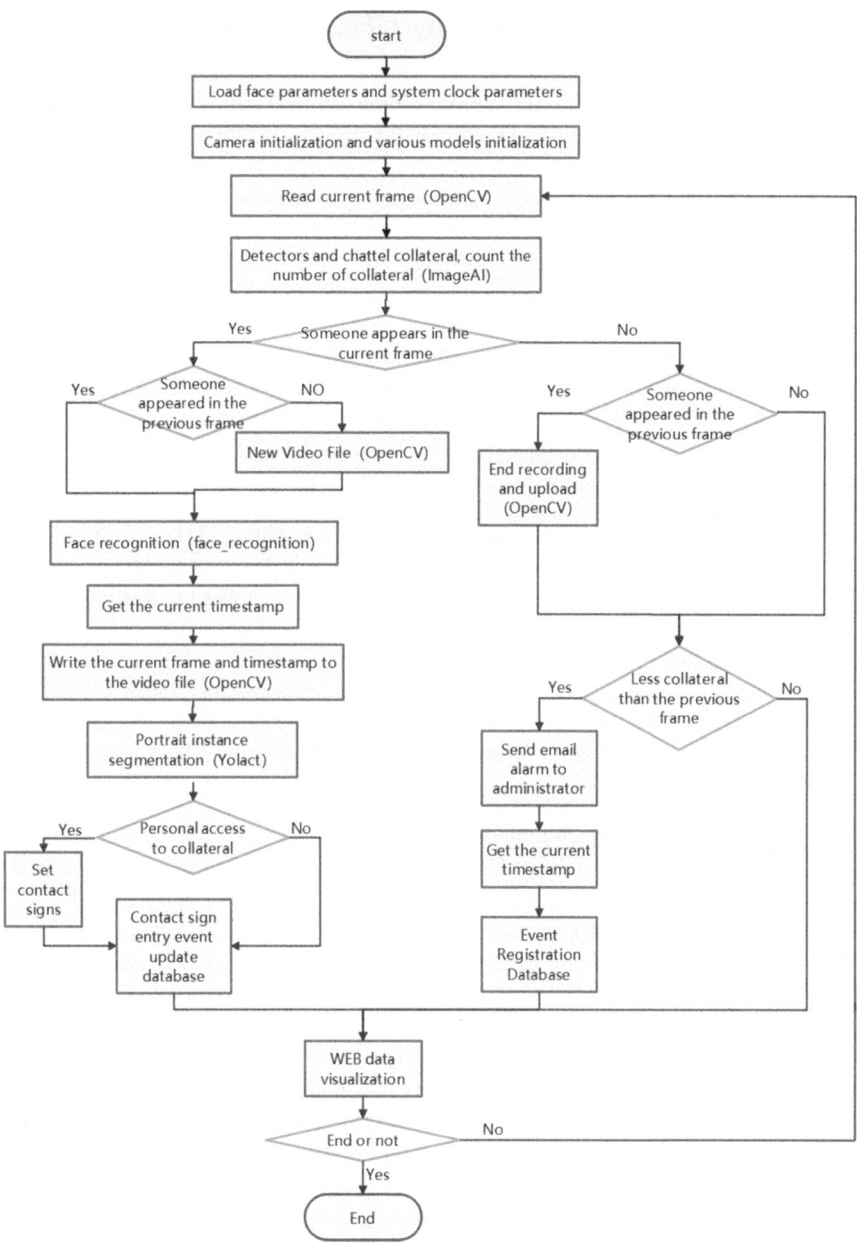

Fig. 8.2 Flowchart of this example

8.2.1 Computer Vision Processing Library: OpenCV

In the field of computer vision, OpenCV is almost universally known as an open-source library and occupies a very prominent position in the industry. OpenCV is an open-source computer vision library from Intel under the Apache 2.0 open-source license. It consists of a series of C functions and C++ classes that implement many common algorithms in image processing and computer vision. OpenCV has a very rich cross-platform medium and high-level API, can run on Linux, Windows, Android, and MacOS operating systems, while providing Python, Ruby, MATLAB, C#, Ruby, GO, and other language interfaces. The operation of OpenCV is completely independent, does not rely on other external libraries, does not need to add new external support can be fully compiled link generation execution program, these features make it become a lot of visual development work first choice.

The OpenCV Logo was designed by Israeli Adi Shavit, a member of the OpenCV community, as shown in Fig. 8.3. Its meaning is as follows:

1. The three notched rings are derived from the "keyhole" Logo, a registered trademark of the open-source initiative. The keyhole Logo is a letter O with a notch in the shape of a keyhole, implying that the key to software development is Open. The OpenCV Logo is made up of three.
2. The opening directions of the three rings are different from each other. The three rings visually resemble the letters O, C, and V, representing Open Computer Vision.
3. The three rings use the three primary colors of red, green and blue, symbolizing the most basic color space in computer vision.

The OpenCV project was founded by Gary Bradsky at Intel in 1999, and the first version came out in 2000. In 2005, OpenCV was used in the Stanley model and won the 2005 DARPA Grand Challenge. Later, the project was led by Gary Bradsky and Vadim Pisarevsky with support from Willow Garage. OpenCV support of many related to computer vision and machine learning algorithms, https://opencv.org/ is the official address, code warehouse address for https://github.com/opencv/opencv, the heat of the current 58.1 K star, the latest version for 4.6.0.

Figure 8.4 shows the module structure of OpenCV. The lowest Layer is OpenCV HAL(Hardware Acceleration Layer), which is a hardware acceleration layer and handles hardware-related tasks. The upper layer is the core layer of OpenCV, which includes modules such as image processing and object detection. The next layer is OpenCV Contrib layer, which is code contributed by other developers and contains high-level API packages such as faces and text. The next layer up is the language interface, method examples, applications, and solutions.

Fig. 8.3 Logo of OpenCV and the open-source initiative

8.2 Development Libraries and Frameworks

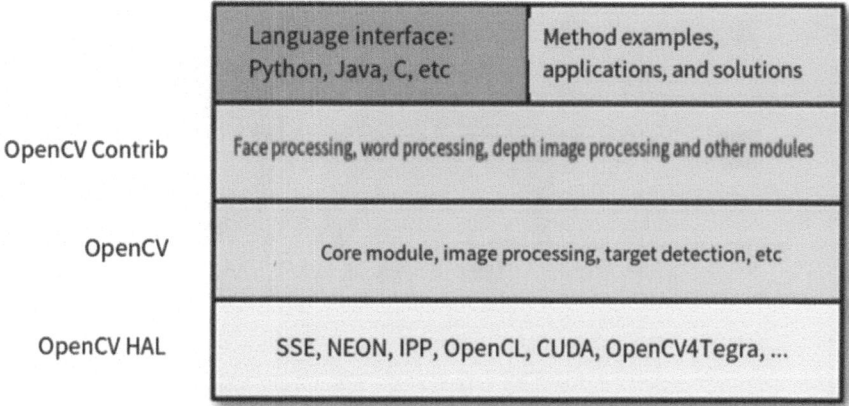

Fig. 8.4 Module structure of OpenCV

This case uses Python API provided by OpenCV 4.4.0.42 to complete video capture, image editing, video saving and other work.

8.2.2 Open-Source Library for Face Recognition: Face_Recognition

Face_Recognition, acclaimed as the world's most efficient open-source face recognition library, adheres to the MIT open-source license. It offers dual interaction methods: a Python code interface and a command line invocation. The library excels in tasks such as face detection, facial landmark extraction, face feature encoding, and face comparison and recognition, boasting an impressive accuracy rate of 99.38%. Built upon the robust C++ visual library dlib, Face_Recognition supports both the Histogram of Oriented Gradients (HOG) model and the CNN model for face detection. These models encode facial contours, unique facial features, and additional biometric information into 128-dimensional feature vectors. These vectors are subsequently compared against those in a pre-existing database to achieve precise face recognition. Face_Recognition's high-level encapsulation and efficient code interfaces make it an optimal tool for rapid development of face recognition applications. Developers can quickly implement sophisticated facial analysis functionalities by leveraging this library. The project's code repository can be accessed at https://github.com/ageitgey/face_recognition, with the latest version tagged as 1.2.2.

Face detection serves as a critical component, determining the presence of a face in a given image via an algorithm. Upon detecting a face, the algorithm identifies and marks the location and size of the facial area. Typically, the face detection algorithm outputs the smallest enclosing rectangle that contains all detected faces. An

Fig. 8.5 Face detection

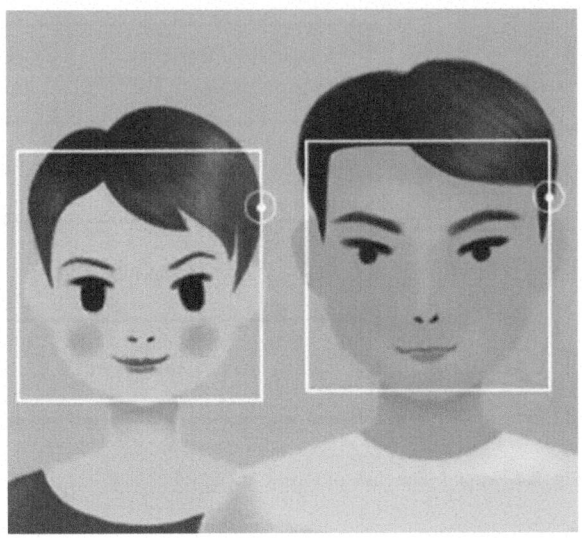

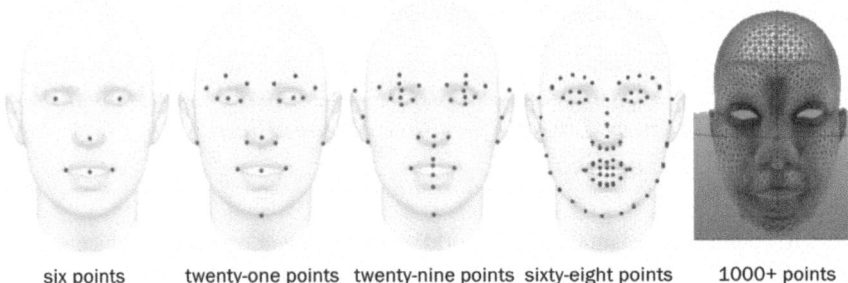

six points   twenty-one points   twenty-nine points   sixty-eight points   1000+ points

Fig. 8.6 face key point detection

illustration of this process can be seen in Fig. 8.5, where the algorithm accurately delineates the boundaries of detected faces.

Face key point detection refers to locating feature key points such as contour points and corner points in contour areas such as eyebrows, eyes, nose, mouth, and face, given a face image. Face key point detection is the basic technology of face pose alignment, face beauty, face expression analysis, and other tasks. For 2D face images, there are usually 6 points, 21 points, 29 points, 68 points, 106 points, 186 points, and so on. For 3D face images, the number of key points can reach a density of more than 1000 points, Face++, Meitu and other enterprises have realized the technology. Face_Recognition uses 68-point detection technology. Figure 8.6 shows the detection effect of face key points.

Face feature vector extraction is to extract facial features of multiple dimensions on the basis of face key point detection. These features are not simple length, width and proportion, but relatively stable features obtained after complex operations (such

8.2 Development Libraries and Frameworks

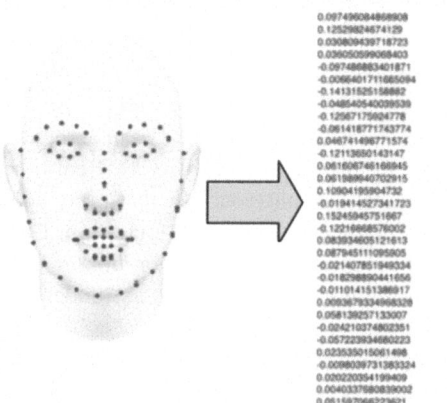

Fig. 8.7 Face feature vector extraction process of Face_Recognition

Fig. 8.8 Face recognition effect of Face_Recognition

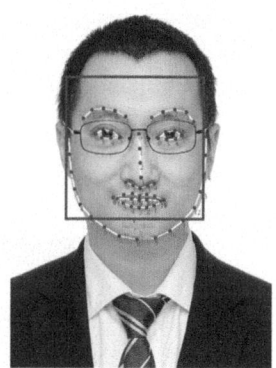

as normalization). Face_Recognition uses 128 floating point numbers to represent the feature vector of a face, as shown in Fig. 8.7.

Face feature vector is the "face" expressed by the algorithm with data. Therefore, once the feature vector is determined, it can determine who the face is. If the data of the two eigenvectors are very close, it means that there is a high probability that the two faces are the same person. By comparing the feature vector of the face to be recognized with that of the known face database, the best matching database face can be found to complete the face recognition task. The recognition effect is shown in Fig. 8.8.

This case uses Face_Recognition to detect and recognize each person in the field of view, and judge whether this person is a registered permitted person, thus providing input for risk assessment.

8.2.3 *Instance Segmentation Open-Source Library: Yolact*

The IEEE International Conference on Computer Vision (ICCV), along with the Computer Vision Pattern Recognition Conference (CVPR) and the European Conference on Computer Vision (ECCV), are the three leading conferences in computer vision. The algorithm Yolact, introduced in the paper "YOLACT: Real-time Instance Segmentation" accepted at ICCV 2019, stands out for its speed with 33 FPS and 30 mAP, making it the fastest instance segmentation algorithm. Currently, the Yolact algorithm is encapsulated in an open-source Python library with over 4k stars on GitHub, available at https://github.com/dbolya/yolact under the MIT open-source license.

In image processing, identifying pixels of specific objects in an image, known as image segmentation, is crucial. This technology is divided into semantic segmentation, instance segmentation, and panoramic segmentation, as shown in Fig. 8.9.

As illustrated in the preceding diagram, semantic segmentation classifies various objects of identical type into a unified class of pixels; in contrast, instance segmentation differentiates each object of the same type into distinct pixel groups. However, instance segmentation only targets the specified object and disregards other entities. Furthermore, panoramic segmentation operates on the entirety of the image, building upon the principles of instance segmentation.

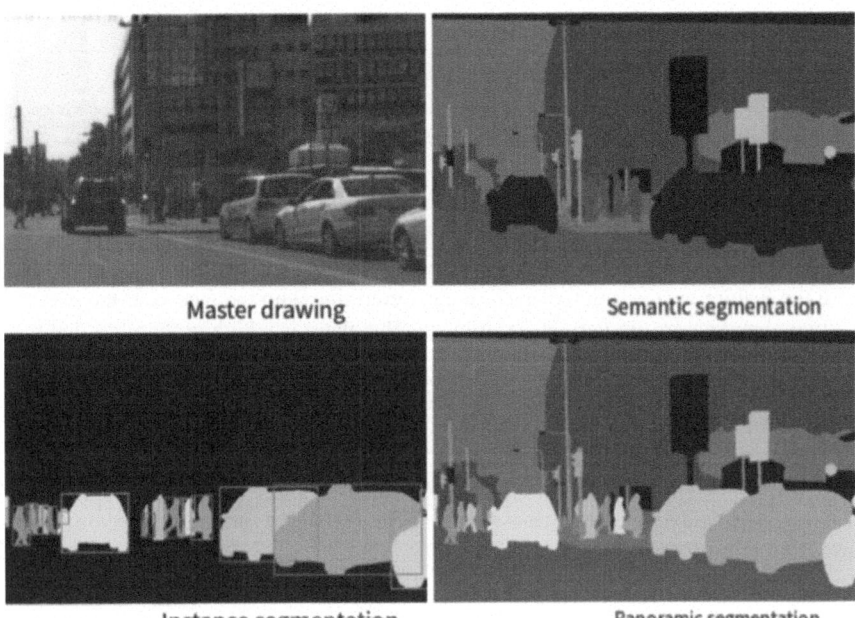

Fig. 8.9 Comparison of image segmentation effect

8.2 Development Libraries and Frameworks

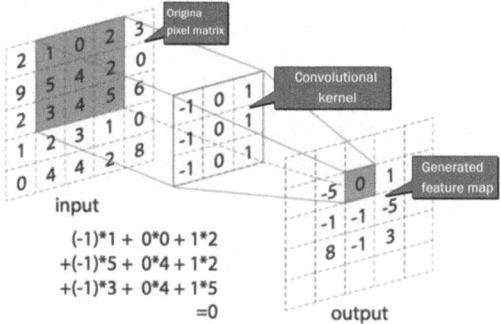

Fig. 8.10 Schematic diagram of a convolution operation

Typically, instance segmentation technology is bifurcated into two-stage and one-stage methodologies. The two-stage approach adheres to the "detect first, segment later" ideology. It initially identifies the boundaries of the target object and subsequently segments it within those borders. In simpler terms, it separates detection from segmentation, with the Mask R-CNN algorithm exemplifying this technique. Conversely, the one-stage approach amalgamates detection and segmentation into a singular comprehensive process. Generally, two-stage algorithms boast superior accuracy but exhibit slower execution speeds. In contrast, one-stage algorithms, while slightly less accurate, offer enhanced speed, with the Yolact algorithm being an exemplar of this category.

To segment a given graph by instance, Yolact employs a unique strategy—it "differentiates each entity based on semantic segmentation." Before delving into this methodology, it is imperative to grasp the underlying concepts of convolution, deconvolution, and fully convolutional networks. Figure 8.10 elucidate the convolution process in detail.

The convolution operation creates a new feature map through the multiplication of the original pixel values within the convolutional kernel matrix. When this kernel function aims to extract a specific feature from an image, the resulting feature map encapsulates that feature's information. For instance, using a convolution kernel to extract an eye results in a feature map highlighting the eye, while a similar operation for the nose results in a nose feature map. In CNNs, distinct kernels function at different levels. The low-level convolutional kernels focus on pixel-level features such as edge textures. Middle-level kernels identify semi-semantic features that span multiple pixels. High-level kernels extract features that closely align with the recognizability of semantic attributes.

Figure 8.11 illustrates the progression of feature extraction capabilities from the first to the fifth convolutional layer. The left side displays feature maps and the right side shows the original image. Analysis reveals that the second layer primarily detects texture edges. The third layer maps out larger contours, including the head and torso. By the fifth layer, only prominent features of the subject, such as the main facial area, are captured. This progression underscores that high-level convolutional layers enhance the representation of semantic features, which are crucial for distinguishing one subject category from another. Furthermore, if the image contains

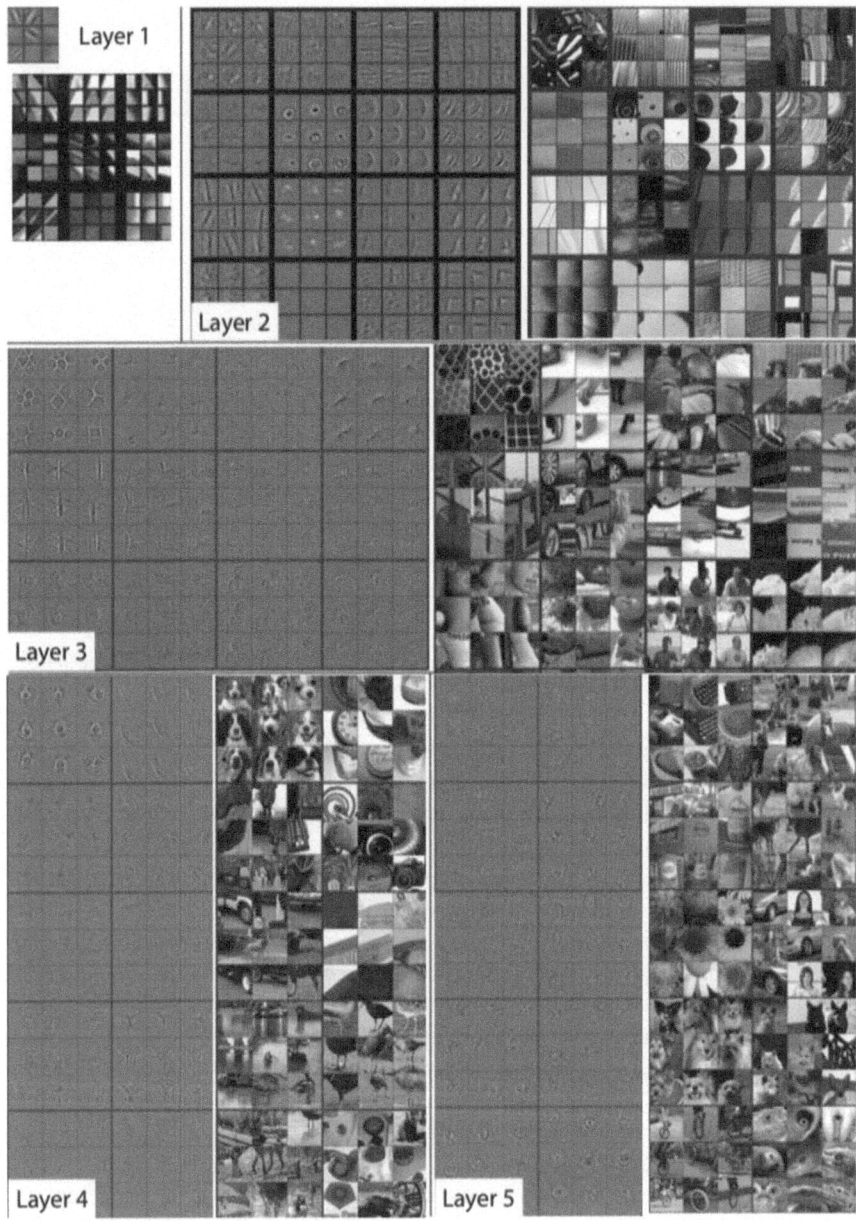

Fig. 8.11 Convolutional kernels at different levels extract feature maps at different levels

multiple objects, the individual features of each object are extracted. This indicates that convolution operations excel in handling the spatial coherence of diverse objects, thereby enabling the differentiation between various subjects.

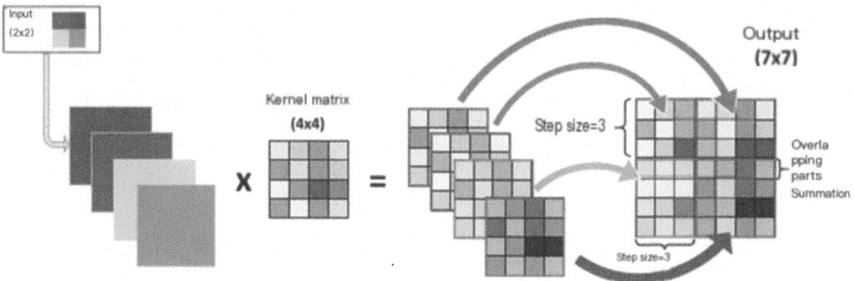

Fig. 8.12 Schematic diagram of semantic segmentation realized by deconvolution operation

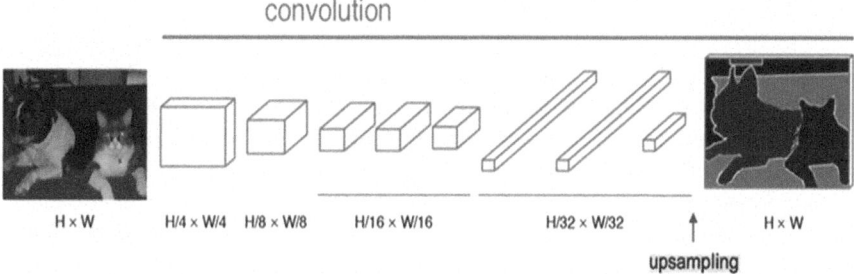

Fig. 8.13 Schematic diagram of semantic segmentation of the full convolutional network

Then look at deconvolution, also known as transposed convolution, upsampling, about equal to the inverse operation of the convolution operation. In the forward propagation process of convolutional neural network, since the input image is extracted by convolutional kernel, the size of the output feature map often becomes smaller, how to restore the image to the original size? In this way, deconvolution operation is needed to realize the mapping from small resolution to large resolution. The convolutional check is used to carry out multiplication operation for each pixel, then size reduction is performed according to the corresponding position, step size and filling, and the overlapping part is added and summed, and finally the feature map of the original size is obtained (note: it is not exactly equivalent to the original image), as shown in Fig. 8.12.

Fully convolutional networks (FCNs) are neural networks in which all the layers are convolutional, hence the name. FCN accepts the input of any size, and then upsamples the feature map of the last convolutional layer through deconvolution to restore it to the input image size, thus generating a category prediction for each pixel, while retaining the spatial information of the input image, thus obtaining the semantic segmentation prediction graph. To put it simply, FCN is good at generating semantic hierarchy division, that is, objects with the same features can get the same category prediction after deconvolution, which is the opposite of the feature of convolution operation, as shown in Fig. 8.13.

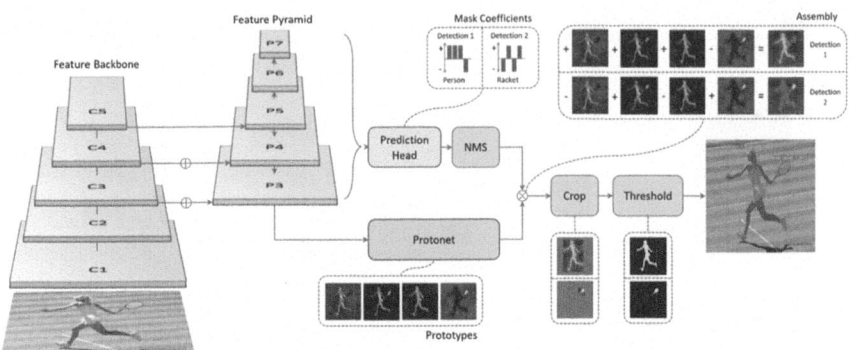

Fig. 8.14 Yolact architecture schematic

After understanding the concepts of convolutional and full convolutional, we return to Yolact algorithm, whose architecture principle is shown in Fig. 8.14.

Yolact divides the case segmentation problem into two parallel parts. The first part uses the full convolutional network to generate "mask coefficient," that is, semantic vector representation, at the semantic level, without targeting any instance. The second part uses the convolutional layer to generate a "prototype mask" at the spatial coherence level. It can be simply understood as distinguishing between different categories and distinguishing between different objects in the same category. Deformable convolution is added to the backbone network to improve the feature sampling capability of the backbone network for instances with different shapes. Set up an object detection branch, add additional headers to predict the mask, and weight the prototype mask for specific instances. These are the key techniques of the Yolact library.

In this case, Yolact is used to pick the pixels of each person in the field of view. This gives a more accurate portrait range than target detection. If the portrait range overlaps with the collateral detection range, the person may have touched the collateral (regardless of spatial depth).

8.2.4 Deep Learning Image Processing Library with Target Detection Migration Learning: ImageAI

ImageAI is a deep learning image processing open-source library designed for programmers without a background in machine learning. It is developed and maintained by the Olafenwa brothers, Moses Olafenwa, and John Olafenwa. The library offers high-level encapsulations for tasks such as image recognition, object detection, video detection, and video object tracking. It uses TensorFlow as the underlying deep learning framework, supports various deep learning algorithms and models, and allows easy creation of custom models. The library is known for its simplicity, strong code readability, and ease of maintenance. ImageAI follows the

8.2 Development Libraries and Frameworks

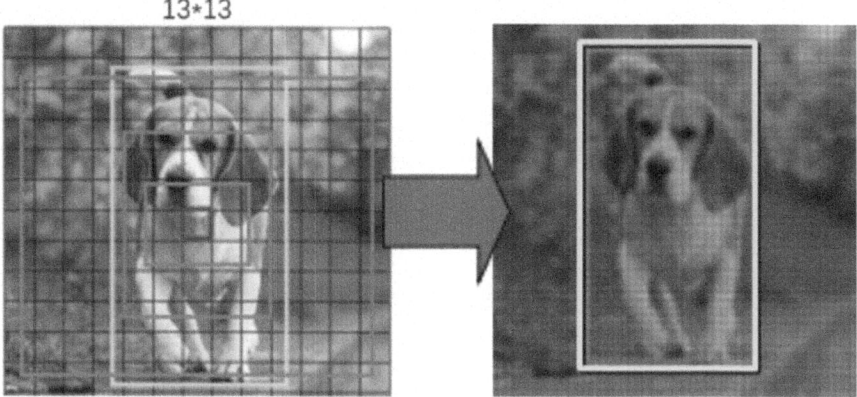

Fig. 8.15 Schematic diagram of the Yolov3 algorithm

MIT open-source license agreement, has gained popularity with 6.7 K stars on Github, and the latest version is 2.1.6. The code repository can be found at https://github.com/OlafenwaMoses/ImageAI. A valuable aspect is that ImageAI provides comprehensive Chinese documentation, available at https://imageai-cn.readthedocs.io/zh_CN/latest.

In this case, ImageAI is used to detect personnel and movable collateral and to count the collateral. Of course, this is done within the limited field of view of the camera. This requires the use of an object detection algorithm, the Yolov3 algorithm used in this case. This algorithm splits the input image into a 13 × 13 grid and uses the neural network to calculate the confidence of the bounding box of the contained object as well as the probability that the contained object belongs to a particular category. Yolov3 uses a non-maximum suppression algorithm to determine the detection region. In simple terms, this algorithm is to continuously eliminate the low confidence bounding boxes surrounding the target until the multiple high confidence bounding boxes are eliminated to only one, as shown in Fig. 8.15.

Figure 8.16 The network structure of Yolov3.

Given that this book primarily focuses on practical development, the underlying principles of algorithms are not discussed in depth. Readers with an interest in these principles may consult other resources. This case utilizes the Yolov3 pre-trained model, which includes a built-in person detection feature. However, for this model, movable property collateral represents a novel concept. How can we develop a new model capable of recognizing both persons and movable property collateral? Fortunately, ImageAI offers functions for custom object detection and transfer learning. Custom object detection involves collecting a substantial number of photos of the movable property collateral (at least 200 images). Then, we use ImageAI to train a brand-new object detection network to identify the collateral. Starting anew to train a fresh model is not only time-consuming and computationally expensive but also yields predictions for only one category. Transfer learning, however, provides a shortcut. It allows us to build a custom small dataset with a limited number of

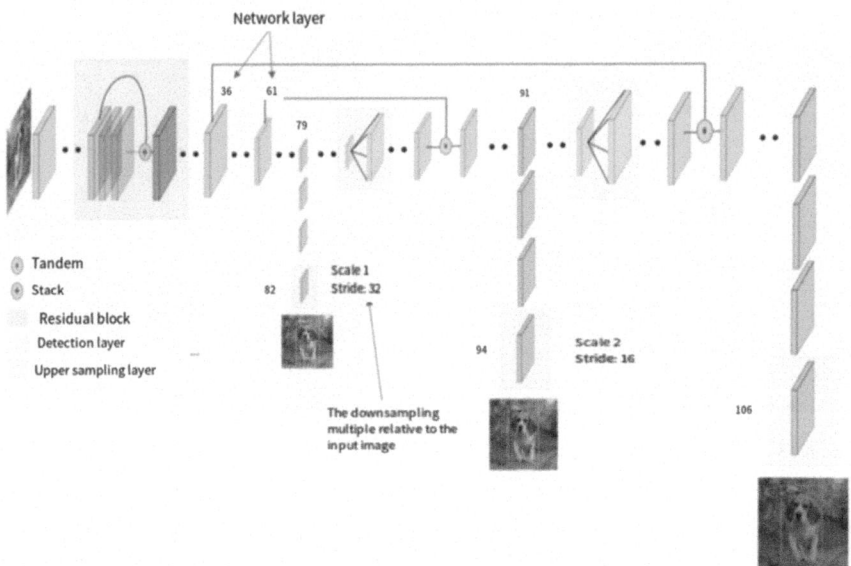

Fig. 8.16 Network structure of Yolov3

collateral photos. We can then perform secondary training on this small dataset using a large pre-trained model as a base. The benefit of this approach is that the new model inherits the performance capabilities of the pre-trained model while also acquiring the ability to detect custom objects. This way, it can identify both persons and collateral without needing to be trained from scratch, thus saving time and computational power.

Transfer learning can be aptly compared to "standing on the shoulders of giants." The large pre-trained model, created using extensive datasets, serves as the giant. Transfer learning allows for fine-tuning the network so that even with a small amount of additional training, we can achieve a high-performance model. This technique is frequently employed in practical engineering applications for its efficiency and efficacy.

8.2.5 Framework and Pyecharts Data Visualization Library: Django

Django is a robust web framework built with Python. It provides extensive modules, enabling developers to efficiently create web applications with minimal coding. As an open-source framework under the BSD license, its repository resides at https://github.com/django. Django employs a modular, plug-in-friendly MVC architecture, dividing the system into Model, View, and Controller. This structure improves maintainability, scalability, and reusability. The Model (M) maps business objects

8.2 Development Libraries and Frameworks

Fig. 8.17 Software architecture of Django

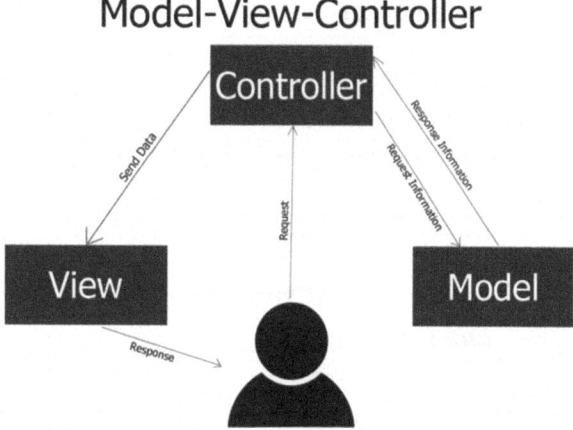

to the database via ORM. The View (V) manages user interactions and page rendering. The Controller (C) handles user requests and routing. Figure 8.17 illustrates Django's software architecture.

Apache ECharts is an open-source data visualization software from Baidu, with clever chart design and good interactivity. Pyecharts is an Echarts chart class library written in Python, providing a variety of Python interfaces for visualization charts, with the following features:

1. Concise API design, easy to use, support chain call;
2. Support more than 30 chart types;
3. Support mainstream Notebook environment, Jupyter Notebook and JupyterLab;
4. Can be easily integrated into Flask, Sanic, Django, and other mainstream Web frameworks;
5. Highly flexible configuration items, can easily match beautiful charts;
6. Provide detailed documentation and examples to quickly develop projects;
7. Support map display, more than 400 map files, support native Baidu map.

This case needs to show the collateral monitoring results to the background managers, including the identification of personnel entering and leaving the collateral and the statistics of personnel touching the collateral, etc. Django is used to build a Web access application, and Pyecharts is integrated to show the statistics visualization.

8.3 Case Practice

This section describes the development and operation environment construction, open-source framework application, code writing and project operation of this case. After reading this section, you will understand the details of the project implementation process.

8.3.1 Hardware and Software Environment Setup and Case Runs

Since this case requires running the neural network, a substantial number of matrix operations are necessary. The CPU's computing speed falls short of requirements, making a GPU essential. Two options exist here. First, embedded edge computing devices like Jetson Nano, Jetson TX2, Xavier NX, and AGX Xavier. These compact devices feature embedded GPUs and support NVIDIA CUDA and CUDNN for accelerating neural network tasks. They can pair with high-definition night vision cameras and deploy the model on a dedicated Linux system. Figure 8.18 details the hardware specifications. Alternatively, Scheme two employs an HD night vision camera, connects via USB port, and supports model deployment on Linux or Windows. This section introduces the actual process using Scheme two.

Hardware configuration: I9-10900K, RTX2080 Super 8G graphics card, 32GB memory, 1T SSD.

Software environment: Windows 10 64-bit, Anaconda, Cuda 10.0, Cudnn7.6.4, TensorFlow-GPU1.13.1, Pytorch1.2.0. See Sect. 1.3.1 of this book for Anaconda installation, Sect. 2.5.1 of this book for Cuda and Cudnn installation. Run the command pip install tensorflow-gpu==1.14.0 to install tensorflow. Run the command pip install torch==1.2.0 torchvision==0.4.0 to install pytorch.

Download the source code from the WeChat public account for this book. Run the monitor.bat command in the "AI client" directory to execute the example program. For ease of explanation, the program reads an image named input.jpg from the current directory to complete the reasoning process. You can modify the comment line in monitor.py to process each video frame from the camera. The collateral here is the keg, and detection prompts "KEG." After executing the example program, a model inference result graph will appear. It will display results for face recognition model loading, target detection, and instance segmentation. The program also determines if the person touches the collateral and if the collateral reduces by comparing the current frame's processing result with the previous frame's, as illustrated in Figs. 8.19 and 8.20.

The program will automatically save the video recording during the monitoring period. It will frame and label the identified face and confidence level on the individual and the identified wine barrel. Upon detecting someone in the field of view, it will export the video, annotating it with the date and time. It also

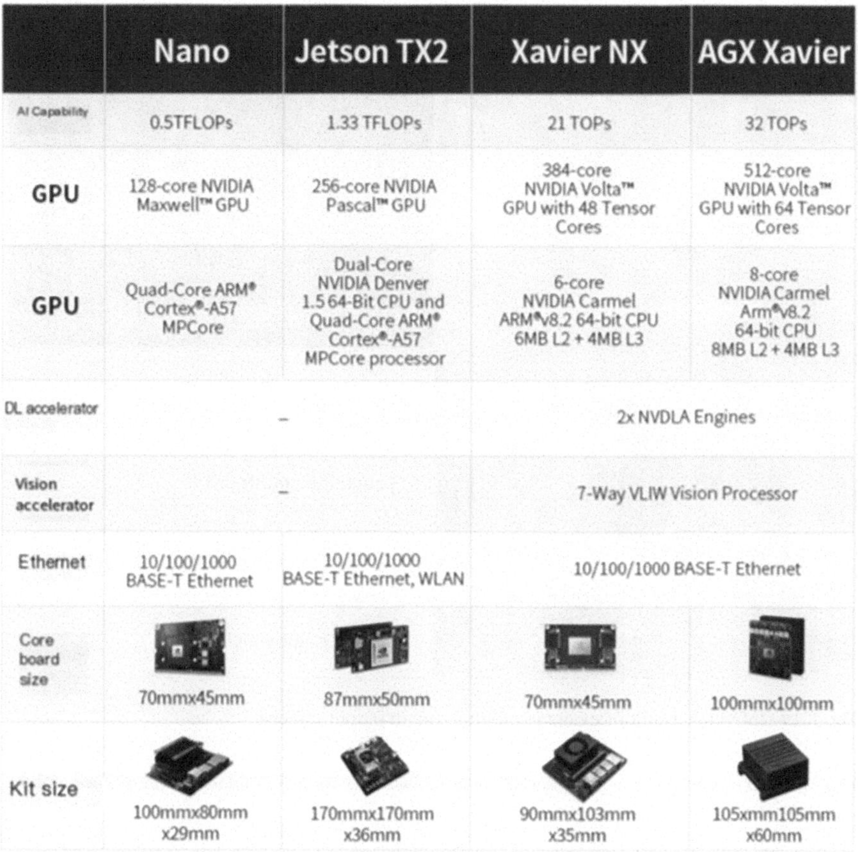

8.3 Case Practice

	Nano	Jetson TX2	Xavier NX	AGX Xavier
AI Capability	0.5TFLOPs	1.33 TFLOPs	21 TOPs	32 TOPs
GPU	128-core NVIDIA Maxwell™ GPU	256-core NVIDIA Pascal™ GPU	384-core NVIDIA Volta™ GPU with 48 Tensor Cores	512-core NVIDIA Volta™ GPU with 64 Tensor Cores
GPU	Quad-Core ARM® Cortex®-A57 MPCore	Dual-Core NVIDIA Denver 1.5 64-Bit CPU and Quad-Core ARM® Cortex®-A57 MPCore processor	6-core NVIDIA Carmel ARM®v8.2 64-bit CPU 6MB L2 + 4MB L3	8-core NVIDIA Carmel Arm®v8.2 64-bit CPU 8MB L2 + 4MB L3
DL accelerator	–	–	2x NVDLA Engines	
Vision accelerator	–	–	7-Way VLIW Vision Processor	
Ethernet	10/100/1000 BASE-T Ethernet	10/100/1000 BASE-T Ethernet, WLAN	10/100/1000 BASE-T Ethernet	
Core board size	70mmx45mm	87mmx50mm	70mmx45mm	100mmx100mm
Kit size	100mmx80mm x29mm	170mmx170mm x36mm	90mmx103mm x35mm	105xmm105mm x60mm

Fig. 8.18 Nvidia Embedded GPU series devices

assesses risk: whether the person touches the collateral or if the collateral diminishes. The four small figures in Fig. 8.20 represent the original input, portrait instance segmentation matting, face recognition and target detection, and business reasoning results, respectively. In Fig. 8.21, "cuda=True" signifies that the program utilizes GPUs. "Loading face feature data" means retrieving the face library's feature vector from the preprocessed pkl encoded file (see Sects. 8.2 and 8.3.2 for details). "person" and "keg" refer to the result labels detected by ImageAI, with the subsequent numbers indicating confidence levels. Confidence is expressed as a decimal percentage, detailing the detection model's certainty in assigning a specific label to a target. The subsequent list denotes the pixel range of the target detection formatted as [$x1$, $y1$, $x2$, $y2$]. "cutouts" signifies that Yolact marks the pixel range of the region where the person pixel is located in the instance segmentation, formatted as [$y1$, $y2$, $x1$, $x2$]. The image above shows two list elements for cutouts, indicating Yolact detected two individuals.

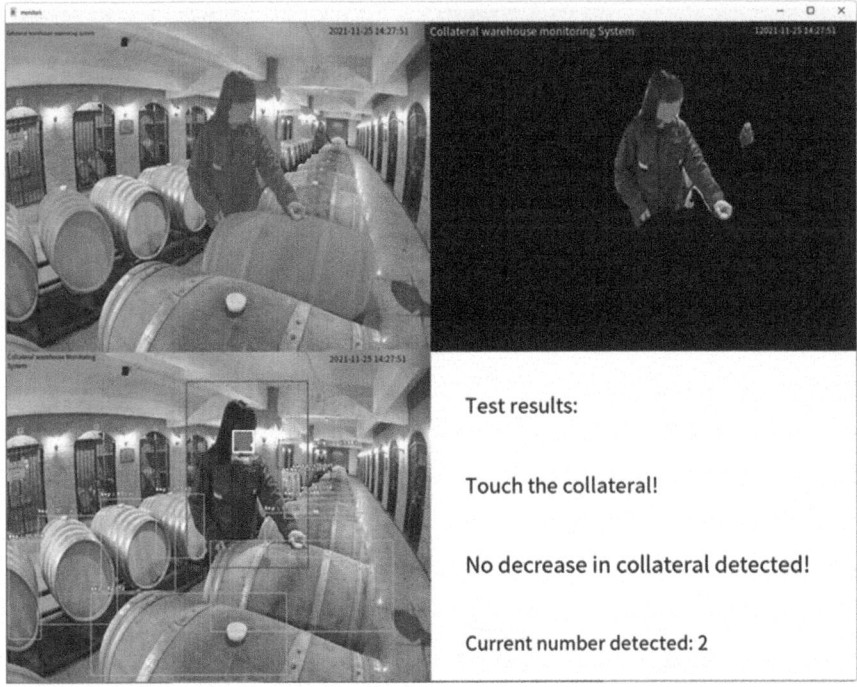

Fig. 8.19 Inference result of example program (main output screen)

Fig. 8.20 Example program inference result (console)

```
loading annotations into memory...
Done (t=0.50s)
creating index...
index created!
cuda= True <class 'bool'>
person  :  96.16777300834656  :  [546, 86, 911, 630]
keg  :  83.12153816223145  :  [617, 550, 1167, 815]
keg  :  82.95161128044128  :  [252, 703, 847, 941]
keg  :  83.66404175758362  :  [880, 411, 1060, 444]
keg  :  85.94346046447754  :  [841, 427, 1068, 485]
keg  :  91.36657118797302  :  [783, 470, 1094, 597]
keg  :  83.09018611907959  :  [283, 428, 511, 697]
keg  :  85.77929735183716  :  [94, 476, 334, 792]
person  :  93.44937801361084  :  [928, 281, 981, 374]
keg  :  84.08395051956177  :  [866, 350, 886, 415]
keg  :  75.39458870887756  :  [883, 347, 900, 411]
keg  :  78.32701802253723  :  [836, 357, 865, 431]
keg  :  79.3161392211914  :  [860, 358, 878, 420]
keg  :  97.19732403755188  :  [453, 414, 598, 600]
keg  :  95.35842537879944  :  [5, 550, 102, 835]
cutouts= [[132, 616, 543, 904], [287, 371, 933, 978]]
box= [617, 550, 1167, 815]
box= [252, 703, 847, 941]
box= [880, 411, 1060, 444]
```

8.3 Case Practice

Fig. 8.21 Description of coordinates

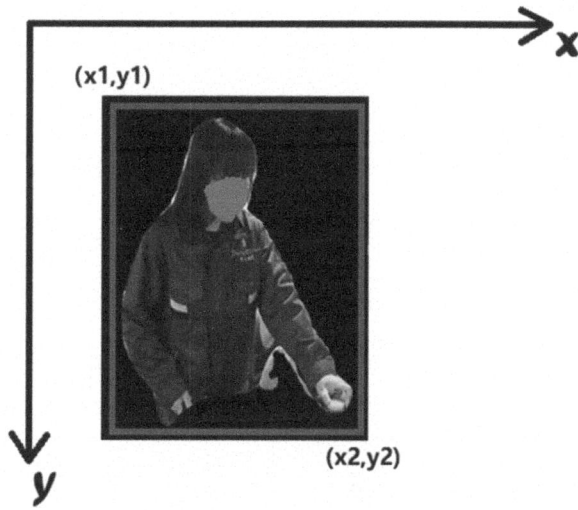

It's worth noting that both ImageAI and Yolact use the same coordinate system, with $(x1,y1)$ representing the top-left coordinate and $(x2,y2)$ representing the bottom-right coordinate, but the order of their vectors is different. ImageAI uses the order of $[x1,y1,x2,y2]$. Yolact adopts the order of $[y1,y2,x1,x2]$. The coordinates are shown in Fig. 8.21.

During the run, use the GPU-Z software to observe the GPU usage, as shown in Fig. 8.22. Just now we have done a frame of reasoning and can see that there is a peak usage of video memory and GPU computing core.

You can also run the nvidia-smi command delivered with the graphics card driver to view the GPU usage, as shown in Fig. 8.23. The circles in the command show the usage of video memory and computing core.

Now, let's start the data visualization report. Go to the "Data Visualization" directory of the download code in this book, and run the command line: python manage.py runserver 127.0.0.1:8081 to start the Django service, as shown in Fig. 8.24. 127.0.0.1 is the IP address of the WEB service, and 8081 is the port of the WEB service. You can modify it as required.

Type https://127.0.0.1:8081/monapp/login in the browser end, according to the following login screen, the user name input 01, 1 password input, as shown in Fig. 8.25.

The home screen is displayed, as shown in Fig. 8.26.

In Monitoring Browse, enter a date range, and click Query to display visual data statistics, as shown in Fig. 8.27.

At this point, the case has been quickly built and can actually run.

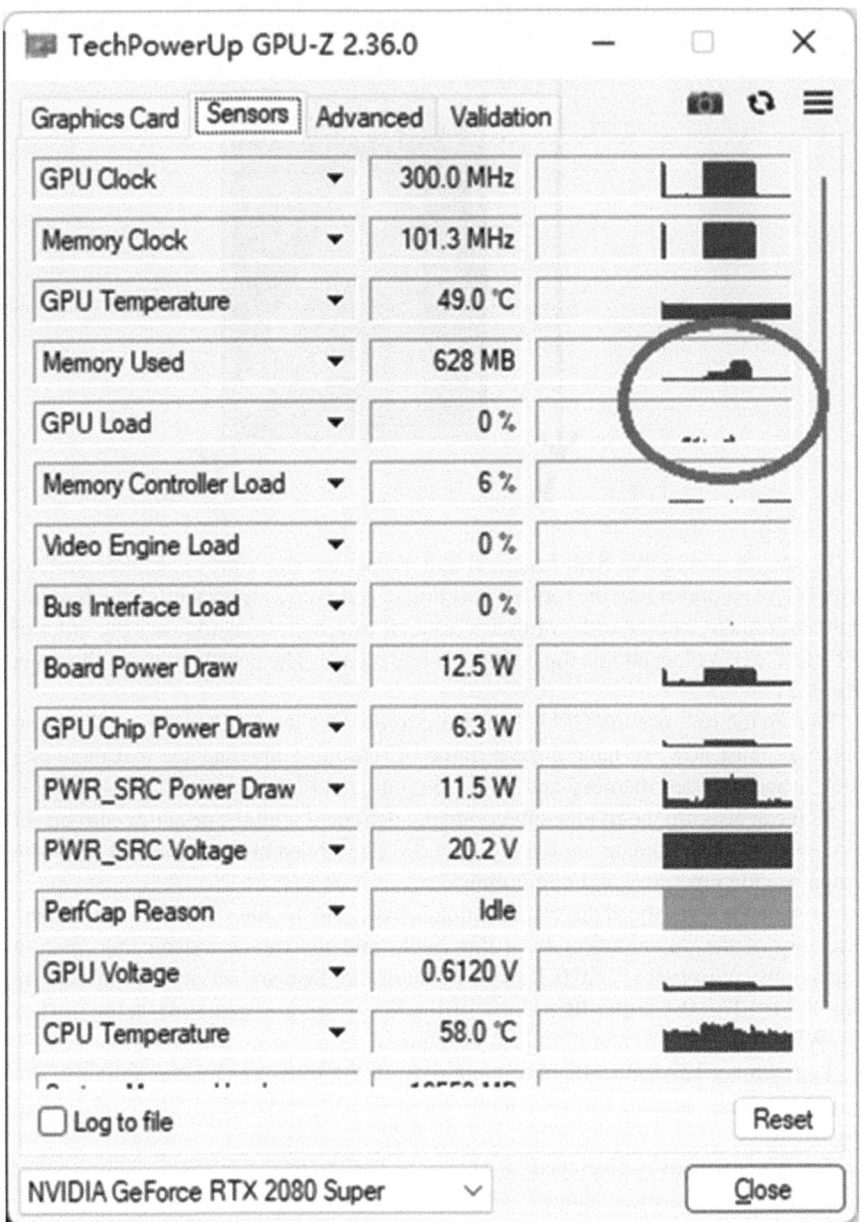

Fig. 8.22 GPU usage during model run

8.3 Case Practice

```
Thu Nov 25 11:46:56 2021
+-----------------------------------------------------------------------------+
| NVIDIA-SMI 471.96       Driver Version: 471.96       CUDA Version: 11.4     |
|-------------------------------+----------------------+----------------------+
| GPU  Name            TCC/WDDM | Bus-Id        Disp.A | Volatile Uncorr. ECC |
| Fan  Temp  Perf  Pwr:Usage/Cap|         Memory-Usage | GPU-Util  Compute M. |
|                               |                      |               MIG M. |
|===============================+======================+======================|
|   0  NVIDIA GeForce ...  WDDM | 00000000:01:00.0  On |                  N/A |
| N/A   55C    P2   63W /   N/A |  4034MiB /   8192MiB |    28%       Default |
|                               |                      |                  N/A |
+-------------------------------+----------------------+----------------------+

+-----------------------------------------------------------------------------+
| Processes:                                                                  |
|  GPU   GI   CI        PID   Type   Process name                  GPU Memory |
|        ID   ID                                                   Usage      |
|=============================================================================|
|    0   N/A  N/A       428    C+G   ...bbwe\PaintApp\mspaint.exe         N/A |
|    0   N/A  N/A      1728    C+G   Insufficient Permissions             N/A |
|    0   N/A  N/A      2580    C+G   C:\Windows\explorer.exe              N/A |
|    0   N/A  N/A      5384    C+G   ...ge\Application\msedge.exe         N/A |
|    0   N/A  N/A      6308    C+G   ...054.29\msedgewebview2.exe         N/A |
|    0   N/A  N/A     10404    C     Insufficient Permissions             N/A |
|    0   N/A  N/A     11364    C+G   ...y\ShellExperienceHost.exe         N/A |
|    0   N/A  N/A     11396    C+G   ...lPanel\SystemSettings.exe         N/A |
|    0   N/A  N/A     13316    C+G   ...artMenuExperienceHost.exe         N/A |
|    0   N/A  N/A     13340    C+G   ...nlh2txyewy\SearchHost.exe         N/A |
|    0   N/A  N/A     15244    C+G   ...ekyb3d8bbwe\YourPhone.exe         N/A |
|    0   N/A  N/A     16668    C+G   ...wekyb3d8bbwe\Video.UI.exe         N/A |
|    0   N/A  N/A     17880    C+G   ...2txyewy\TextInputHost.exe         N/A |
|    0   N/A  N/A     17896    C+G   ...3d8bbwe\CalculatorApp.exe         N/A |
+-----------------------------------------------------------------------------+
```

Fig. 8.23 GPU usage during model operation

Fig. 8.24 Starting the Django service framework

```
Watching for file changes with StatReloader
Performing system checks...

System check identified no issues (0 silenced).
November 28, 2021 - 09:08:45
Django version 3.2.9, using settings 'monitor.settings'
Starting development server at http://127.0.0.1:8081/
Quit the server with CTRL-BREAK.
```

Fig. 8.25 Case management login page

pledge

Warehouse monitoring system

Username: 01

Secret Code:

Log In

Fig. 8.26 Project management screen

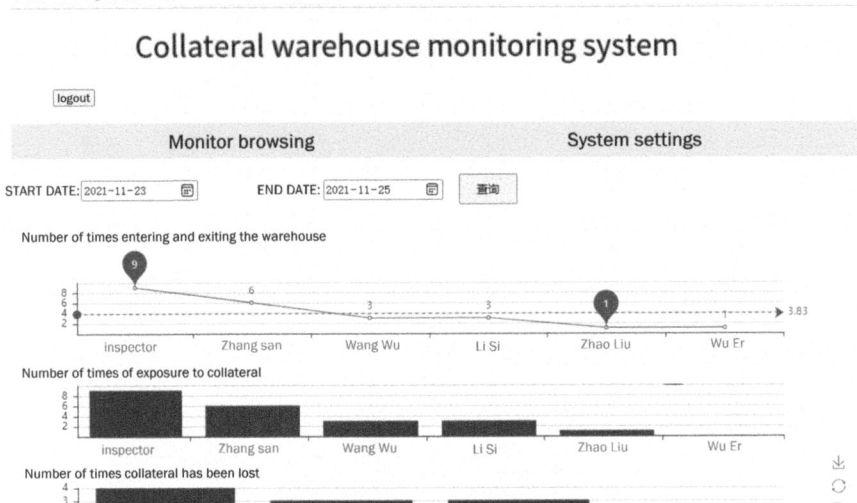

Fig. 8.27 Data visualization page of this project

8.3.2 Code Practice

ImageAI Transfer Learning Practice

First, we need to prepare the data set. In this case, we take photos from a real warehouse. In view of data privacy security, this book uses example photos to

8.3 Case Practice

illustrate the problem. To train a deep learning model, you have to tell the model in the data set which regions are pixels of which labels. This requires image labeling. We use labelImg to do the annotation work. Download the labelImg.exe file from the WeChat public account of this book, and double-click it to run it. It is worth noting that labelImg.exe cannot be stored in a path containing Chinese characters. Figure 8.28 shows the labelImg annotation screen.

After entering the labelImg interface, click Open to open a picture, click CreateRectBox to annotate the picture, and mark a label behind the picture frame. In this example, keg and person are marked. The right side displays the type and number of labels. Click Save to save the tag result. The software will generate an xml file with the same name, as shown in Fig. 8.29.

Create a keg_data directory to annotate more than 200 photos, and create two subdirectories under this directory to store the training set and validation set respectively, and two directories to store the annotation xml file and image file respectively for annotations and images, respectively. Figure 8.30 shows the structure of the directory.

Import the modules used by the project from the ImageAI library.

```
from imageai.Detection.Custom import DetectionModelTrainer
```

Create a target detector training.

```
trainer = DetectionModelTrainer()
```

Specify target detection model type as Yolov3 and data set directory.

Fig. 8.28 Annotation screen of labelImg

```xml
<annotation>
    <folder>images</folder>
    <filename>img1.jpeg</filename>
    <path>E:/imageai/keg_data/train/images/img1.jpeg</path>
    <source>
        <database>Unknown</database>
    </source>
    <size>
        <width>800</width>
        <height>533</height>
        <depth>3</depth>
    </size>
    <segmented>0</segmented>
    <object>
        <name>person</name>
        <pose>Unspecified</pose>
        <truncated>0</truncated>
        <difficult>0</difficult>
        <bndbox>
            <xmin>289</xmin>
            <ymin>6</ymin>
            <xmax>441</xmax>
            <ymax>528</ymax>
        </bndbox>
    </object>
    <object>
        <name>keg</name>
        <pose>Unspecified</pose>
        <truncated>0</truncated>
        <difficult>0</difficult>
        <bndbox>
            <xmin>448</xmin>
            <ymin>205</ymin>
            <xmax>536</xmax>
            <ymax>304</ymax>
        </bndbox>
    </object>
</annotation>
```

Fig. 8.29 Annotation screen of labelImg

```
trainer.setModelTypeAsYOLOv3()
trainer.setDataDirectory(data_directory="keg_data")
```

The object_names_array parameter specifies the tag sequence as keg and person, and the model is trained to recognize the target of these two types of tags. The batch_size parameter specifies the number of batch samples used during training, the num_experiments parameter specifies the number of epochs, and the train_from_pretrained_model parameter specifies which pre-trained model to start the migration from.

8.3 Case Practice

```
>> train      >> images       >> img_1.jpg  (shows Object_1)
              >> images       >> img_2.jpg  (shows Object_2)
              >> images       >> img_3.jpg  (shows Object_1, Object_3 and Object_n)
              >> annotations  >> img_1.xml  (describes Object_1)
              >> annotations  >> img_2.xml  (describes Object_2)
              >> annotations  >> img_3.xml  (describes Object_1, Object_3 and Object_n)

>> validation >> images       >> img_151.jpg (shows Object_1, Object_3 and Object_n)
              >> images       >> img_152.jpg (shows Object_2)
              >> images       >> img_153.jpg (shows Object_1)
              >> annotations  >> img_151.xml (describes Object_1, Object_3 and Object_n)
              >> annotations  >> img_152.xml (describes Object_2)
              >> annotations  >> img_153.xml (describes Object_1)
```

Fig. 8.30 Directory structure of data set

```
trainer.setTrainConfig(object_names_array=["keg","person"],
batch_size=4, num_experiments=100,
train_from_pretrained_model="yolo.h5")
```

With the above work done, you can execute the following code to start the training!

```
trainer.trainModel()
```

See train.py in the imageai Transfer Learning directory for the above code. Run python train.py to start the training process and you will see the interface shown in Fig. 8.31.

Ensure the above code runs on TensorFlow 1.13.1 for proper functionality. During training, it first outputs parameters, then iterates through each epoch, observing a consistent decrease in the loss value. After each iteration, the model optimizes anchor box predictions and saves the model files in the models directory. Iterations conclude upon reaching the num_experiments parameter limit. There are four directories in keg_data: cache, json, logs, and models. Only copy files from the json directory and the latest model files in the models directory up one level. This book only shows a sample data set and the corresponding trained example model (the actual application case by the author has a much smaller model loss than the example weight). The example model's weight file is detected at detection_model-ex-098--loss-0014.018.h5 under the models directory. The json file mapping the data set and anchoring the model object is detection_config.json in the json directory.

Face_Recognition Face Coding Acceleration Practice

When operating a project, extracting the feature vector from each photograph in the facial database individually and subsequently comparing it to the face needing recognition will substantially increase processing time as the number of photos in

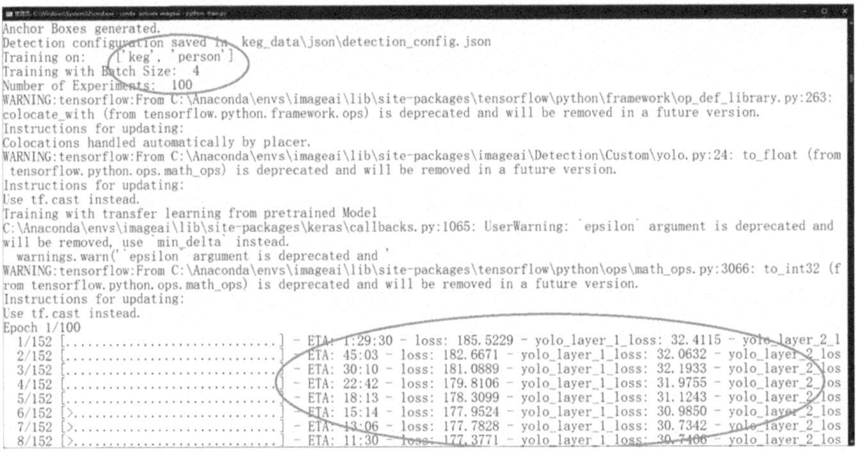

Fig. 8.31 ImageAI transfer learning training process

the database grows. This increase can become exceedingly prohibitive. To address this issue, we developed a suite of programs designed to extract feature vectors from every photograph in the database, store these vectors in a binary file, and then load all feature vectors into memory during project initialization. This approach ensures that during runtime, only vector similarity calculations are performed to achieve face recognition. This methodology significantly reduces operational time and, consequently, enhances the project's overall execution speed. Python's pickle module facilitates this process by enabling the storage of any in-memory data into a binary file and the retrieval of that data back into memory, offering a user-friendly solution.

First prepare the face database, put the photos in the know_face directory, the file name is the name, the file suffix does not matter. Because of face data privacy concerns, this book does not show the data set. Import the library package module:

```
import pickle
import face_recognition
```

To initialize variables:

```
time_start=time.time()
path = os.getcwd()+'/known_face'
os.chdir(path)
images_file = os.listdir('.')
know_names = []
know_paths = []
know_encodings = []
```

To create a list of names and file paths from the photo dataset:

8.3 Case Practice

```
for each in images_file:
name = os.path.splitext(each)[0]
know_names.append(name)
image_path = path+'/'+each
know_paths.append(image_path)
```

Extract the face feature vector for each photo:

```
count = 1
```

for each_path in know_paths:

```
    img = face_recognition.load_image_file(each_path) #
face_recognition framework loads photos
    face_locations = face_recognition.face_locations(img,
model='cnn') # Use a convolutional neural network (cnn) model for face
location
    encoding = face_recognition.face_encodings(img, face_locations,
num_jitters=1, model='large')[0] # There is only one face per photo in
the face photo dataset, Extract the feature vector of the first face
    know_encodings.append(encoding) # Forms a list of face database
feature vectors
    count = count+1
```

Here we use the pickle module to save the extracted list of face feature vectors and the list of names to the binary file.

```
  pickle_encoding_file = open('../models/face_encodings.pkl','wb')
  Pickle.dump (know_encodings,pickle_encoding_file) # writes the memory
list directly to the binary file using the pickle module
  pickle_encoding_file.close()
  pickle_name_file = open('../models/face_names.pkl','wb')
  Pickle.dump (know_names,pickle_name_file) # writes the memory list
directly to the binary file using the pickle module
  pickle_name_file.close()
```

For details, see gen_facecode_file.py. Run python gen_facecode_file.py. Two pkl files are generated in the models directory, as shown in Fig. 8.32.

Main Code Development Practice

Exciting moment has finally arrived! After preparing for the migration of the model and face feature vector files, we can now start writing the main code of the project.

| face_encodings.pkl | 2021-11-25 20:19 | PKL 文件 | 6 KB |
| face_names.pkl | 2021-11-25 20:19 | PKL 文件 | 1 KB |

Fig. 8.32 Generating the face feature vector file

Download the YOLACT code from https://github.com/dbolya/yolact, rename the project evaluation code eval.py to monitor.py, and import the required libraries (refer to the monitor.py file in the download package for specifics).

First, since ImageAI relies on tensorflow-gpu, and the TensorFlow framework by default occupies the entire GPU memory, in order to avoid conflicts with subsequent code, it is necessary to limit the GPU memory allocation for TensorFlow by writing the following code:

```
os.environ["CUDA_VISIBLE_DEVICES"] = "0" # Use the first graphics card to run the deep learning task
import tensorflow as tf
from keras.backend.tensorflow_backend import set_session
config = tf.ConfigProto()
config.gpu_options.per_process_gpu_memory_fraction = 0.1 # Set the video memory ratio of the tensorflow framework
set_session(tf.Session(config=config))
```

Then load the ImageAI object detection model. Note that the migration model we trained in Sect. 8.3.2.1 is used here, just specify the model file and json file:

```
print(' Load object detection model... ')
detector = CustomObjectDetection() # Create target detection instance from class
The detector.SetModelTypeAsYOLOv3() # designated use Yolov3 detection model
Detection.setmodelpath("./weights/detection_model-ex-098--loss-0014.018.h5") # Load the custom model trained in Sect. 8.3.2.1
Detect.setjsonpath("./weights/detection_config.json") # Load the json result file from the training in Sect. 8.3.2.1
detector.loadModel() # Load custom model
```

Next, we write the main logic code. Create an instance of the lact class in the domain() function, load the model weight file, and set the model to inference mode:

```
net = Yolact() # Instantiate the Yolact class
net.load_weights(args.trained_model) # Load the pre-trained models that come with the Yolact framework
net.eval() # Set the network to inference mode
```

8.3 Case Practice

To speed up the instance splitting, you need to put the Yolact network into video memory and write code:

```
if args.cuda:
    device=torch.device("cuda:0") # Specifies the serial number of the graphics card device, where 0 indicates the first graphics card on the host
    net = net.to(device) # puts the Yolact instance network into video memory
```

Use the pickle module to load the list data from the face feature vector list file and the person name list file into the memory list.

```
print(' Load face feature data... ')
pkl_encodings = open('models/face_encodings.pkl','rb')
all_face_encodings = pickle.load(pkl_encodings) # load the data from the binary to the memory list using the pickle module
pkl_encodings = open('models/face_names.pkl','rb')
all_face_names = pickle.load(pkl_encodings) # load the data from the binary to the memory list using the pickle module
known_faces_encodings = []
known_faces_names = []
for each_name in all_face_names:
  known_faces_encodings.append(all_face_encodings[all_face_names.index(each_name)])
   known_faces_names.append(each_name)
print(' Face feature data loaded.')
Connect to the database to prepare for the next action:
conn = sqlite3.connect('./database/dypdb.db')
cursor = conn.cursor()
```

Set two variables to hold the number of people and the number of collateral detected in the previous frame:

```
person_count_last = 0
keg_count_last = 0
Use Opencv to read the video frames that the camera feeds to the host in real time:
capture=cv2.VideoCapture(0) #0 means to use the first camera and read frames from the video file if replaced with XX.mp4
  while capture.isOpened(): # if the camera is readable
  ref, frame=capture.read() # Reads a frame from the camera
```

```
...
capture.release() # Release the camera handle
```

Add the following main code at the ellipsis. Start with object detection:

```
dete_frame, detections = detector.detectObjectsFromImage
(input_image=frame, input_type='array', output_type='array',
minimum_percentage_probability=70)
```

It is worth noting that after input_type and output_type are specified as matrices, the detectObjectsFromImage() function will return two values, one is the output image matrix dete_frame marked with the detection flag box and confidence. The other is a list of detections. The minimum_percentage_probability parameter means ignore the detection if the confidence is less than 70%.

If no personnel are detected in the current frame, and personnel are detected in the previous frame, it means that the entrance personnel have left the scene. In this case, it is necessary to save the current frame in the recorded video, then stop saving the video, and update the departure time recorded in the database to the current time:

```
if person_count == 0:
    if person_count_last > 0:
        videoWrite.write(frame) # Write a frame to the video file
        videoWrite.release() # Releases the handle to the video file
        cursor.execute("update monitor_data set leave_datetime=datetime
(CURRENT_TIMESTAMP,'localtime') where id=(select max(id) from
monitor_data)") # enrolls in the last record in the database
```

If the amount of collateral detected in the current frame is less than the amount detected in the previous frame, the collateral may have been stolen or obscured, and an alarm is required, either by SMS (SMS modem or SMS gateway required) or by email (free of charge). In this case, email alarm is used:

```
if keg_count < keg_count_last:
    goods_loss = True # Set the number of collateral loss flag
    cursor.execute("update monitor_data set goods_loss=1 where id=
(select max(id) from monitor_data)")
    Send_smtp_mail1 (server = "XXX", the port = 25, PSW = 'XXX', sender =
'XXX', receivers = 'XXX', cc = "", the BCC =" ", the subject = 'detection
to reduce collateral, body = now +' Collateral decrease detected ',
ssl='no',from_name='XXX')
```

If personnel are detected in the current frame and no personnel are found in the previous frame, it indicates that there is an intruder in the current field of view. At this time, it is necessary to start the video immediately and save the video during the

Fig. 8.33 Schematic histogram of directional gradient

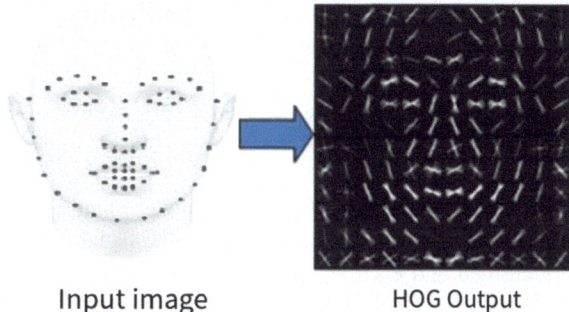

Input image   HOG Output

entry period automatically. Here, we use the Xvid encoding format to create the video file. Xvid is an open-source MPEG-4 codec based on OpenDivX that supports multiple encoding modes, quantization methods, motion detection, curve balance allocation, and many other encoding technologies. When Xvid encoding is specified, the video file size will be compressed to a reasonable range, and the uncompressed video file size will be very large if the encoding is not specified.

```
fourcc = cv2.VideoWriter_fourcc(*'XVID') # Specify video encoding
format
videoWrite = cv2.VideoWriter('./record/'+now2+'.avi', fourcc,
10, (1280,960)) # Create a video recording video file using Opencv at
10 frames per second with a resolution of 1280*960 per frame
videoWrite.write(frame) # Write the current frame to the recorded
video at 1280*960 resolution
```

Once inside, we need to identify who he is. The general steps of facial recognition include face detection (i.e., locating), face encoding (i.e., extracting feature vectors from the located area), and face matching (i.e., comparing the to-be-recognized feature vector with the database feature vectors to identify the name). Face_recognition offers two facial detection technologies, one being HOG and the other being CNN. The main idea behind HOG is that the distribution of edge directions can effectively represent the outline of the target, as shown in Fig. 8.33.

Compared to the CNN method, the HOG is faster but slightly less accurate. In this case, the camera position may be far away from the target, so the recognition accuracy is more important than the speed, so we choose the CNN method to detect faces:

```
face_locations = face_recognition.face_locations(frame,
model='cnn')
```

Face encoding for Face_recognition, available in both large and small ways. large uses a large model, which is more accurate and slower, while small uses a small model, which is faster but easier to ignore encoded faces. In order to pursue the face recognition effect, we use large mode and pass the aforementioned detected face region matrix as a parameter:

```
face_encodings = face_recognition.face_encodings(frame,
face_locations, num_jitters=1, model='large')
```

Next comes the process of identifying the name through database matching. We will calculate a distance value between the facial feature vector of the individual to be identified and each vector in the face database. When this distance value is less than a certain threshold, it indicates a high degree of similarity, suggesting that they are the same person. Here's a small detail: if multiple faces to be recognized in the same frame have distances smaller than the threshold to a certain individual, the one with the smallest distance is selected as the final recognition result.

```
  for face_encoding in face_encodings: # Iterate over every face in the face database
    face_distances = face_recognition.face_distance
(known_faces_encodings, face_encoding) # Pass the face vector to be recognized and the database face vector, and return the similarity distance
    best_match_index = np.argmin(face_distances) # Best matching database face index
    best_name = known_faces_names[best_match_index] # best matching database name
    if face_distances[best_match_index] < args.face_match_distance: # Face is considered to be a person only when the feature vector distance is less than the threshold
      if best_name in face_names: # If two celebrities are identified in the same frame, take the closest one and leave the name of the other blank
        if face_dis[face_names.index(best_name)] > face_distances[best_match_index]:
          face_dis[face_names.index(best_name)] = 1
          face_names[face_names.index(best_name)] = ""
          name = known_faces_names[best_match_index]
          dis = face_distances[best_match_index]
        else:
          name = ""
          dis = 1
      else:
        name = known_faces_names[best_match_index]
        dis = face_distances[best_match_index]
    face_names.append(name)
    face_dis.append(dis)
```

8.3 Case Practice

With this done, we are ready to split the portrait instance. Since we've placed the Yolact network instance on video memory before, the data entering the network needs to be placed on video memory as well:

```
frame2 = torch.from_numpy(frame).cuda().float()
```

Perusing Yolact's code, the prep_display() function implements an instance partition of the input image matrix, calling it as follows:

```
cutouts, croppeds, frame3, person_img = prep_display(preds, frame2,
None, None, undo_transform=False)
```

Because this case wants to get everyone's matting pixels and their range of coordinates, you need to transform the code for the prep_display() function. Line 342 of Monitore.py, img_gpu represents the image matrix, which needs to be copied so as not to interfere with subsequent tasks:

```
img_gpu_old = img_gpu
```

Set the xys list to save everyone's Matting box range in format [$y1,y2,x1,x2$]; Set the croppeds list to save everyone's matting pixels. In Yolact, multiply the image matrix with the mask to filter out other pixels that are not yourself for matting:

```
img_gpu = img_gpu_old * mask_t
```

On line 311, classes, scores, boxes = [x[idx].cpu().numpy() for x in t[:3]] This code already has each instance's label category, confidence level, and matting box, so naturally you get each person's matting pixel:

```
img_numpy = (img_gpu * 255).byte().cpu().numpy()
x1, y1, x2, y2 = boxes[j, :]
cropped = img_numpy[y1:y2, x1:x2] # Crop coordinates
```

For interested readers, I've reworked the code on lines 342–410. After that, back to the main flow code, we need to mark the name of the face we recognize, just write the name in Chinese on the bottom edge of the face box. Here we use the ft21 module, which I wrapped as the put_chinese() function:

```
for (top, right, bottom, left), name in zip(face_locations,
face_names):
    cv2.rectangle(frame3, (left, top), (right, bottom), (0, 0, 255), 2) #
Frame
        frame3 = put_chinese(frame3, name, (left, bottom-16),
```

Fig. 8.34 Comparison between the matting box and the detection box

```
(255,255,0), 20, 'data/simkai.ttf') # Display the name on the bottom
edge of the face detection box
```

For cases where the current frame finds the person and the previous frame did not, the name needs to be registered in the database:

```
cursor.execute("insert into monitor_data values (……
```

If Yolact's matting box overlaps with the target detection box of the collateral, the person may have touched the collateral. Why not tell by the overlap between the portrait target detection box and the collateral target detection box? Because the portrait matting box has a more accurate pixel range than the target detection box, the former is to fit the portrait, the latter is to surround the portrait. As shown in Fig. 8.34, the image on the left is the matting box and the image on the right is the detection box.

Given that we retain the video footage upon entry for detailed manual review, the algorithm's task is simplified to approximating the probability of an individual interacting with the collateral rather than precisely determining physical contact. This approach significantly lowers project complexity. Here is an alternative proposition. We could develop a specialized instance segmentation dataset focusing on collateral and personnel. This would involve gathering a substantial collection of images, manually segmenting each one, annotating them meticulously, and then employing the Yolact framework to train a custom model for instance segmentation of both personnel and collateral. The criterion for assessing contact hinges on whether the minimum distance between the personnel mask and the collateral mask falls below a predetermined threshold. This methodology necessitates extensive labeling efforts, thus it will be presented as a subject for further investigation.

8.3 Case Practice

Fig. 8.35 Schematic diagram of determining rectangles' coincidence

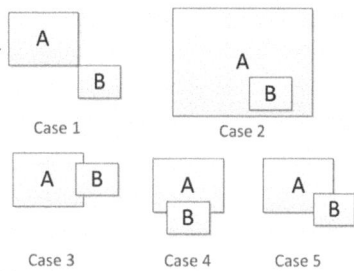

In this case, box *A* represents the personnel matting box, and box *B* represents the chattel collateral matting box. We roughly judge whether the personnel touch the collateral by whether the two boxes coincide. As can be seen from Fig. 8.35, if two rectangles intersect, the distance between the two center points must be less than half of the sum of the length of side AB. This judgment holds in both width and height directions.

Therefore, encapsulate this judgment as the is_rect_intersect() function, enter the $x1,y1,x2,y2$ coordinates of the two rectangles, and return whether there is an overlap area. After completing the above work, we can splicing face recognition graph, portrait matting graph, target detection graph and final judgment graph into a large graph. numpy library provides this matrix operation:

```
htich = np.hstack((frame1,frame2))
htich2 = np.hstack((frame3,frame4))
vtich = np.vstack((htich, htich2))
```

First horizontally concatenate pairs of images and then vertically concatenate the concatenated horizontal images. As for the email sending code, I have encapsulated it in the function send_smtp_mail1(), utilizing the libraries smtplib and email. Interested readers can study it on their own. Due to space limitations, further details are not discussed. The aforementioned code can be found in the "AI Client" directory under monitor.py.

The last part is the presentation of Web data visualization. Execute the django framework installation command: pip install django. Then run the django-admin startproject monitor command to create a project named monitor. A series of directories and files will be created. The project structure is shown in Fig. 8.36.

Run the cd monitor command to go to the monitor directory and run the python manage.py startapp monapp command to create the application named monapp. This will create the directory named monapp. Add the monapp app name to the INSTALLED_APPS settings in the monitor/ Settings.py file. Set ALLOWED_HOSTS = ['*'] to allow all addresses to access this project, and set X_FRAME_OPTIONS = 'ALLOWALL https://127.0.0.1' to allow pages to be displayed in frames from the specified source. Go to TEMPLATES and specify 'DIRS':[os.path.join(BASE_DIR,'templates')] to set the templates directory and

Fig. 8.36 Django project structure

```
└─monitor
    │  manage.py
    └─monitor
          asgi.py
          settings.py
          urls.py
          wsgi.py
          __init__.py
```

create the templates directory under the monitor directory. Add the following code in the urls.py file in the monitor directory to specify the foreground page access route:

```
from django.urls import path, include
urlpatterns = [
  path('admin/', admin.site.urls),
  path('monapp/', include('monapp.urls')),
]
```

The above code says that the monapp/XXX type url passed in by the browser will be forwarded to the monapp/urls.py file for routing. Therefore, add the full routing information of the project to the monapp/urls.py file as follows:

```
from django.urls import path, re_path
from . import views
urlpatterns = [
  path('login', views.login),
  path('do_login', views.do_login),
]
```

The above code says that the monapp/login url request will be handled by the login() function of the monapp/views.py file; the url request for monapp/do_login will be handled by the do_login() function of the monapp/views.py file. In this case, the interaction between the front and back ends is as follows: the html form routes the request to Django's url, and in the corresponding view.py handler returns the page to the front end via the render() function, and so on. At this point, the Django project controller part is set up.

Next, develop the data visualization section. Write the following code in the get_query() function. First get the parameter information for the front-end post:

```
  date1 = request.POST.get('date1') + ' 00:00:00'
  date2 = request.POST.get('date2') + ' 23:59:59'
Connect to sqlite database:
conn = sqlite3.connect('../codes/database/dypdb.db')
```

To get data from a database using a cursor:

8.3 Case Practice

```
cursor = conn.cursor()
cursor.execute("select person,count(*) as count from monitor_data
where enter_datetime>=datetime('"+date1+"') and
enter_datetime<=datetime('"+date2+"') group by person order by count
(*) desc")
res = cursor.fetchall()
```

To save data to memory list:

```
list_person = []
list_count = []
for i in range(len(res)):
    list_person.append(res[i][0])
    list_count.append(res[i][1])
```

Fill the data into the pyecharts chart:

```
from pyecharts import Line
line = Line(" Number of times in and out of warehouse ")
line.add("", list_person, list_count, mark_point=["max", "min"],
mark_line=["average"],legend_pos="1%",is_label_show=True,
xaxis_rotate=30)
```

In the above code, instantiate the line chart class, specify the chart title as "times of entering and leaving the warehouse," add() function sets the X axis and Y axis data list, mark_point parameter indicates the maximum and minimum values, mark_line parameter indicates the average line, legend_pos parameter specifies the legend position, is_label_show parameter sets whether the data is displayed, xaxis_rotate=30 sets the label rotation of X axis by 30 degrees. In the same way, draw the bar chart:

```
bar1.add(' number of collateral contacts ', list_person,
list_contact, legend_pos="75%", legend_top="27%", xaxis_rotate=30)
#x axis rotated by 30 degrees
```

When drawing a pie chart, the center parameter sets the pie position, the radius parameter sets the inner diameter and the outer diameter size, and the legend_orient parameter sets the direction displayed by the legend:

```
Pie. Add (" ", list_person list_loss, radius = [6], center = [50] 30,
legend_pos = "20%", legend_top = "70%", legend_orient="vertical",
is_label_show=True)
```

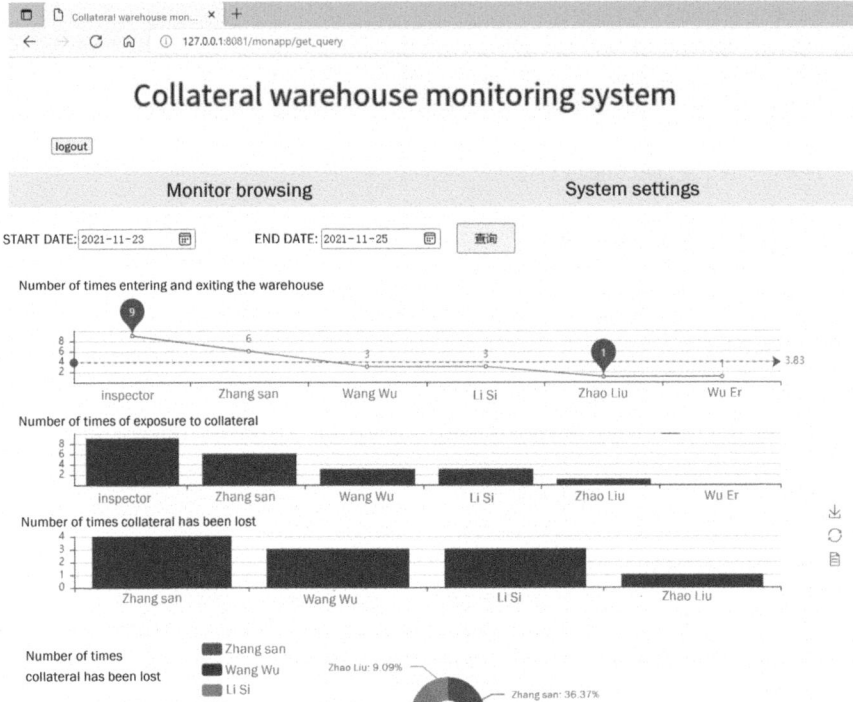

Fig. 8.37 Visualization of collateral warehouse monitoring data

The Pyecharts library provides the Grid module, which allows visual charts to be displayed in grid form. The following code concatenates several visual charts into an html file and specifies the location of each chart inside the Grid:

```
grid = Grid(height=600, width=1000)
grid.add(line, grid_left="5%", grid_bottom="80%")
grid.add(bar1, grid_left="5%", grid_bottom="60%", grid_top="30%")
grid.add(bar2, grid_left="5%", grid_bottom="40%", grid_top="50%")
grid.add(pie, grid_left="5%", grid_bottom="10%", grid_top="75%")
grid.render('./templates/page/get_data.html')
```

Run the above code, and you're done! The effect is shown in Fig. 8.37.

It is worth noting that the visualizations generated by pyecharts are interactive. Moving the mouse over a chart area can display the data values and labels in this area, which has a good use experience. See the download file "Data Visualization" for the code.

8.4 Case Summary

This scenario employs advanced computer vision technology to facilitate the monitoring of commercial bank collateral warehouses. The procedures outlined here oversee movable assets, including automobile collateral and raw wine collateral, across various warehouse environments. In practical application, this technology significantly reduces the need for on-site security personnel in commercial banks, thereby lowering operating costs, enhancing risk control efficiency, and contributing positively to the protection of state-owned asset values and the prevention of asset loss.

Moreover, image understanding technology proves beneficial in diverse banking contexts. In branch monitoring, it can alert to emergencies, monitor customer crowding, detect the destruction of financial equipment, and identify unauthorized personnel accessing bank offices at unusual times. Additionally, it aids in bank conference management by recognizing issues such as mobile phone usage and the presence of non-participants. It also ensures that service actions by bank branch staff meet operational standards. Overall, image understanding technology significantly enhances the safety and operational quality within banking institutions.

The present case illustrates the deployment of artificial intelligence (AI) technology within the risk management frameworks of commercial banking institutions. However, the prevailing technical methodology remains limited to frame-by-frame image processing, lacking comprehensive contextual understanding over sequential video data. This shortcoming necessitates enhancement. For an optimized solution, it is advisable to integrate the SlowFast model, developed by He Keming's research team, which secured the top position in the AVA Video Detection Challenge at the CVPR2019 conference. The SlowFast model introduces an innovative approach to video semantic analysis, effectively emulating the operational principles of retinal neural mechanisms in primate vision. It simultaneously extracts pertinent information from videos at disparate frame rates—slow and fast. This model processes static background elements via the slow channel while addressing dynamic objects through the fast channel. Subsequently, it integrates the data from both channels through lateral connections, facilitating a cohesive understanding of events within the continuous video frames. Compared to other methodologies, SlowFast demonstrates significantly reduced computational complexity and enhanced accuracy in video analysis. Practitioners are encouraged to adopt these advanced techniques in analogous scenarios, particularly within state-owned asset management contexts. Doing so can contribute to constructing a more harmonious society and adhering to the political and social obligations incumbent upon financial institutions.

Chapter 9
Personal Loan Delinquency Prediction Project: Bayesian Network Techniques

Personal loan businesses hold significant importance both at the national and banking levels. Nationally, personal loans enhance the purchasing power of both urban and rural populations. This boost in purchasing power stimulates consumer demand, cultivates a thriving market, and supports production and industrial growth, thereby promoting overall economic development. At the banking level, personal loans serve to adjust the credit structure, thereby improving the quality of credit portfolios. They also increase interest income and generate additional revenue through various financial services. Given the extensive number of individual borrowers, personal loans also help diversify the credit risk faced by commercial banks. They reduce operational costs and bolster the stability of banks by ensuring a steady flow of income. Consequently, the sustainable and healthy growth of the personal loan sector is pivotal for the efficient and resilient functioning of commercial banks.

Determining the likelihood of personal loan delinquencies in a scientific and effective manner is a critical issue for commercial banks. A frequently employed practice involves leveraging statistical machine learning or deep learning methodologies to analyze and fit a range of characteristics pertaining to loan delinquency, such as occupation, billing history, and income levels. Nonetheless, this approach suffers from significant drawbacks, primarily its lack of interpretability. The models created in this process often act as mere reflections of the dataset, raising concerns about whether they genuinely capture the underlying factors of interest. The interpretability challenge amplifies with increased model complexity. For instance, deep learning neural networks commonly function as opaque black boxes, making it difficult to discern whether their predictions stem from genuine data insights or coincidental patterns. This creates substantial communication barriers when conveying predictive results to business personnel responsible for implementing risk control measures. Without a clear understanding of the model's decision-making process, these personnel are likely to question the model's reliability, asking, "Why should I trust your model?" Such skepticism can lead to reduced engagement from business staff, hampered integration of technical solutions with industry practices, and vague risk predictions. To address these issues, it is essential to develop models that not only deliver accurate predictions but also provide sufficient transparency and interpretability to foster trust and collaboration between technical and business teams.

To address the challenge of predicting personal loan delinquency, we must design a highly interpretable model. Within the domain of machine learning, interpretability is crucial. Techniques like data rules, decision trees, and linear models are known for their high interpretability. However, the loan repayment process involves complex factors such as customer quality, customer credit ratings, the value of collateral, total customer assets, and liabilities. These factors exhibit numerous dependencies and inherent uncertainties, which complicate their suitability for traditional methods like data rules, decision trees, and linear models. Instead, due to these complexities and probabilistic relationships, Bayesian networks provide a more appropriate framework for constructing predictive models.

9.1 Introduction to Bayesian Networks

This section introduces the basic concepts of Bayesian learning in plain language, which led to the development of Bayesian network techniques to model relationships between multiple complex events.

9.1.1 Bayesian Learning Concepts

Machine learning entails deriving inferences about unobserved or prospective data (events) based on analyzed data (events), incorporating an intrinsic degree of uncertainty. Fundamentally, uncertainty pervades all aspects of the world. A learning system leverages observable data to quantify and model these uncertainties, thereby enabling predictive analytics concerning future (data) events. Within the machine learning field, the Frequency school and the Bayesian school represent two predominant paradigms. Frequencists employ statistical machine learning methodologies to model uncertainty by fitting mathematical models to empirical data. Conversely, Bayesians utilize probabilistic programming techniques to model uncertainty through probabilistic inference. Probability, universally acknowledged as the mathematical framework for characterizing and managing uncertainty, diverges in interpretation between these schools. Frequencists conceptualize probability as the empirical frequency of an event's occurrence over historical instances; meanwhile, Bayesians regard probability as an evolving subjective expectation of an event, which adjusts dynamically in light of new evidence. This Bayesian perspective hinges on two fundamental constructs: "prior probability," representing the initial expectation, and "posterior probability," representing the updated expectation, as formalized by Bayes' theorem, introduced in 1763.

9.1 Introduction to Bayesian Networks

$$P(B_i|A) = \frac{P(B_i)P(A|B_i)}{\sum_{j=1}^{n} P(B_j)P(A|B_j)}$$

In the formula, $P(B_i|A)$ is the conditional probability of the occurrence of the event B_i under the condition of the occurrence of the event A, that is, the posterior probability. $P(B_i)$ is people's prior probability of the event B_i, which is generally a summary of experience. $P(A|B_i)$ is the conditional probability of the event A occurring under the condition of the event B_i occurring, also known as likelihood. The denominator $\sum_{j=1}^{n} P(B_j)P(A|B_j)$ is called the total probability formula, which is actually equal to $P(A)$, that is, the probability of the event A happening is equal to the sum of the product of the probability of the event B_i happening and the conditional probability of the event A happening when B_i happens, which is the sum of the global equal to the local. The Bayes formula can also be written in a symmetric esthetic form:

$$P(A|B)P(B) = P(B|A)P(A)$$

This makes sense because $P(A, B) = P(A)P(B|A) = P(B)P(A|B)$, the joint probability of two events occurring at the same time is equal to the product of the probability of one event occurring with the probability of the other event occurring under that event. Bayesians use a probability distribution to represent all uncertain or unobservable random variables in a model and their relationships.Bayesian inference involves transitioning from a prior probability distribution, which represents subjective judgment before data observation, to a posterior probability distribution, a synthesized judgment after data observation. This transition utilizes mathematical probability theory to construct learning models from the acquired data. To infer unobserved data from observed data effectively, this probabilistic method is indispensable. Bayes' formula underpins this process, where the prior probability encapsulates initial expectations, the likelihood captures the implicit model generating observed phenomena, and the posterior probability reflects these expectations post-adjustment through continuous observation. Bayes' formula thus provides the foundational criteria for Bayesian inference, facilitating the mutual transformation of conditional probabilities. When considering A and B in the formula as observational data and model parameters respectively, we can formally express this relationship in precise mathematical terms.

P(model parameter | observed data)
$= P$(model parameter)$^{*}P$(observed data | model parameter)$/P$(observed data)

The provided formula highlights the objective of maximizing "P(model parameters|observed data)." This implies that, given the observed data, the specific model parameters that maximize the likelihood of that data represent the optimal parameters we seek. This concept lies at the heart of Maximum a Posteriori estimation (MAP), which focuses on maximizing the posterior probability of the model parameters, conditioned on the observed data samples. This aligns with the essential principles of probabilistic programming, where the goal is to identify the most suitable model for explaining the observed data. In this framework, an observed dataset could fit several potential models, and the uncertainty inherent in these models is managed through probabilistic methods. In a Bayesian learning system, both the observed data and the model parameters are treated as random variables. Given their random nature, the task involves determining the optimal values of these random variables, typically by maximizing the posterior probability (MAP). This optimization relies on probabilistic integration techniques. In contrast, the frequentist approach distinguishes observed data as random variables but views model parameters as fixed but unknown constants. Therefore, frequentists approximate these constants using mathematical methods, defining a loss function and utilizing various mathematical optimization techniques such as gradient descent or Newton's method to find the best model parameters. These contrasting perspectives underscore the fundamental differences between statistical machine learning and Bayesian learning. Each approach presents unique modeling methodologies, highlighting the diverse strategies within the field of artificial intelligence and statistical analysis.

9.1.2 From Bayesian Learning to Bayesian Networks

In practical scenarios, occurrences are typically influenced by the interplay of several stochastic variables. The conditional probability distributions of these variables are interrelated, enabling their combination to form more comprehensive, all-encompassing, and intricate models. These advanced mathematical frameworks are known as probabilistic graphical models. A probabilistic graphical model delineates the conditional independence of multiple stochastic variables through a graphical structure, simplifying and clarifying probabilistic computations. These models find extensive applications across various domains. Probabilistic graphical models are categorized into Bayesian networks and Markov networks (or Markov random fields). The former employs directed edges to explicate the relationships between the variables, while the latter uses undirected edges. Bayesian networks, popularized by Turing Award laureate Professor Judea Pearl in 1988, have risen to prominence within artificial intelligence research. Often referred to as Belief networks, these structures comprise a directed acyclic graph coupled with a probability table. Each vertex signifies a random variable, and directed edges between vertices depict the derived associations among these variables. The acyclicity of the network ensures the absence of loops, maintaining the integrity of the model. Within a Bayesian network, the tail of a directed edge (observation node A) serves as the progenitor (parent) to the head of the directed edge (result node B). Consequently, node B represents the offspring (child) of node A. When a directed pathway extends from node A to node B, node A is termed an ancestor of B, while B is designated as a descendant of A.

9.1 Introduction to Bayesian Networks

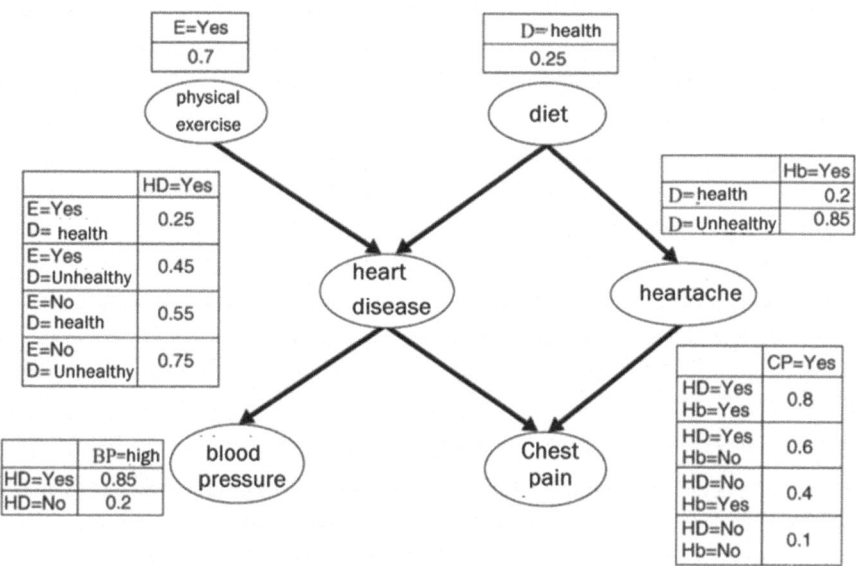

Fig. 9.1 Example of Bayesian network

In a Bayesian network, the probability table for a non-root node (a node that has a parent) is a conditional probability table, determined by the values of its parent. If a node has more than one parent, it is expressed as a combined conditional probability table of multiple parent nodes. For a root node (a node that has no parent), its probability table is expressed in terms of a prior probability (that is, the estimated probability before the observation). For example, in Fig. 9.1, we consider the probability of someone exercising to be 0.7 (prior probability) and his probability of healthy eating to be 0.25 (prior probability), based on the combination of values of these two nodes, we can infer the probability of heart disease (usually derived from the statistical sample), both exercise and diet nodes are the parents of the heart disease node, The heart disease node is their child node, and the probability of heart disease is determined by the different values of the exercise and diet nodes. How healthy a diet is leads to heart pain, which, together with heart disease, leads to chest pain, and heart disease leads to high-blood pressure. We connect the relationship between the random variables of each node with directed edges, and label the conditional probabilities under various combination conditions to obtain the Bayesian network as shown in Fig. 9.1.

We can do two major tasks on Bayesian networks: parameter learning and structure learning. Parametric learning refers to calculating the probability distribution of each node, given the observed data and the structure of the network, which is calculated using the joint probability distribution of the Bayesian network and the Bayesian formula. As an example, for a simple four-node Bayesian network, as shown in Fig. 9.2.

Fig. 9.2 Four-node Bayesian network

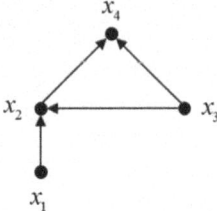

The joint probability distribution of the four random variables is $P(x_1, x_2, x_3, x_4) = P(x_1)P(x_3)P(x_2|x_1, x_3)P(x_4|x_2, x_3)$. As you can see, because there is a derivation relationship between the variables, the conditional probability of each node is determined by all of its parent nodes. The joint probability distribution of a Bayesian network can then be written as $(X) = \prod_{i=1}^{n} P(x_i|\pi_i)$, where n is the number of nodes and π_i is the set of parent nodes of the nodes x_i. And because Bayes' formula can be written as $P(A, B) = P(A)P(B|A) = P(B)P(A|B)$, the conditional probability distribution for each node can be calculated based on these two formulas. Structure learning means that if the structure of the Bayesian network is unknown, the structure of the network can be learned from the observed data.

Bayesian networks offer significant advantages in three primary areas of application. Firstly, they employ graphical techniques to represent the interrelationships among variables, ensuring clarity and ease of comprehension due to their explicit semantics and high interpretability. Secondly, Bayesian networks are robust in probabilistic modeling; they maintain the capability to construct comprehensive probabilistic relational frameworks even in the absence of certain data variables, a feat that traditional machine learning methods find challenging when confronted with incomplete datasets. Thirdly, Bayesian networks adeptly integrate prior knowledge with observational data. This dual-faceted approach allows for a thorough representation of relationships among variables. Here, prior knowledge pertains to the prior probability of root nodes (nodes without parent nodes), while observational knowledge relates to the conditional probabilities of intermediate nodes, often derived from extensive statistical samples. This probabilistic mechanism proves highly effective, particularly in scenarios where data is sparse or not readily accessible. In essence, Bayesian networks extend Bayesian learning to multivariable contexts, making them suitable for representing and analyzing uncertain events. They proficiently manage decisions involving multiple control factors and facilitate inferences from incomplete, imprecise, or uncertain data.

Bayesian networks represent the foundational framework of causal science, which forms the core science endowing artificial intelligence with "true intelligence." The nodal derivation relationships in Bayesian networks illustrate dependencies between observed variables and outcome variables, encompassing both causation and correlation. By facilitating the learning of causal relationships among variables, Bayesian causal networks underpinned Professor Judea Pearl's development of a novel paradigm of causal inference. This paradigm has significantly advanced causal science into what is now recognized as "intelligent science" for artificial intelligence, demonstrating substantial future potential. The maturation of Bayesian networks, reflecting Professor Pearl's earlier research, has established the groundwork for causal science to emerge as an independent discipline and has laid a pivotal foundation for the future progress of artificial intelligence. For those seeking further insight, Professor Pearl's comprehensive works on causal science, including "Why" and "Causality: Models, Reasoning, and Inference," are highly recommended.

9.2 Probability Graph Calculation Library: Pgmpy

In this section, Pgmpy, the Python probability graph computing library for Bayesian machine learning, is introduced in a very short space.

Pgmpy is an open source probability graph computing library, which can handle a series of tasks such as Bayesian model, Bayesian network, naive Bayes, Markov chain, Markov random field, and factor graph. It can complete the two tasks of parameter learning and structure learning. Among them, parameter learning is a directed acyclic graph (DAG) given a set of data samples and the dependency between random variables, predicting the (conditional) probability distribution of a single variable. Structural learning is a DAG that, given a set of data samples, predicts the dependencies between random variables.

Pgmpy released under the MIT open source license, code hosting warehouse for https://github.com/pgmpy/pgmpy, the official web site is http://pgmpy.org/, Sample site is https://github.com/pgmpy/pgmpy/tree/dev/examples.

9.3 Case Practice

This section describes how to set up the development and operation environment of the case and how to use Pgmpy library to develop the case model.

9.3.1 Environment Setup and Case Runs

You can refer to Sect. 1.3.1 of this book to install Anaconda, activate the virtual environment, and run the following command to install the library package in the Internet environment:

```
pip install pgmpy.
```

Command line run the following commands to train the model and do sample reasoning:

```
python Bayesian Network - train.py.
```

Run the following commands to build a Bayesian Network and do sample reasoning:

```
python Bayesian Network - Specify probabilities.py.
```

9.3.2 Code Practice

In this particular analysis, we selected the following data from outstanding loan customers as observational metrics: credit rating, number of delinquencies in the past 2 years, number of loan approval inquiries in the past 3 months, number of currently outstanding loans, long-term loan ratings, and whether loans are presently delinquent. These data points provide multifaceted insights into the overdue indicators from varied perspectives. The credit rating encompasses evaluations based on factors such as age, occupation, identity, industry, and other fundamental information. It also incorporates an assessment of credit default history, income liabilities, and overall financial conditions. The number of delinquencies over the past 2 years aggregates the frequency of overdue amounts on credit cards and personal loans, derived from records of various banks and lending institutions. In the past 3 months, the count of loan approval inquiries reflects the frequency with which financial institutions have evaluated the customer's creditworthiness during loan applications. This metric indicates both the volume of loan applications submitted by the customer and the urgency of their financing needs. Additionally, the current number of outstanding loans details the cumulative loan liabilities held by the customer across all financial institutions, depicting their existing financial commitments. This comprehensive data selection provides a thorough understanding of the customer's credit and financial behavior, crucial for precise credit risk assessment.

The higher the customer's credit rating, the lower the likelihood of delinquencies, and vice versa; The more delinquents a customer has in the last 2 years, the higher the likelihood that their current loan will be delinquent, and vice versa; The more likely customers are to have long loans, the more likely they are to be past due, and we can infer long loan ratings from two indicators: the number of loan approval inquiries in the last 3 months and the number of loans currently outstanding. According to the above derivation, the Bayesian network structure constructed in this case is shown in Fig. 9.3.

It is worth noting that the figure above only labels the prior probability table of the root node and does not label the conditional probability table of the intermediate

9.3 Case Practice

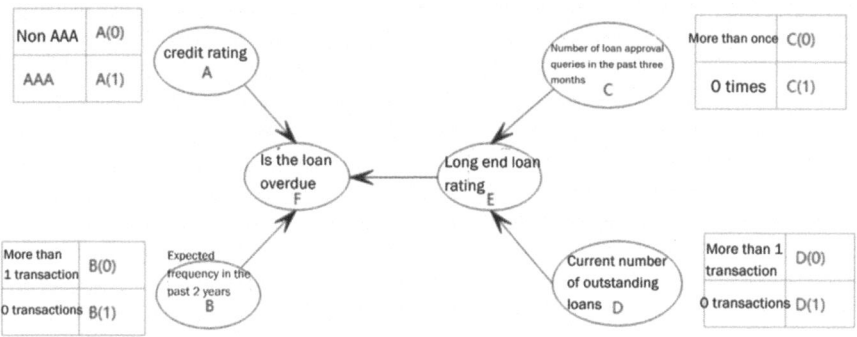

Fig. 9.3 Bayesian network structure in this case

node, because this is the part that needs to be solved. After determining the network structure, we extract the data set from the background database in the format of:

A	B	C	D	E	F
0	1	0	0	0	0
0	1	0	0	0	1
0	1	1	0	1	1
......					

The ABCDEF field of the data set corresponds to each node in the network, note that the data set should contain all the node data in the network, the data range is the index of the node value (0 or 1), and each customer generates one row of data. The data set file is data.xlsx. Below, we write the program code (Bayesian Network - train.py). First import the relevant library package:

```
import pandas as pd
from pgmpy.factors.discrete import TabularCPD
```

The pgmpy library provides an algorithm wrapper for Bayesian networks, using the following code to import the algorithm package.

```
from pgmpy.models import BayesianNetwork
from pgmpy.estimators import MaximumLikelihoodEstimator, BayesianEstimator
```

Load data into DataFrame structure from data set file with pandas module:

```
df = pd.read_excel('data.xlsx')
```

Fig. 9.4 Sample set of node data

	A	B	C	D	E	F
0	0	1	0	0	0	0
1	0	1	0	0	0	1
2	0	1	1	0	1	1
3	0	1	0	0	0	1
4	0	1	0	0	0	1
...	..	..	..	..	..	..
44210	1	0	0	0	0	1
44211	1	0	0	0	0	1
44212	1	0	0	0	0	1
44213	1	0	1	0	1	1
44214	1	0	0	0	0	1

df data is a two-dimensional table form (index and ABCDEF node data values), as shown in Fig. 9.4.

Next, use the Bayesian network module of the pgmpy library to describe the node names and network structure of the Bayesian network. Here you only need to specify the node pairs for each edge, which are written in the form "(start node, end node)." It is worth noting that older versions of the pgmpy library use the BayesianModel module instead of the BayesianNetwork module, and the author installed pgmpy version 0.1.16, which already supports the BayesianNetwork module.

```
overdue_model = BayesianNetwork([('A', 'F'),
                                 ('B', 'F'),
                                 ('C', 'E'),
                                 ('D', 'E'),
                                 ('E', 'F')])
```

The BayesianNetwork function produces a model handle identified as overdue_model. After defining both the node structure and the network framework, the next step involves inputting the dataset into the model to begin the training of the Bayesian network. This training phase utilizes algorithms to determine the prior probability distribution of the root node and the conditional probability distribution of intermediate nodes automatically. The goal is to ensure that the resultant node probability distributions are optimally aligned with the data samples provided. This is consistent with the fundamental principle of probabilistic programming, as outlined in Sect. 9.1.1, which conceptualizes a learning system as a mechanism to derive the most suitable model for interpreting observed data. In this scenario, the maximum likelihood estimation algorithm is employed to fit the sample dataset effectively. The algorithm is specified through the estimator parameter, ensuring precision in the training process:

```
overdue_model.fit(df, estimator=MaximumLikelihoodEstimator)
```

The so-called maximum likelihood estimation is to use the known information of the sample results to invert the values of the model parameters that are most likely (maximum probability) to cause these sample results. This is a method of solving the

9.3 Case Practice

CPD of A:

A(0)	0.00169626
A(1)	0.998304

CPD of F:

A	A(0)	A(0)	A(0)	A(0)	A(1)	A(1)
B	B(0)	B(0)	B(1)	B(1)	B(1)	B(1)
E	E(0)	E(1)	E(0)	E(1)	E(0)	E(1)
F(0)	1.0	0.0	0.21818181818181817	0.0	0.08292798110979929	0.05639934216890781
F(1)	0.0	1.0	0.7818181818181819	1.0	0.9170720188902007	0.9436006578310921

CPD of B:

B(0)	0.0521542
B(1)	0.947846

Fig. 9.5 Node probability table obtained by training

best model parameters given the observed data under the condition of "model determined and parameters unknown." After the training is complete, look at the probability table for each node:

```
for cpd in overdue_model.get_cpds():
    print("CPD of {variable}:".format(variable=cpd.variable))
    print(cpd)
```

You will get the output shown in Fig. 9.5.

It is worth noting that the root node probability table is expressed directly in the form of "index–probability" and the intermediate node probability table is expressed in the form of "index combination of all its parent nodes–probability." After the training is completed, a complete Bayesian network is obtained, as shown in Fig. 9.6.

The following uses the trained Bayesian network to reason about a sample of data. Here we just need to input the observations of the root node into the model. First import the Bayesian inference module:

```
from pgmpy.inference import VariableElimination
```

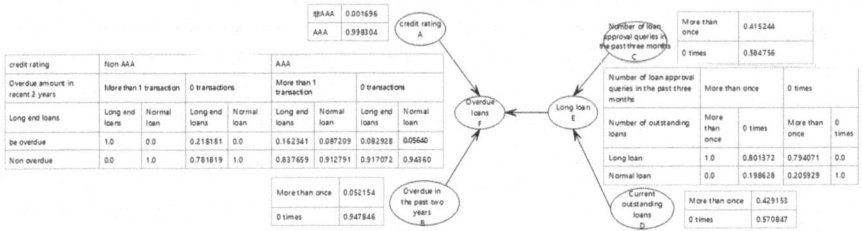

Fig. 9.6 Bayesian network obtained through training

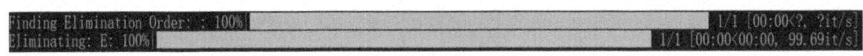

Fig. 9.7 Reasoning process progress

To do Bayesian inference on a trained network:

```
overdue_infer = VariableElimination(overdue_model)
```

If A customer sample, his credit rating is "non-AAA" ($A = 0$), the number of overdue loans in the last 2 years is 0 ($B = 1$), the number of loan approval inquiries in the last 3 months is more than 1 ($C = 0$), and the number of outstanding loans is 0 ($D = 1$), we want to predict his loan overdue probability, we only need to input the observed value of the root node into the inference engine. We can return the inference result:

```
prob_F = overdue_infer.query(
     variables=['F'],
     evidence={'A':0, 'B':1, 'C':0, 'D':1})
```

The program will display the inference process progress, as shown in Fig. 9.7. Show the inference results:

```
print(prob_F)
```

Print (prob_f) 3 Get the inference result shown in Fig. 9.8.

This means that the probability of this customer being late is 17.48% and the probability of not being late is 82.52%. pgmpy's calculations are so fast that the authors trained the network on 44,000 data points in just 2 s. We can update the data set and train the model every day. Readers can also design more complex network structures, such as adding nodes to the network such as collateral depreciation and customer involvement. In the case of more complex and larger data sets, we can also use statistical methods to determine the probability table of each node, just specify

9.3 Case Practice

Fig. 9.8 Network inference result of the sample

F	Phi(F)
F(0)	0.1748
F(1)	0.8252

the probability value in the form of parameters. See Bayesian networks–Specifying probabilities.py for more details. The key code is interpreted as follows:

Define the conditional probability table of the intermediate node. variable_card specifies the number of the value of the node; values specifies the value of the probability table; evidence specifies the parent node; evidence_card specifies the number of the parent node. Note that the order of values, evidence, and evidence_card must be consistent.

```
# Conditional probability distribution for long loans
E_cpd = TabularCPD(
   variable='E', # node name
   variable_card=2, # Number of node values
   values=[[1,0.801371,0.794071,0.0], # Table of probabilities for this node
   [0.0, 0.198628, 0.205929, 1.0]].
   evidence=['C', 'D'], # Parent of this object
   evidence_card=[2, 2] # Number of parent node values
)
```

Define the prior probability table of the root node, and directly fill in the number of values and probabilities.

```
# Credit rating prior probability distribution
A_cpd = TabularCPD(
     variable='A',
     variable_card=2,
     values=[[0.001696], [0.998304]]
)
```

Define the ABCDEF node probability table in turn, then add the nodes to the network model with the add_cpds() function:

```
overdue_model.add_cpds(A_cpd, B_cpd, C_cpd, D_cpd, E_cpd, F_cpd)
```

In this way, we have completed all parameter designations of Bayesian network and can directly call the inference module to predict the sample. The code of this chapter can be found in "BayesianNetwork - training.py," "BayesianNetwork—Specifying probability.py," and the data can be found in download file data.xlsx.

9.4 Case Summary

In our current operational practice, we stratify each customer's overdue probability. We give precedence to customers exhibiting a high likelihood of delinquency, ensuring timely risk management to implement pre-emptive risk mitigation measures to some extent. This case study elucidates both theoretical and practical methods for predicting customer loan delinquency, enabling readers to deploy and utilize these methods directly within their professional environment. Such implementation holds significant value in averting financial risks. Notably, this case study focuses solely on the parameter learning aspect of the Bayesian network, serving as a foundational comparison framework. The scenario of predicting loan delinquency can be expanded to include structural learning among multiple interrelated nodes. For instance, a downturn in the real estate industry may influence customers' liabilities, income, mortgage values, repayment intentions, and credit ratings. The Bayesian network can model these structural relationships based on varying circumstances. Furthermore, as time progresses, key characteristics and overdue behaviors of customers will evolve, albeit interrelated. Utilizing dynamic Bayesian networks can process more extensive datasets over different timeframes, construct more complex and comprehensive models, and yield more actionable overdue prediction insights.

Bayesian networks currently represent a significant area of focus within artificial intelligence research and are anticipated to spearhead future advancements in causal science. This sophisticated technology excels in managing predictions that are uncertain and multifactorial in nature. Within the realm of banking operations, numerous scenarios benefit from these networks, including, but not limited to, predicting risk transmission in supply chain loans and co-guarantee loans, as well as determining the root causes behind performance improvements stemming from customer marketing activities. The utilization of probability graph models exemplified by Bayesian networks, along with Bayesian inference methodologies, finds extensive and effective application in the banking sector. Professionals within the industry are encouraged to study the techniques detailed in this chapter and apply them in their daily operations to ensure the sustainable and robust growth of the banking industry.

Part III
Intelligent Operation

Chapter 10
Enterprise WeChat Private Traffic Customer Cold Start Program: Automated Control Technology

In the contemporary era of mobile Internet, the establishment and advancement of online and mobile ecosystems have transformed the landscape of retail banking. Channels such as mobile banking, online banking, and WeChat banking now predominantly handle the majority of retail banking operations. Moreover, electronic payment systems, credit card installment plans, and Internet lending services have seamlessly integrated into daily Internet usage scenarios. Notably, these environments are increasingly controlled by Internet companies rather than traditional banks. Consequently, the distinctive advantage that banks once held through their widespread offline branch networks is diminishing. Recent trends in the banking industry reveal a continued decline in the number of customers visiting physical branches to conduct business. Instead, branches are progressively serving as venues for high-end clientele, in-person contract signings, and specialized services. Conversely, the Internet has emerged as the principal platform facilitating interactions between banks and a vast demographic of ordinary customers. A significant proportion of retail banks' deposits and assets under management (AUM) are contributed by these ordinary customers, whose engagement critically influences the overall performance of retail banks. Hence, it is imperative to develop strategic models for the online management and operation of retail customer segments to sustain and enhance their contributions to the retail banking sector.

During the initial phase of the Internet's development, the period known as the traffic dividend era emerged. At that time, the correlation between web traffic and sales was remarkably direct. Companies widely invested in acquiring public traffic from centralized platforms such as WeChat and headlines. However, as the benefits of public traffic began to dwindle and the acquisition costs surged rapidly, this dynamic induced significant anxiety among enterprises. The necessary shift from a traffic acquisition mindset to a user management approach became evident as a strategic adaptation. The fragmentation of mobile Internet disrupted the monopoly of concentrated public domain traffic. Consequently, traffic centralization began transitioning toward decentralization. This market evolution signaled a need for returning to business rationality. There was a noticeable reorientation from reliance

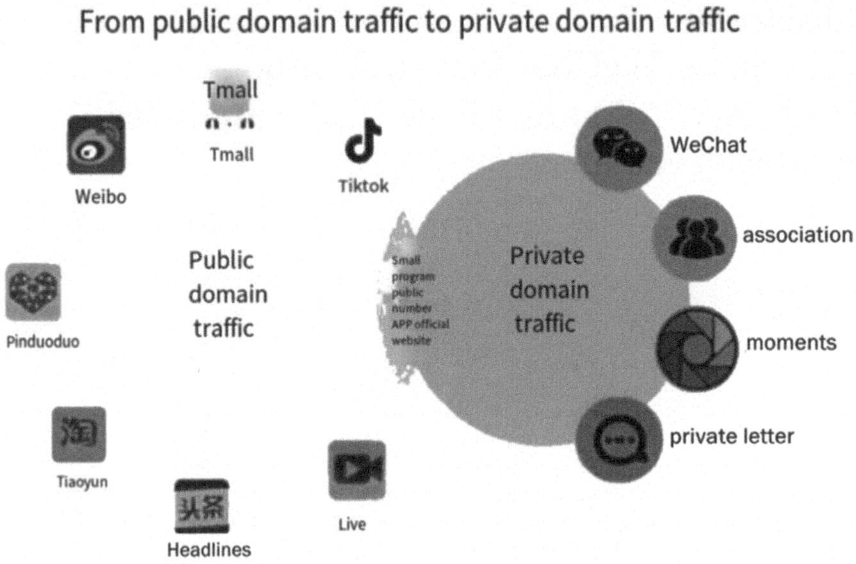

Fig. 10.1 From public domain traffic to private domain traffic

on public domain traffic toward cultivating private domain traffic, a trend illustrated in Fig. 10.1.

Since 2018, the concept of private domain traffic has garnered increased attention. This type of traffic is diverted from public domains (the Internet), other domains (such as platforms, media channels, and partners), to their own private domains (official websites and customer lists), as well as traffic generated within the private domain itself (visitors). Private domain traffic is cost-free and can facilitate multiple customer interaction points, sales, and other marketing activities. These customers can be directly contacted at any time and frequency through various channels, including media, user groups, and micro-signals. In recent years, the operation of private domain traffic has evolved into a ubiquitous business model. From large enterprises to small vendors, the practice is widespread. The market has also seen the emergence of private domain SAAS technology service providers (such as Youzan and Weimeng), private domain operation service providers, and private domain education and training institutions (such as Orange). This maturation of the private domain business is evident. According to the "2020 China Digital Marketing Trend" data, up to 62% of advertisers indicated that their own traffic pools were the most significant form of digital marketing in 2020, second only to social marketing. Private traffic, akin to domain names, trademarks, and goodwill, constitutes a company's private digital assets. Currently, private domain traffic has solidified as a prominent business model. It is now commonplace for individual businesses to construct their own private domain traffic WeChat customer groups and integrate them with live video, microblogging, public accounts, and other marketing strategies.

Fig. 10.2 Private traffic is approximately equal to the WeChat ecosystem

WeChat stands out as the most optimal platform for generating private domain traffic for several reasons. Initially, as a nationally recognized application, WeChat remains the preeminent social software with the highest frequency of use, extensive daily engagement time, and the greatest user reach. Within the WeChat ecosystem, each individual acts as a traffic node, and this ecosystem boasts an unparalleled richness in traffic resources. Moreover, the social connections on WeChat continually expand, enabling users to forward and share content effortlessly. This capability transforms a single traffic node into multiple nodes, effectively driving exponential traffic growth for enterprises. Additionally, the WeChat ecosystem integrates a comprehensive suite of functionalities, including customer contact management, customer interaction, video marketing, and enterprise verification. These features collectively provide a seamless and enhanced experience for managing private domain traffic. Consequently, in the modern retail landscape, leveraging private traffic essentially means integrating into the WeChat ecosystem, as depicted in Fig. 10.2.

According to The Times, retail banks increasingly adopt integration into the WeChat ecosystem as part of their strategic approach to effective customer acquisition. This practice considers factors such as operating costs, customer retention, and overall customer management. Bank account managers now actively use WeChat to communicate directly with customers, promote various products and activities through official enterprise accounts, and seamlessly integrate life scenarios via WeChat mini programs. Additionally, they leverage live video broadcasting and interest exploration within the WeChat ecosystem. These practices have become standard in enhancing online customer management for retail banks. A critical challenge to address is the "cold start" problem, which pertains to efficiently attracting and onboarding customers during the initial establishment phase of the bank's presence on WeChat. Essentially, this involves finding effective methods to quickly accumulate a substantial customer base on enterprise WeChat. Successful resolution of this issue is crucial for subsequent marketing strategies and overall customer engagement. In this context, the author discusses the development of an automated program designed to facilitate the bank's enterprise WeChat operations.

This program enables automated sending of customer invitations, along with the ability to modify customer names and assign marketing labels. This initiative represents a foundational step in the broader strategic transformation of retail banking, setting the stage for more advanced customer relationship management and targeted marketing efforts.

10.1 Program Design

Using enterprise WeChat as the platform, retail banks have implemented a dual-tiered customer service system comprising "enterprise-employee" and "employee-customer" interactions. At the enterprise level, management oversees employees and sets key performance indicators (KPIs), ensuring alignment with organizational goals. Employees, in turn, maintain relationships with customers, engaging in targeted marketing activities. Given the vast number of retail customers, employees typically need to add substantial customer lists to the enterprise WeChat system. This involves manually entering customer data, sending connection invitations, awaiting customer acceptance, and updating customer identification details. These manual tasks are both time-intensive and arduous, often leading to significant workloads. Figure 10.3 illustrates the customer maintenance model employed by retail banks.

In this case, we engineered two distinct plug-in applications, named "Batch Customer Invitation Robot" and "Batch Customer Tag Robot," to replicate all operational functions performed by employees on the enterprise's WeChat platform. Specifically, the "Batch Customer Invitation Robot" facilitates the execution of automated tasks under defined resolution parameters. Once an employee logs in to the enterprise's WeChat with their individual account credentials, this program emulates various mouse and keyboard activities in a loop. Subsequently, it interfaces with the WeChat user interface to dispatch invitations to every customer managed by that employee. Conversely, the "Batch Customer Tag Robot" performs batch updates. It systematically alters the names and marketing labels of all customers who have accepted the invitations on the enterprise's WeChat after they receive said invitations. The comprehensive functionalities of both programs are depicted in the logical architecture diagram found in Fig. 10.4.

This case uses Pywin32 library to call various API functions of Windows operating system, including positioning enterprise WeChat window, adjusting screen resolution, operating clipboard, etc. Use pillow and OpenCV library to complete a simple image classification task, so as to identify whether the label in the enterprise WeChat customer information interface has been marked; using Pyqt5 library to write the interface of external program; and use the Pandas library to handle bulk customer data. The workflow of the batch customer invitation robot for this case is shown in Fig. 10.5.

Figure 10.6 shows the workflow of the batch customer marking robot in this case.

10.2 Development Library

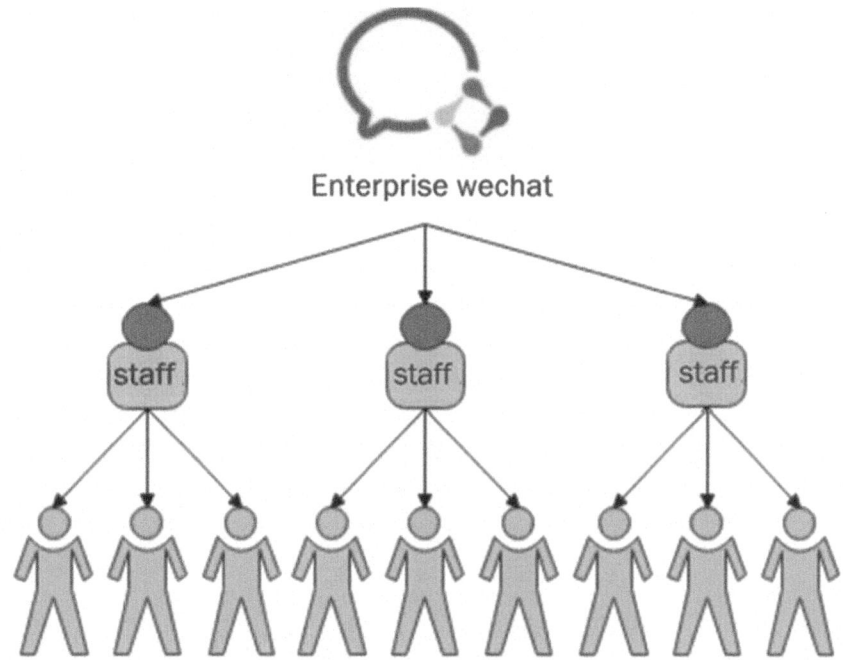

Fig. 10.3 Two-layer structure of enterprise WeChat customer service system

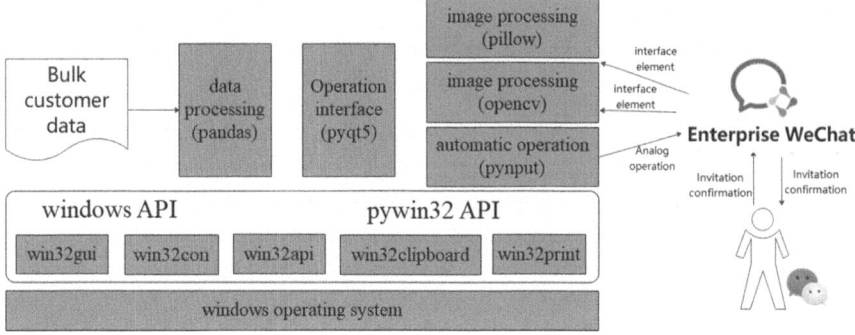

Fig. 10.4 Logical architecture of this case

10.2 Development Library

This section introduces the automatic control, image processing, data processing related technologies, and open source libraries used in this case.

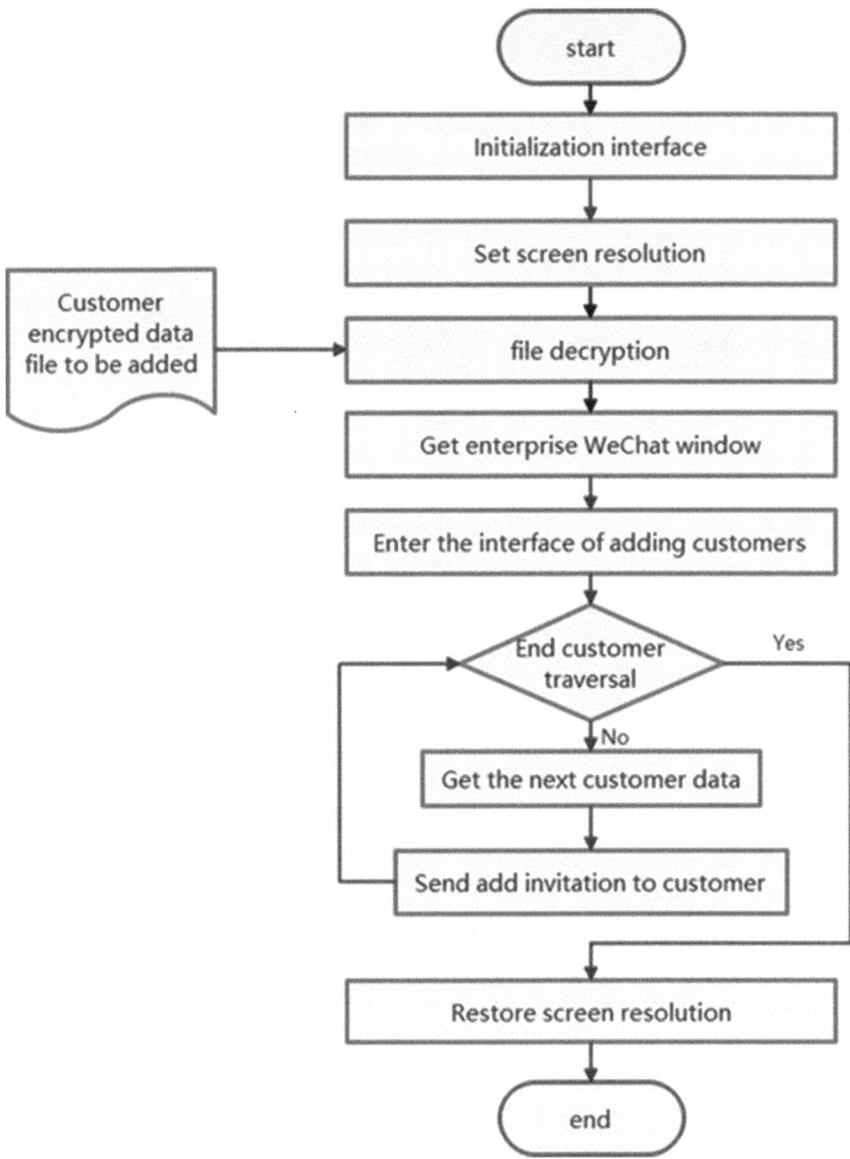

Fig. 10.5 Batch customer invitation robot workflow

10.2.1 Underlying Interface Library: Pywin32

On the Windows platform, Python modules are usually over-packaged for ease of use, and some functions cannot be flexibly used. In special cases, Windows API must be directly invoked to achieve. To achieve this goal, there are two ways, one is

10.2 Development Library

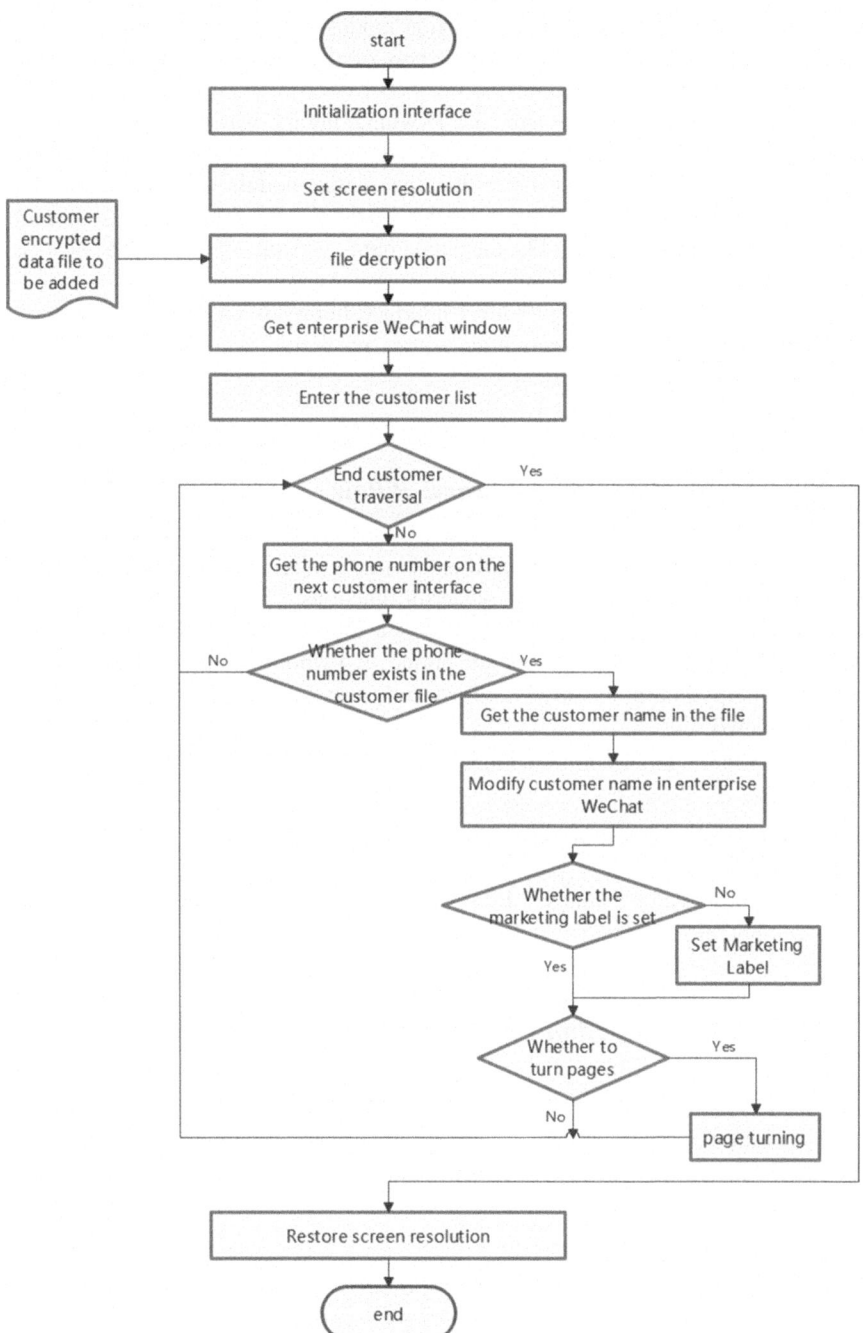

Fig. 10.6 Workflow of the batch customer marking robot

to use C language to write Python extension modules or is to write ordinary DLL through Python Ctypes to call, but this part of the sacrifice of Python's rapid development, compile-free features.

In order to solve this problem, the Pywin32 library was born. Pywin32, or "Python For Win32," wraps almost all Windows apis and can be easily called directly from Python. Another major function of this module is to use Python for COM programming. Pywin32 open source code escrow warehouse for https://github.com/mhammond/pywin32, the maximum version of 303, the heat is 3.5 K star.

In this case, the Pywin32 low-level interface library is used to call Windows functions such as window positioning, screen resolution acquisition and setting, clipboard reading and writing.

10.2.2 Image Processing Library: Pillow

Python Image Library (PIL) is a third-party image processing library for Python, which is widely used because of its powerful functions. The official home page for http://pythonware.com/products/pil/. PIL has a long history, early versions only support Python 2.x, later versions are ported to Python 3, named Pillow, home page is http://python-pillow.org/, Code custody warehouse for https://github.com/python-pillow/Pillow.

Pillow mainly provides two kinds of functions: image archive and image processing.

1. Image archiving: batch processing of images, generating image preview, image format conversion, etc.
2. Image processing: image basic processing, pixel processing, color processing, etc.

10.2.3 Computer Vision Processing Library: OpenCV

The OpenCV library is described in Sect. 8.2.1 of this book. This case uses OpenCV to process the histogram of the screenshot of the button element on the enterprise WeChat, and judge whether the button is valid.

10.2.4 Data Processing Library: Pandas

Pandas (Python Data Analysis Library) is a NumPy-based data analysis library that provides functions and methods for importing and exporting, cleaning, processing, calculating, and other data structures such as Series, Time Series, DataFrame, and

Panel. Originally developed by AQR Capital Management in April 2008 and open sourced in late 2009, Pandas continues to be developed and maintained by the PyData team as part of the PyData project. Pandas was originally developed as a financial data analysis tool, and as such, Pandas provides excellent support for time series analysis. Pandas takes its name from Panel Data and Python Data Analysis. Nowadays, Pandas is widely used in scientific computing, data processing, and other fields. This case uses Pandas to read batch user information.

10.2.5 Pynput Library

Pynput is a cross-platform third-party library for controlling and monitoring input devices, available for Windows, MacOS, and Linux. By control, we mean allowing input actions to be sent to the operating system through code, which is equivalent to a user tapping a keyboard or clicking a mouse. Monitoring means that once the user's mouse and keyboard input triggers the monitoring condition set by the code, the event is captured by the program and the response code is executed. In pynput, two classes are provided, Pynput.mouse and Pynput.keyboard, to control and monitor the mouse and keyboard. Pynput follows the LGPL 3.0 open source license, its code warehouse for https://github.com/moses-palmer/pynput, the official document at https://pynput.readthedocs.io/en/latest/.

In this case, Pynput is used to control mouse and keyboard input with code, simulating various operations of users in enterprise WeChat.

10.3 Case Practice

This section introduces the case development and operation environment construction and code implementation. After reading this section, you will understand the specific implementation process of this case in detail.

10.3.1 Environment Setup and Case Runs

This case is built in Windows 10 environment. Connect to the Internet, create a command line, and enter the virtual environment, execute:

```
conda create -n qywx python=3.7 -y & conda activate qywx
```

Full name	Cell-phonenumber
Customer 1	XXXXXX...
Customer 2	XXXXXX...
Customer 3	XXXXXX...
Customer 4	XXXXXX...
Customer 5	XXXXXX...
Customer 6	XXXXXX...
... ...	... ...

Fig. 10.7 Format of batch customer data

Perform the following steps to install the dependency library packages in turn. Wait until the execution is complete and no error is reported to complete the environment setup:

```
pip install pandas pywin32 pynput pyqt5 pillow opencv-python
```

Command line execution activation environment:

```
conda activate qywx
```

Fill in data.xlsx, this is the bulk customer list data in the format shown in Fig. 10.7.

To ensure data security, we encrypt this data. To do so:

```
python batch customer data file encryption.py
```

Select data.xlsx and the encrypted file encrypt.xlsx will be generated in the current directory.

Start enterprise WeChat after installation. The computer will pop up the QR code, open the mobile enterprise WeChat app to scan the code (note that it is not WeChat), and use the computer to log in to the enterprise WeChat. Be careful not to minimize the window. This is the "initial state" of the enterprise WeChat. At this point, execute on the command line:

```
python bulk Client Invite bot.py
```

10.3 Case Practice

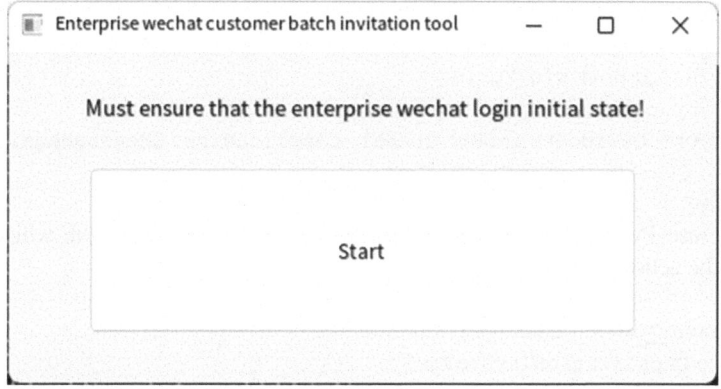

Fig. 10.8 Batch Customer Invitation robot screen

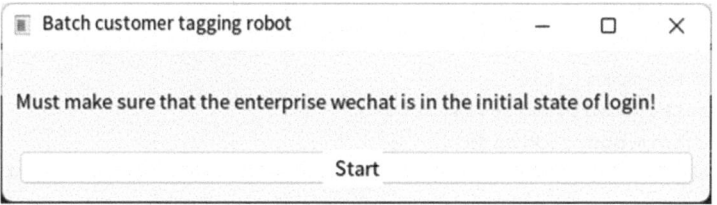

Fig. 10.9 Batch customer tag robot interface

The Batch Customer Invitation robot main screen is displayed, as shown in Fig. 10.8.

Click the "Start" button, select the batch customer data encryption file encrypt. xlsx, and you can enjoy a wonderful scene: the mouse and keyboard seem to be endowed with a soul, moving by themselves, and operating continuously in the enterprise WeChat. After a while, all the customers in encrypt.xlsx have received the enterprise WeChat invitation.

Run the command line: python Batch customer tag robot.py, the batch customer tag robot interface is displayed, as shown in Fig. 10.9.

Select the batch customer data encryption file encrypt.xlsx and you will see the program automatically modify customer names and set marketing labels. At this point, we have finished setting up the environment and running the program.

10.3.2 Code practice

Batch Customers Invite Robot Code Combat

First, import the necessary library package for this case:

```
import win32gui,win32con
import win32api
from win32 import win32print
from pynput import mouse, keyboard
from PyQt5.QtWidgets import QLabel, QApplication, QMessageBox,
QWidget, QPushButton, QFileDialog
```

Then, use PyQt5 to instantiate a custom interface class, MainWin, which will display the action interface:

```
if __name__ == '__main__':
    app = QApplication(sys.argv)
    ex = MainWin()
    sys.exit(app.exec_())
```

In the MainWin class, define an initUI() function to initialize interface elements:

```
# Define the form title
self.setWindowTitle(self.title)
# Define form position size
self.setGeometry(self.left, self.top, self.width, self.height)
# Create a text control inside the form and sets its display content and coordinate position
textLabel1 = QLabel(" Must ensure that the enterprise WeChat is in the initial state of login!" , self)
textLabel1.move(50, 30)
# Create button object inside form
button = QPushButton(" Start ", self)
# Define the message that appears when the mouse hovers over the button
button.setToolTip(" Select customer data to start batch adding powder ")
# Set button position size
button.move(50, 70)
button.resize(340, 100)
# Set the response function for clicking the button
button.clicked.connect(self.begin)
self.show()
```

Firstly, we define the callback function for the button interaction. Considering that the "pynput" library references mouse coordinates based on screen resolution (beginning from the top left at 0,0), the mouse positioning is intrinsically linked to the display settings. Thus, identical screen locations may yield different coordinate values under varying resolutions. Consequently, a fixed resolution is imperative for accurately locating the mouse position when interacting with the enterprise WeChat interface. Our approach involves initially capturing the current screen resolution,

10.3 Case Practice

adjusting it to the desired resolution for the operation, and ultimately restoring it to the original resolution upon program termination. We achieve this by defining the "get_real_resolution()" function to ascertain the current resolution, utilizing the "win32gui" and "win32print" modules from the "pywin32" library.

```
def get_real_resolution():
    # Get the handle to the current screen
    hDC = win32gui.GetDC(0)
    # Get landscape resolution
    w = win32print.GetDeviceCaps(hDC, win32con.DESKTOPHORZRES)
    # Get vertical resolution
    h = win32print.GetDeviceCaps(hDC, win32con.DESKTOPVERTRES)
    return w, h
```

We define the set_resolution () function to set the screen resolution:

```
def set_resolution(w, h):
    # Get graphic display Settings
    dm = win32api.EnumDisplaySettings(None, 0)
    # Set the number of pixels in both the high and wide directions
    dm.PelsHeight = h
    dm.PelsWidth = w
    # Set the number of bits per pixel
    dm.BitsPerPel = 32
    # Set display mode
    dm.DisplayFixedOutput = 0
    # Apply the display parameter change to the current screen
    win32api.ChangeDisplaySettings(dm, 0)
```

We use the atexit module to set the function to be called when exiting the program, which should restore the original screen resolution, that is, the return value of get_real_resolution() w, h:

```
atexit.register(recovery_resolution, w, h)
```

This means that before exiting the program, first execute the recovery_resolution () function, passing the w, h arguments, and then reset the original resolution in recovery_resolution():

```
def recovery_resolution(w, h):
    set_resolution(w, h)
```

Then, select the Batch Customer Encrypted data file, which is a two-column excel table containing the name and phone number:

```
Filename, filetype = QFileDialog.GetOpenFileName (self, "select file",
"c:\\", "All Files (*);; Excel Files (*.xlsx);; Excel Files 97 (*.xls)")
```

Decrypt the file and read it into the DataFrame.

```
decrypt_file(filename,'decrypt.xlsx','key')
df = pd.read_excel('decrypt.xlsx')
```

Find the form handle of enterprise WeChat through pywin32's win32gui module and specify the title of the form. If the enterprise WeChat is not open, return 0:

```
hwnd = win32gui.FindWindow(None, 'enterprise WeChat ')
```

In order to fix the position of operation elements, we maximize the form of enterprise WeChat, and set the form prefix:

```
win32gui.ShowWindow(hwnd, win32con.SW_SHOWMAXIMIZED)
win32gui.SetForegroundWindow(hwnd)
```

It takes a certain amount of time to manipulate the form, we wait 0.3 seconds before continuing:

```
Time.Sleep(0.3)
```

Next, we use pynput's mouse controller and keyboard controller to manipulate the elements on the enterprise WeChat interface. First create the controller handle:

```
m_mouse = mouse.controller() # Create a mouse Controller
m_keyboard = keyboard.controller() # Create a keyboard Controller
```

Because we set the enterprise WeChat form front, so to manipulate this form element, just move the mouse hotspot to the corresponding position and click. When developing, it is necessary to spend more effort to align the coordinate position. For example, click the enterprise WeChat address book, and then click "New customer":

```
 m_mouse.position = (20, 147) # Move the mouse to the address book position
 m_mouse.click(mouse.button.left) # Click the left mouse Button
 Time.Sleep(0.3)
 m_mouse.position = (100, 100) # Move the mouse to the "new customer" position
 m_mouse.click(mouse.button.left) # Click the left mouse Button
```

10.3 Case Practice

It should be noted that the above operation needs to wait for the enterprise WeChat app to respond, and we set a waiting time of 0.3 seconds. In the place where you need to enter the keyboard (such as entering the mobile phone number of each customer), we will operate the m_keyboard handle:

```
for i in range(len(df)): # Loop an invitation to add to each customer
  m_keyboard.type(str(df['phone number'][i])) # Enter the phone number in the customer data file
  m_keyboard.press(keyboard.Key.enter) # hit Enter
  m_keyboard.release(keyboard.Key.enter) # Release the Enter key
  ...
```

In this way, we have completed the development of the mass customer invitation robot. See "Batch Customer Invitation Bot.py" for the code.

Batch Customer Tag Robot Code Practice

The subsequent step involves developing the code for a batch customer tagging robot. The implementation's function entails: once the customer consents to the invitation addition, their nickname is updated to reflect their actual name, and marketing labels are applied in bulk. This requires manually selecting each customer in the enterprise WeChat customer list, as depicted in Fig. 10.10. We extract the registered mobile phone number from each customer's enterprise WeChat account, cross-reference this data with the batch customer records, retrieve the corresponding name, and subsequently alter the nickname accordingly.

The key codes are as follows:

```
for i in range(......):
  if i<17: # Customer List The number of customers on the home page is less than or equal to 17
      m_mouse.position = (140, 230+i*30) # Move the mouse to a customer location
      m_mouse.click(mouse.button.left) # Click the left mouse Button
    else: # not the home page, page_counts is the number of customers on the current page
      m_mouse.position = (140, 90+page_counts*30) # Move the mouse to a customer location
  m_mouse.click(mouse.button.left) # Click the left mouse Button
```

How do I get the phone number in the screen? We mimic the user operation: double click the phone number text control, press ctrl+c to copy it to the clipboard, and then get it from the clipboard:

Fig. 10.10 Enterprise WeChat customer list

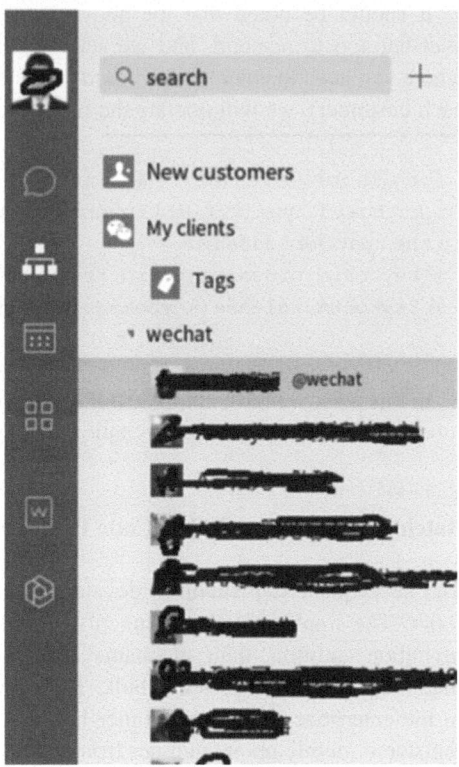

```
m_mouse.position = (570, 260) # Move the mouse over the phone number
m_mouse.click(mouse.button.left) # Click the left mouse Button
Time.Sleep(0.2)
# Click the left mouse button again, since the two clicks are only 0.2 s
apart, will trigger the double click event, causing the phone number
contents to be selected
m_mouse.click(mouse.Button.left)
# Press ctrl+c to copy the phone number to the clipboard
m_keyboard.press(keyboard.Key.ctrl) # press ctrl
m_keyboard.press(keyboard.KeyCode(char='c')) # press c
m_keyboard.release(keyboard.KeyCode(char='c')) # release c
m_keyboard.release(keyboard.Key.ctrl) # release ctrl
telnum = get_clip() # copies the phone number in the clipboard to the
variable
# Using pywin32's win32clipboard module, get the clipboard contents
function
  import win32clipboard as wb
  def get_clip():
    wb.OpenClipboard() # Open the clipboard
```

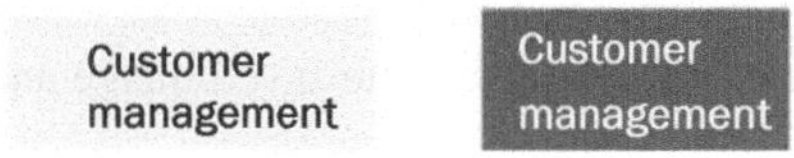

10.3 Case Practice

| Customer | Customer |
| management | management |

Fig. 10.11 Unlabeled vs. labeled

```
# Gets the clipboard contents as text, returning bytes type data
    d = wb.GetClipboardData(win32con.CF_TEXT)
    wb.CloseClipboard() # Close the clipboard
    return d.decode('GBK') # decodes the bytes type to GBK-encoded str type
```

Next, find the real name corresponding to the phone number in the bulk customer data DataFrame:

```
Name = df.Loc[(df) mobile phone number.Astype(' int64) == int
(telnum)), 'name'].Values[0]
```

Delete the original remarks of the customer, here directly press backspace continuously:

```
for j in range(20):
    m_keyboard.press(keyboard.Key.backspace) # Press backspace
    m_keyboard.release(keyboard.Key.backspace) # Release backspace
```

Then enter the customer's name and click "OK" to change the customer's real name in the enterprise WeChat:

```
m_keyboard.type(name)
m_mouse.position = (520, 540) # Move the mouse to the "OK" position
m_mouse.click(mouse.button.left) # Click the left mouse Button
```

Below, we set the marketing TAB for "custodial customer." Click on the label bar and you will see all of the marketing labels for that customer. If a label is already set for the customer, its color is blue (black and white illustrated in this book, dark gray), and if it is not set, it is gray, as shown in Fig. 10.11. In the process of customer polling, there are two situations: set and unset.

How do you classify your customers' marketing label buttons? This requires designing an algorithm to classify the images. We reserve the button picture with the blue label, then capture the button picture from the interface, calculate the similarity with the reserved picture, and judge whether the current button is gray or blue according to the similarity score. Use PIL module to capture the image area of the current button:

```
grb = ImageGrab.grab(bbox=(660,310,743,340))
```

Calculate the similarity score between the saved image and the captured image:

```
grb.save('grab.jpg')
img1 = cv2.imread('button.jpg')
img2 = cv2.imread('grab.jpg')
degree = classify_gray_hist(img1,img2)
```

We use the histogram intersection algorithm to measure how similar two graphs are. A histogram is a statistical distribution plot where the horizontal axis represents the number of bins and the vertical axis represents the statistics for each bin. In this example, the value range of the horizontal axis is 0–255, representing 256 values of each pixel in the grayscale plot, and the vertical axis is the number of pixels corresponding to the pixel value. A histogram of two images, the more pixels they overlap in the same pixel values, the more similar the images are. The intersection distance of the histograms is usually used to measure their similarity, let M, N be two histograms with K BIN, their components are $M(i)$, $N(i)$, where $i = 1, 2, 3, 4,...$ k whose formula is:

$$D(M,N) = \frac{\sum_{i=1}^{K} MIN(M(i), N(i))}{\sum_{i=1}^{K} M(i)}$$

where the numerator is the number of overlapping pixels in each component of the two histograms, divided by the total number of pixels in a certain graph to obtain the normalized result in the range (0, 1). In this example, the index of "average coincidence degree" is used to measure the image similarity, that is, the average of the coincidence ratio on each component is taken to establish the similarity function:

```
def classify_gray_hist(image1,image2,size = (256,256)):
    # Scale the input image to 256 by 256 pixels
    image1 = cv2.resize(image1,size)
    image2 = cv2.resize(image2,size)
```

The arguments to the #calcHist() function are: Enter the image matrix, channel number, mask (value 1 means to process part of the image, none means to process the whole image), how many bin to use (256 means 256 bin, each bin represents a pixel value), the value range of pixels (0–255), return an array, where each component represents the number of pixels on each bin.

```
Hist1 = cv2.CalcHist([image1], [0], None, [256], [0.0, 255.0])
Hist2 = cv2.CalcHist([image2], [0], None, [256], [0.0, 255.0])
```

10.3 Case Practice

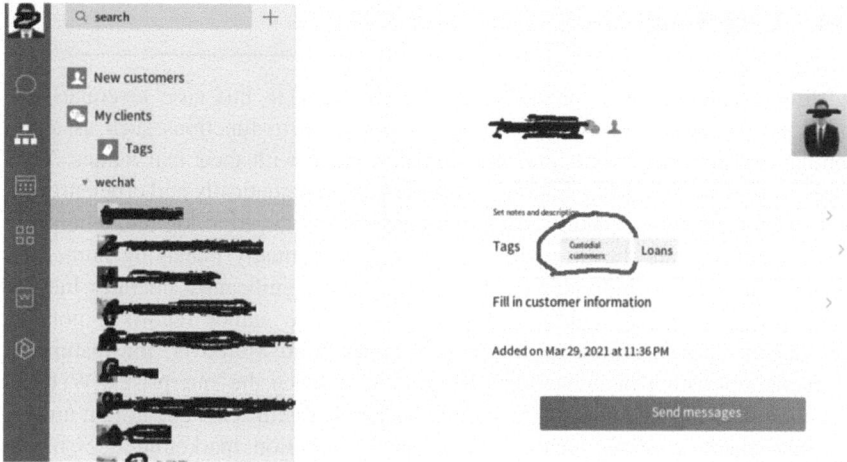

Fig. 10.12 Setting a marketing label for the customer

```
#degree Indicates the average coincidence of the two histograms
degree = 0
for i in range(len(hist1)):
    if hist1[i] != hist2[i]:
        # Sum the overlap proportions on each bin
        degree = degree + (1 - abs(hist1[i]-hist2[i])/max(hist1[i],hist2[i]))
    else:
        degree = degree + 1
    # Divide the sum result of the coincidence ratio by the total number of
bins to get the average degree of coincidence
    degree = degree/len(hist1)
    return round(float(degree),2)
```

If the current TAB button is not checked, its gray image has a smaller degree (lower degree of coincidence) than the saved blue image, or a larger degree (higher degree of coincidence) if it is checked. Therefore, we set a threshold to distinguish whether the marketing label is selected or not:

```
if degree < 0.7: # Marketing label not selected
    Complete Marketing label selection...
```

After completing the above work, the marketing label set can be seen in the customer information of the enterprise's WeChat. As shown in Fig. 10.12.

The code is detailed in the download file "Bulk Customer Tag Bot.py." At this point, we have completed the development of "Batch Customer Invitation Bot" and "Batch Customer Tag Bot."

10.4 Case Summary

The implementation of automated control technology in this case leverages the capabilities of enterprise WeChat to seamlessly perform functions such as auto-sending customer invitations and marking customers with their real names. This process ensures that fixed operations are carried out systematically and efficiently by the system itself. Consequently, bank employees merely need to log into enterprise WeChat and activate the program, as opposed to manually managing numerous customer interactions individually. This automation significantly liberates human resources and enhances operational productivity. The author reports a notable threefold increase in efficiency in practical application. Moreover, the feature of real name annotation facilitates data integration between the enterprise's WeChat marketing interface and the bank's internal backend system. This integration underpins subsequent processes such as performance recognition, marketing assessment, and the development of customer business strategies.

This case study represents an initial step in deploying a customer-centric online business strategy within the emerging retail landscape and plays a critical role in the evolution of commercial banks' retail private domain traffic business model. This case offers both practical applications and areas for optimization. For instance, utilizing Pynput requires resolution adjustments, with each operational coordinate being static. Consequently, any layout change in enterprise WeChat necessitates program coordinate repositioning. Future enhancements can consider transitioning this system to a robotic process automation (RPA) model, enhancing its general applicability. Nonetheless, this case illustrates a comprehensive automatic control framework compatible with various applications and web pages. Numerous automatic control scenarios exist within banking operations, providing a foundation for readers to extrapolate and innovate new applications inspired by this case study.

In this scenario, the case can be converted into an automated online system for managing private domain traffic customers. Specifically, this involves gathering customer data on browsing, orders, and payments from a personalized marketing website on WeChat. Subsequently, one needs to analyze this data to identify customer interests, construct a customer preference analysis model, and disseminate tailored marketing content to individual customers through automated control technology. The feedback, based on customer interactions, aids in generating new interest data. This feedback loop is vital for the continuous optimization of the model and the retention of private domain traffic, thereby achieving a closed-loop system for the automated management of private domain customers.

Chapter 11
Intelligent Inspection Robot for Commercial Bank Data Centers: Computer Vision Technology

Information technology risks can instantly paralyze a bank. Previously, this was a topic in the banking qualification exam. When IT risks occur, they can lead to business disruptions or casualties, causing negative societal sentiment. Lessons from such events are painful. The People's Bank of China and the China Banking Regulatory Commission always prioritize IT risks in commercial banks' regulatory supervision. During critical periods, like political events and major online shopping festivals such as Double 11, ensuring bank information system stability and business continuity is crucial. IT risks mainly revolve around operational risks in commercial banks' data centers. Controlling these operational risks is essential for the banks' IT departments. Recently, bank data centers have expanded, with more cloud computing clusters, cloud storage, GPU clusters, network equipment, and connections. System security and electrical safety risks are growing challenges. Traditional patrols face several issues: they can't operate 24/7, can't inspect multiple devices at once, overlook issues, delay alarms, restrict access during emergencies, and can't analyze data from paper-based patrols.

Addressing system security and electrical safety risks through technological innovation is a logical step forward. Can we develop a system that emulates the process of manual inspection while continuously monitoring critical risks? Monitoring the operating status of equipment and electrical safety are crucial factors in mitigating IT operational risks. The status of power indicator lights, hard drive indicator lights, IPMI interface indicator lights, network indicator lights, health indicator lights, and UID indicator lights on the equipment panel represents the operational state of the equipment. Due to the significant power consumption in data centers, local power consumption limits can be exceeded, and aging UPS battery wires can lead to fire hazards. Therefore, risk management focuses on evaluating the normal operational state of devices such as UPS units, power distribution cabinets, and servers by observing the equipment panel indicator lights. Additionally, infrared thermal imaging cameras are utilized to detect potential high-temperature risks at UPS battery terminals, UPS battery surfaces, power distribution cabinet terminals, and critical areas within the computer room. This dual approach enhances the ability to prevent and manage potential hazards efficiently.

11.1 Scheme Design

As previously articulated, the design of an automated inspection robot scheme substantially mitigates numerous challenges associated with personnel inspection processes. This can be approached through two primary methodologies. One involves utilizing the device's out-of-band port to establish a monitoring and management network, enabling real-time monitoring of the devices' operational status by reading its internal chip data. Alternatively, camera vision technology can be employed to observe the device's operational state. Given that the former method is limited to server applications and is not applicable to power supply equipment such as uninterruptible power supplies (UPSs) and distribution cabinets, we adopt the latter approach in practical scenarios. This chapter delineates a scheme where a slide rail is installed at the top of the data center to facilitate the navigation and positioning of the robot. A stepper motor is mounted on the top slide rail of the data center and propels the robot to move accurately to designated inspection points. The motor's displacement distance is programmed to ensure precise positioning of the robot for each inspection task. The stepper motor's control program assigns a scene number to the current inspection, integrating details such as the number and color of panel lights under normal conditions and the database records of flashing lights. This technical endeavor is executed by specialized design and construction firms. For panel light inspections, the robot is composed of a Raspberry Pi and a natural light camera. The visual model, deployed on the Raspberry Pi, continuously reads the panel light status of UPS, power distribution cabinets (PDCs), servers, and other devices, comparing this data against the normal states registered in the database to identify any deviations in operational status. In high-temperature inspection scenarios, particularly in areas prone to fire risks like UPS battery connections and power supply wires, the system also utilizes the slide rail and stepper motor. Here, a Raspberry Pi coupled with an infrared camera deploys a thermal imaging temperature recognition model to detect any regions within the field of vision where temperatures exceed safety thresholds. Should either visual model detect anomalies, an alert is transmitted to the equipment room's monitoring system for prompt intervention.

This inspection system relies on a tri-layered architecture comprising the base layer, operations layer, and application layer. The base layer is tasked with hardware management and drivers, the operations layer handles fundamental image processing algorithms and operations, while the application layer is responsible for image

11.2 Computer Vision Technology

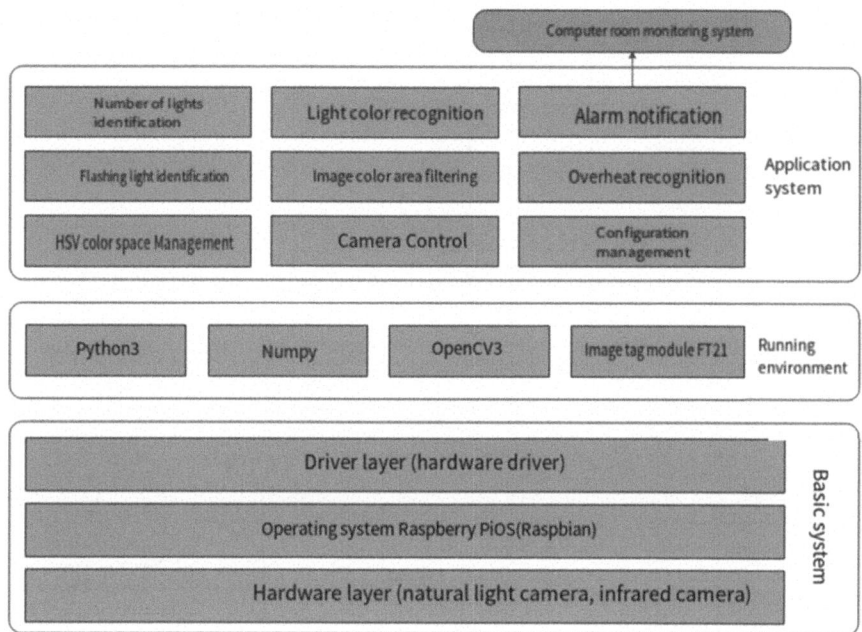

Fig. 11.1 System architecture of this case

recognition, data alarms, and logical processing. The project's architectural framework is illustrated in Fig. 11.1.

In this case, the panel light monitoring module needs to obtain image frames from the video, identifies the position of the panel light through image processing, and compares it with the corresponding position of the previous frame to identify flicker. The processing process is shown in Fig. 11.2.

The thermal imaging monitoring module in this case needs to obtain image frames from the video and identify the heat concentrated area through image processing. Figure 11.3 shows the processing flow.

The basic processing flow of this case is as follows: read the image frames in the video, obtain the panel light area in the image by color, identify the color, quantity, position, blinking, and compare the alarm with the preset normal light color, quantity, position, and blinking.

11.2 Computer Vision Technology

This section introduces the hardware devices, key concepts, and algorithms involved in this case. After reading this section, you will understand the relevant technical implementation of this case.

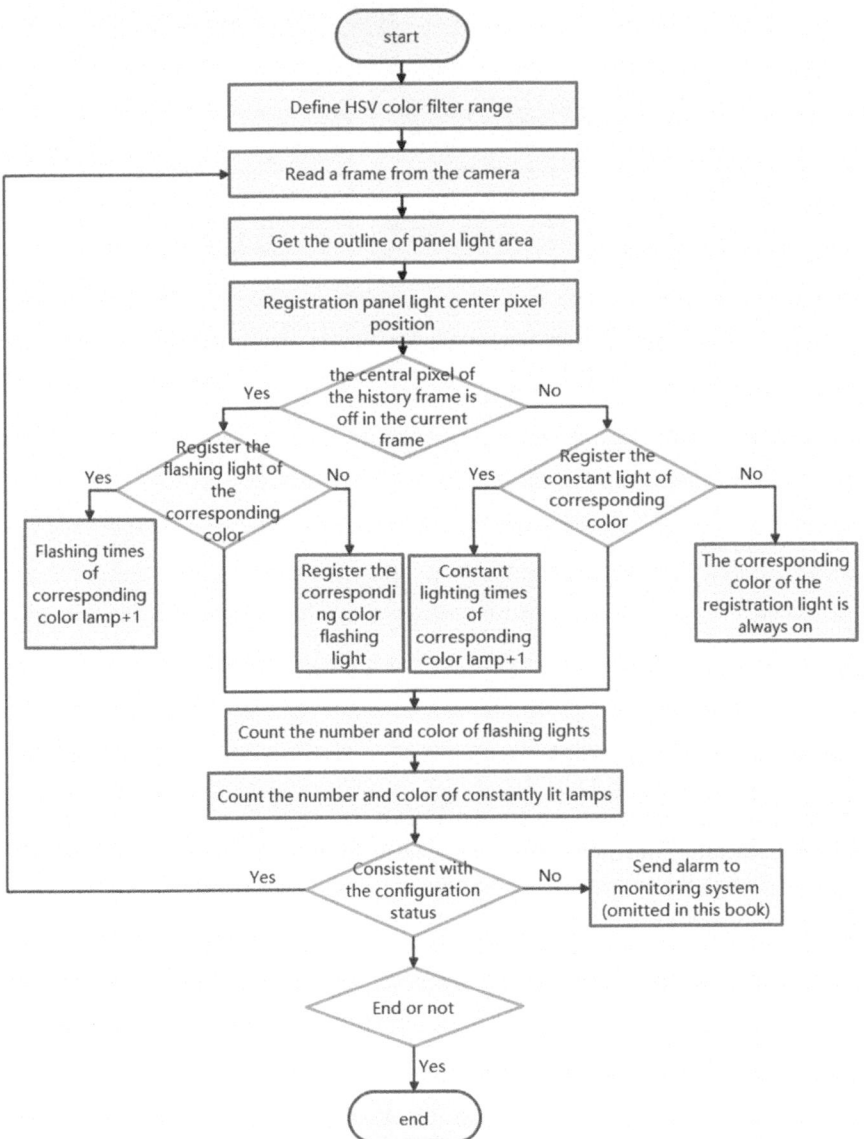

Fig. 11.2 Flow chart of panel light monitoring in this case

11.2.1 Raspberry PI

The Raspberry Pi (RPi or RasPi/RP) is an embedded minicomputer developed by the UK-based Raspberry Pi Foundation, currently evolved into its fourth generation. Designed by Eben Upton, this compact device, comparable in size to a credit card,

11.2 Computer Vision Technology

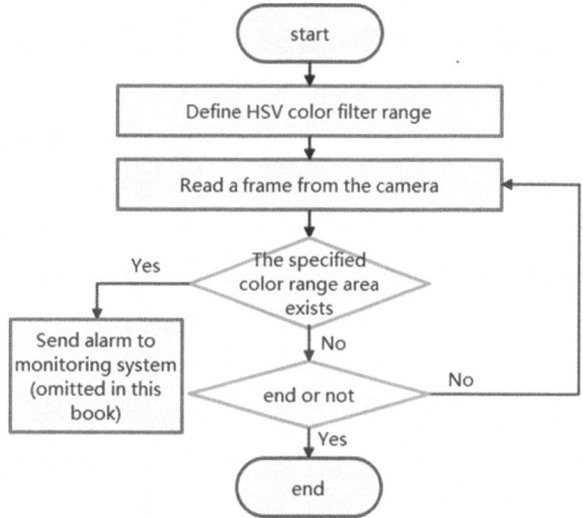

Fig. 11.3 Thermal imaging monitoring flowchart of this example

aims to facilitate computer education in underprivileged regions and operates on a non-profit basis. Despite being a low-cost device, the Raspberry Pi boasts substantial computing prowess, making it highly popular among enthusiasts and educators alike. The Raspberry Pi utilizes an ARM architecture processor, categorizing it as open-source hardware. It supports multiple Linux distributions and Android operating systems, offering robust computational performance, extensive software compatibility, and user-friendly features. The minicomputer includes a network port, USB interfaces, audio and video outputs, and an HDMI interface. It employs a micro-SD memory card as its primary storage medium, effectively functioning as a "hard disk." Users can easily connect external peripherals such as a mouse, keyboard, network, and monitor, rendering it a "small but complete" fully functional computer. Furthermore, the Raspberry Pi is equipped with the capability to interface with various external cameras and peripheral sensors, enhancing its versatility in numerous applications. Figure 11.4 provides an illustration of the Raspberry Pi's comprehensive range of accessories, demonstrating its integrative potential within diverse technological ecosystems.

The specialized application is implemented on the Raspberry PI platform, leveraging its CPU capabilities for visual computation processes. This initiative falls under the realm of embedded systems development. The implementation utilizes the Raspberry PI alongside a natural light camera kit and an infrared thermal imaging camera kit. To successfully complete the development, one must connect the video cable of the camera to the designated interface, as illustrated in Fig. 11.5.

In this case, the Raspberry PI kit slides along a fixed track to inspect devices in the equipment room one by one. We fixed it on the pre-made slide rails, as shown in Fig. 11.6.

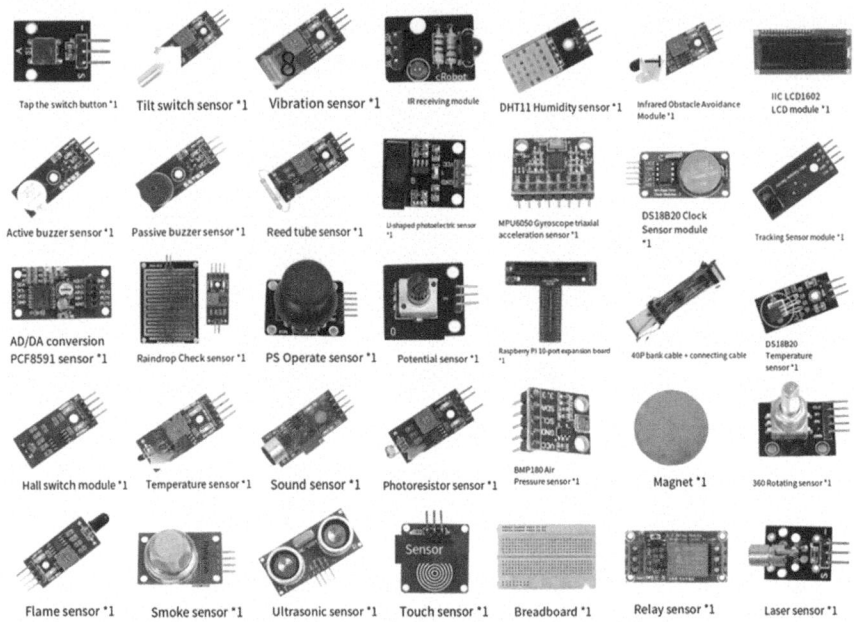

Fig. 11.4 Abundant peripheral hardware of the Raspberry PI

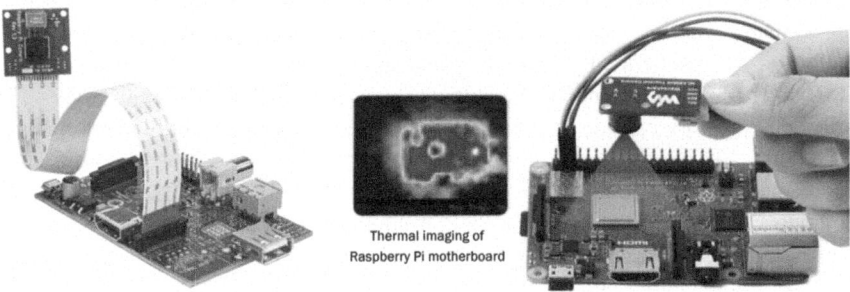

Fig. 11.5 Raspberry PI, natural light camera kit, and infrared thermal imaging camera kit used in this case

11.2.2 HSV Color Space

UPS panel lights, PDC panel lights, and server panel lights in a data center typically show a mix of red, green, and yellow, making them stand out against surrounding pixels. Infrared thermal images highlight high-heat areas with deep red. In both cases, color threshold filtering helps identify areas of interest. This case uses the HSV color space for color range filtering.

The HSV color space, created by A.R. Smith in 1978, is also known as the hexagonal cone model. It consists of three parts: Hue (H), Saturation (S), and Value

11.2 Computer Vision Technology

Fig. 11.6 The Raspberry PI camera track is fixed in this case

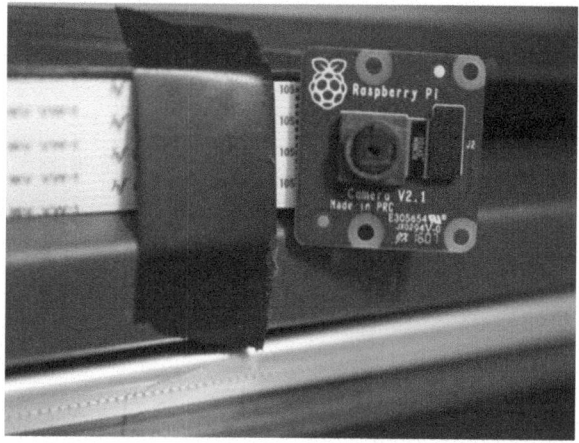

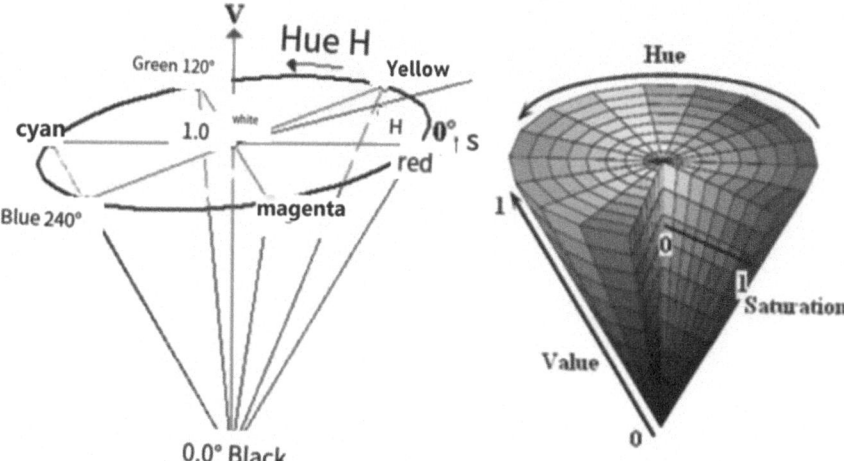

Fig. 11.7 HSV color space concept

(V). As shown in Fig. 11.7, the H parameter represents hue information, or spectral color position, ranging from 0 to 360°. Red, green, and blue are separated by 120° each, with complementary colors being 180° apart. Saturation (S) measures color purity on a scale from 0 to 1, reflecting the ratio between the chosen color's purity and its maximum purity, radial within the cone and grayscale when S = 0. Value (V) indicates color brightness, ranging from 0 to 1, along the cone's exterior in the vertical direction.

It is worth noting that the HSV components are defined with values ranging from H: 0–360, S: 0–1, V: 0–1. But in OpenCV, to match the value of 0–255 for the 8-bit data, the value range of HSV is modified to H: 0–180, S: 0–255, V: 0–255. As shown in Table 11.1.

Table 11.1 Value ranges of HSV components in OpenCV

	Black	Gray	White	Red		Orange	Yellow	Green	Cyan	Blue	Purple
hmin	0	0	0	0	156	11	26	35	78	100	125
hmax	180	180	180	10	180	25	34;	77	99	124	155
smin	0	0	0	43		43	43	43	43	43	43
smax	255	43	30	255		255	255	255	255	255	255
vmin	0	46	221	46		46	46	46	46	46	46
vmaX	46	220	255	255		255	255	255	255	255	255

11.2 Computer Vision Technology

11.2.3 Median Filtering

Median filtering is a nonlinear smoothing technique. It sets each pixel's grayscale value to the median of all pixel values within the pixel's neighborhood window. Essentially, it is a type of statistical sorting filter, including minimum and maximum value filters. It effectively reduces salt and pepper noise and is a common method for image denoising and enhancement.

As shown in Fig. 11.8, a 3 × 3 pixel window (or convolution kernel) moves over the image. It sorts the pixel values within the corresponding ROI, using the median as the output for the center pixel. As the kernel slides, the algorithm resets all centers' values, regenerating the entire image. The kernel size can be adjusted as needed. Note that the kernel size must be an odd number. Larger kernels include more pixels in the median calculation, resulting in a smoother and blurrier image.

Figure 11.9 shows the denoising effect of the median filter. It can be seen that the effect is still good.

This case uses median filtering to de-noise key image areas, which we will explain in detail in Sect. 11.4.2 of Code Parsing.

11.2.4 Edge Computing

Edge computing refers to the practice of providing services at the proximity of the data source using an open platform. Applications are deployed on edge devices, which results in faster response times and reduces the frequency of network

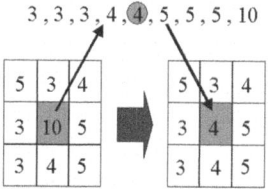

Fig. 11.8 Schematic diagram of the median filtering algorithm

Fig. 11.9 Denoising effect of median filter on image salt and pepper noise

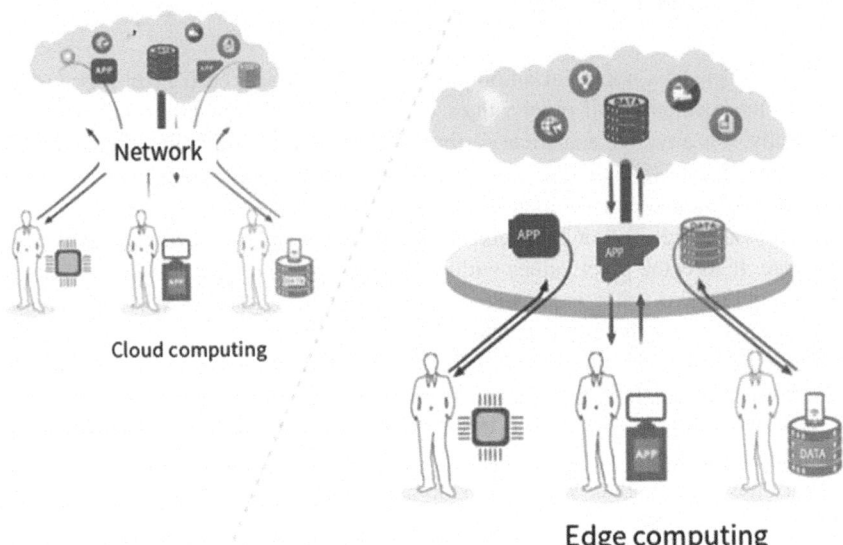

Fig. 11.10 Edge and cloud computing

transmissions. This setup meets the demands for real-time business operations, data security, and privacy protection. Typically, edge computing devices are low-power embedded devices, prominently featuring ARM chip-based devices such as smartphones, NVIDIA Jetson Nano series, and Raspberry Pi, among others. The primary distinction between edge computing and cloud computing lies in their operational focuses. Edge computing leverages the computational power of chips on edge devices, minimizing the need for interaction with cloud servers or, in some cases, eliminating it entirely. In contrast, cloud computing relies on the computational resources of cloud servers, necessitating extensive and frequent data exchanges with them. In essence, edge computing brings the computation closer to the data source, while cloud computing involves higher reliance on remote servers. For a visual representation of the comparison between edge computing and cloud computing, refer to Fig. 11.10.

This case utilizes the edge computing paradigm. The edge computing hardware employed is the Raspberry PI 3B. The application is deployed on this device, leveraging its CPU processing capabilities to perform calculations. Upon detection of anomalies, the computational results are transmitted to the backend monitoring system. Predominantly, the computations occur at the edge, minimizing the need for extensive network communication with the backend. Consequently, this reduces the latency associated with network transmission and facilitates the prompt identification and resolution of faults. This model ensures efficient real-time processing and enhances overall system responsiveness.

11.3 Developing Libraries

This section describes the two development libraries used in this case, OpenCV and Numpy.

11.3.1 Computer Vision Processing Library: OpenCV

See Sect. 8.2.1 for an introduction to OpenCV. This example uses OpenCV3 for related tasks.

11.3.2 Scientific Computing Library: Numpy

NumPy (Numerical Python) represents an open-source extension designed for numerical computation in Python, facilitating the storage and processing of extensive matrices. It surpasses Python's own Nested List Structure in terms of efficiency, effectively supporting multidimensional arrays and a vast array of matrix operations. Moreover, it offers an extensive library of mathematical functions tailored for array manipulations. Originally known as Numeric, it was initially developed by Jim Hugunin, alongside other contributors. In 2005, Travis Oliphant integrated features from Numarray, another homogeneous library, into Numeric, thereby evolving it into NumPy. This library is open source, maintained by various collaborators, and can perform tasks typically executed in languages like C++ and Fortran. NumPy endows users with advanced numerical programming tools, including matrix data types, vector processing capabilities, and a comprehensive arithmetic library. Its primary design caters to rigorous numerical manipulation, making it indispensable in both large financial corporations and eminent scientific computing organizations. Institutions such as Lawrence Livermore National Laboratory and NASA utilize it extensively, especially for applications traditionally associated with MATLAB. For instance, in specific scenarios, NumPy is employed to define HSV (Hue, Saturation, Value) color thresholds, demonstrating its versatility and functionality in diverse computational tasks.

11.4 Case Practice

This section describes the development and operation environment construction, code writing, and operation results of this case. After reading this section, you will understand the implementation process of this case.

11.4.1 Environment Setup and Case Runs

Hardware configuration: Raspberry PI 3B.
  Operating system: Raspbian Buster
  Development environment: Python3.7, OpenCV3.4.

Raspberry PI can be used in two ways. One is to connect the screen directly to the HDMI port, and the mouse and keyboard to the USB port, and use it as a normal computer. The other is to connect the Raspberry PI to a PC with a network cable without the screen, and log in to the Raspberry PI remotely through Putty on the PC. To install openCV on the Raspberry PI, you can download the OpenCV source code and compile it directly, or you can run the sudo pip3 install openCV-python command to install it. The former is prone to problems during compilation, while the latter is slower to download. We download the wheel file of OpenCV whl directly from the following website, https://www.piwheels.org/simple/opencv-python/ opencv_python-3.4.3.18-cp37-cp37m-linux_armv7l.whl, this file adapter arm chips Linux system. In fact, the whl file is essentially a compressed package, which usually contains the py file, the compiled pyd file, and the license file. The advantages of using whl files to install library packages are that you no longer need a compiler to install C extensions, support offline installation, and fast installation. We use winrar to open this whl file, you can see that there are cv2 and opencv_python-3.4.3.18.dist-info two directories, cv2 directory is the OpenCV runtime, there is an __init__.py and a .so dynamic library file, as shown in Fig. 11.11.

First, we install OpenCV's dependency packages by executing the following command:

```
sudo apt-get install libatlas-base-dev
sudo apt-get install libjasper-dev
sudo apt-get install libqtgui4
sudo apt-get install libqt4-test
sudo apt-get install libhdf5-dev
```

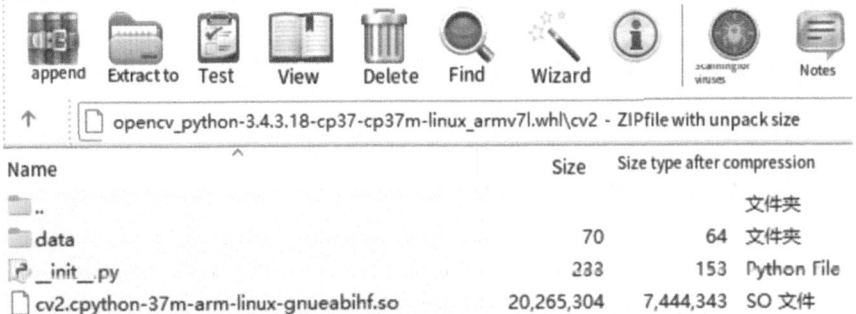

Fig. 11.11 Content of the OpenCV installation package

11.4 Case Practice

```
pi@raspberrypi:~/Downloads $ python3
Python 3.7.3 (default, Apr  3 2019, 05:39:12)
[GCC 8.2.0] on linux
Type "help", "copyright", "credits" or "license" for more information.
>>> import cv2
>>>
```

Fig. 11.12 OpenCV is successfully installed on Raspberry PI

Then execute the following command to install OpenCV3 from the whl file:

sudo pip3 install opencv_python-3.4.3.18-cp37-cp37m-linux_armv7l.whl.

In this case, enter the python3 interpreter and import cv2 successfully, indicating that OpenCV installation is complete, as shown in Fig. 11.12.

Run the following command to install the Numpy library:

sudo apt install python3-numpy.

At this point, the installation is complete.

To run the device panel light monitoring code and thermal imaging monitoring code in this chapter, execute python indicator detection.py and python temperature thermal imaging detection.py, display the real-time reasoning results of the current camera input video, that is, the case runs successfully.

11.4.2 Code Practice

Panel Indicator Light Detection Code Practice

Import related library package:

```
import numpy as np
import ft21 # Chinese font display module
import cv2
```

Set the font:

```
font = cv2.FONT_HERSHEY_SIMPLEX
```

Use the Numpy library to define the value range of the HSV color space vector in red, yellow and green, which defines the color of the panel light:

```
lower_green = np.array([35, 110, 106], dtype = np.uint8) # Green range low threshold
upper_green = np.array([77, 255, 255], dtype = np.uint8) # green range
```

high threshold

```
    lower_red = np.array([0, 127, 128],dtype = np.uint8) # Red range low
threshold
    upper_red = np.array([10, 255, 255],dtype = np.uint8) # red range
high threshold
    lower_yellow = np.array([26, 127, 128],dtype = np.uint8) # Yellow
range low threshold
    upper_yellow = np.array([34, 255, 255],dtype = np.uint8) # Yellow
range high threshold
```

Then, use OpenCV to open the camera, with a value of 0 indicating the first camera to be plugged into the system:

```
cap = cv2.VideoCapture(0) # Open the camera
if (cap.isOpened()): # Video opened successfully
```

Next, initialize some variables to store the panel light data found in each frame:

```
    red_twinkle_center = dict() # Dictionary of the center point where
the red light flashes during the video and the number of times it flashes
    green_twinkle_center = dict() # Dictionary of where and how many
times the green light flashes during a video
    yellow_twinkle_center = dict() # Dictionary of the number of times
the yellow light flashes during the video
    red_dict = dict() # red light location and number of frames found
    green_dict = dict() # green light position and number of frames found
    yellow_dict = dict() # yellow light position and number of frames
found
    red_center = [] # List of red light center points per frame
    green_center = [] # List of green light center points per frame
    yellow_center = [] # List of yellow light center points per frame
```

Each frame of the image is processed below:

```
while (True):
    ret, frame = cap.read() # Read a frame
```

Since OpenCV uses BGR pixel format instead of HSV format by default. The reason is that when early versions of OpenCV were developed, the BGR format was popular with camera providers and software vendors. BGR stands for blue, green, and red, for example # FF0000 for solid blue, because blue is at the maximum FF, while green and red are both at 00. We use the cvtColor() function and COLOR_BGR2HSV identifier to transform the input frame image from BGR (blue, green, and red) to HSV (hue, saturation, brightness):

11.4 Case Practice

```
hsv_img = cv2.cvtColor(frame, cv2.COLOR_BGR2HSV)
```

In the input frame, if the HSV value of a certain pixel is within the previously set panel light threshold range, it outputs 255 (white), otherwise it outputs 0 (black), thus obtaining the mask image. This step is to get the pixel region of the panel light, implemented using OpenCV's inRange() function:

```
mask_green = cv2.inRange(hsv_img, lower_green, upper_green) #
Select the green area by the HSV tri-channel color range
    mask_red = cv2.inRange(hsv_img, lower_red, upper_red) # Press the
HSV three-channel color range to select the red area
    mask_yellow = cv2.inRange(hsv_img, lower_yellow, upper_yellow) #
Select the yellow area by the HSV three-channel color range
```

Due to ambient light, air impurities, and other factors, there is a certain degree of noise in the obtained panel indicator image. These noises may interfere with subsequent tasks, such as panel light contour recognition. We need to use median filtering to remove the noise in order to get a smoother area of panel light pixels, thus making the contour recognition and center point calculation more accurate for subsequent tasks. To get a smoother image, we use a larger size convolution kernel (7 × 7 pixels) to calculate the median filter. OpenCV provides the medianBlur() function, which can easily generate the mask image's median filter denoising:

```
mask_green = cv2.medianBlur(mask_green, 7) # median filter to remove
the green area noise, convolution kernel 7*7
    mask_red = cv2.medianBlur(mask_red, 7) # median filter to remove red
region noise, convolution kernel 7*7
    mask_yellow = cv2.medianBlur(mask_yellow, 7) # median filter to
remove yellow region noise, convolution kernel 7*7
```

With a smooth image of the panel light, we can calculate its center position. The purpose of getting the center position is to mark the position of the light. This can be calculated in three steps, first using the findContours() function to get the outer outline of the lamp, then using the boundingRect() function to get the smallest rectangle that wraps this outer outline, and then calculating the center position of this rectangle:

```
mask_green, contours, hierarchy = cv2.findContours(mask_green, cv2.
RETR_EXTERNAL, cv2.CHAIN_APPROX_NONE) # obtains the outline of the
green light region
```

It is important to note that contours is a list of contours, and we need to work with each one:

```
for cnt in contours: # In the equipment room scenario, the green light
is detected in each green area
    (x, y, w, h) = cv2.boundingRect(cnt) # Get the smallest package
rectangle
    if abs(x-x_old)>80 and abs(y-y_old)>80: # Considering that the
indicator light has many shapes and may be recognized as multiple
objects, consider the object with similar coordinates as one light
        cv2.rectangle(frame, (x, y), (x+w, y+h), (0, 255, 0), 2) # Draws
the smallest wrapped rectangle on the frame image
        cv2.putText(frame, "green", (x, y-5), font, 0.7, (0, 255, 0), 2) #
Displays text at the upper position of the rectangle on the frame image
        green_center.append([(x+w)/2, (y+h)/2])
    x_old, y_old = x, y
```

Next, we need to deal with the event of each frame light on and light off. If the light corresponding to the center point is off in the current frame when processing the previous frame, it means that the light is blinking, register the number of flashes or the position of the flash:

```
for i in range(len(red_center)): # Handle all red light area center
points registered
    if (mask_red[int(red_center[i][0]), int(red_center[i][1])] ==
0): # The center position of the red light area of the red mask map is not
on, and the red light flashes once from on to off
        # Record the location of the center and the number of times it blinks
in the dictionary: {'x,y':times}
        is_in = False
        for key in red_twinkle_center: # traverse the dictionary
            center_x, center_y = getxy(key)
            if abs(center_x-red_center[i][0]) < 80 and abs(center_y-
red_center[i][1]) < 80: # Find this spot blinking in dictionary
                red_twinkle_center[key] += 1 # Number of flashes plus 1
                is_in = True
                break
        if is_in == False: # The flashing point is not in the dictionary, then
add to the dictionary
            key = str(int(red_center[i][0]))+','+str(int(red_center[i]
[1]))
            red_twinkle_center[key] = 1 # Flash 1 time
```

If no lights are found flashing, register the number of frames or the light center location:

11.4 Case Practice

```
        is_in = False
        for key in red_dict: # traverse the dictionary
            center_x, center_y = getxy(key)
            if abs(center_x-red_center[i][0]) < 80 and abs(center_y-
red_center[i][1]) < 80: # Find this spot blinking in dictionary
                red_dict[key] += 1 # Add 1 to the number of frames found
                is_in = True
                break
        if is_in == False: # The red dot is not in the dictionary, then the
dictionary is added
            key = str(int(red_center[i][0]))+','+str(int(red_center[i]
[1]))
            red_dict[key] = 1 # Found 1 time
```

It is worth noting that in the above code, the threshold of 80 is set to prevent interference and increase the stability of the program. After the above work is done, we can count the number of flashing lights and lights. In order to improve the stability of the program, we set 10 thresholds:

```
for key in red_twinkle_center: # Count the number of red blinking
lights
    if red_twinkle_center[key] > 10:
        red_twinkle_num += 1
for key in green_twinkle_center: # Count the number of blinking green
lights
    if green_twinkle_center[key] > 10:
        green_twinkle_num += 1
for key in yellow_twinkle_center: # Count the number of blinking
yellow lights
    if yellow_twinkle_center[key] > 10:
for key in red_dict: # Count the number of red lights
    if red_dict[key] > 10:
        red_num += 1
for key in green_dict: # Count the number of green lights
    if green_dict[key] > 10:
        green_num += 1
for key in yellow_dict: # Count the number of yellow lights
    if yellow_dict[key] > 10:
        yellow_num += 1    yellow_twinkle_num += 1
```

Finally, we display the found statistics on the screen, using the tf21 module to add Chinese characters to the frame image:

Fig. 11.13 Panel indicator monitoring scenario

```
def put_chinese(img, txt, pos, color, text_size):
    ft = ft21.put_chinese_text('msyh.ttf')
    image = ft.draw_text(img, pos, txt, text_size, color)
    return image
```

frame = put_chinese(frame, "Red light blinking :"+str(red_twinkle_num)+" green light blinking :"+str(green_twinkle_num)+" yellow light blinking :"+str(yellow_twinkle_num), (10, 40), (255,255,255), 18) # Displays the detected content in the upper part of the frame image

Use the imshow() function to display the frame image after the operation, because it is a frame-by-frame cycle operation, so the display effect is more coherent:

```
cv2.imshow("Monitor", frame)
```

After obtaining the data of the panel light, compare it with the normal status data registered in the database, and send a warning signal if it is inconsistent. This part of the code, different companies have different specific situations, there is no unified answer, this book is omitted. The following is the effect of our actual work, the upper left corner is the input frame of the Raspberry PI natural light camera, the upper right corner is the mask image after red HSV filtering, the lower left corner is the mask image after green HSV filtering, and the lower right corner is the mask image after yellow HSV filtering. As you can see, the red light is on and the green and yellow lights are blinking, as shown in Fig. 11.13.

11.4 Case Practice

For details about the code in this section, see Indicator Detect.py.

Temperature Thermography Detection Code Practice

The following introduces the code implementation of abnormal high-temperature detection in important locations of the machine room. UPS connectors and wires in the equipment room are important detection areas. Since the infrared camera of Raspberry PI has converted the image into a thermal image, the higher the temperature, the purer the red (or golden yellow), so we only need to judge whether there is an HSV color filter area in the input frame image, if there is, it means that the high temperature exceeds the standard, and send an early warning. We marked the found filter area with a rectangle, and put a text prompt in the output frame image:

```
# Define the HSV range of the high-temperature area, paying attention to
the need to adapt to specific scenes and specific infrared cameras
    lower_red = np.array([0, 127, 128],dtype = np.uint8) # Red Yellow
range Low threshold
    upper_red = np.array([34, 255, 255],dtype = np.uint8) # High
threshold in the red and yellow range
    ...
    mask_red, contours2, hierarchy2 = cv2.findContours(mask_red, cv2.
RETR_EXTERNAL, cv2.CHAIN_APPROX_NONE) # obtains the contours of the red
region, returning multiple objects
    for cnt2 in contours2: # In the equipment room scene, red is detected in
each red region
        (x2, y2, w2, h2) = cv2.boundingRect(cnt2) # Get the minimum package
rectangle
        cv2.rectangle(frame, (x2, y2), (x2 + w2, y2 + h2), (0, 0, 255), 2) #
Draws the smallest wrapped rectangle on the frame image
        cv2.putText(frame, "red", (x2, y2-5), font, 0.7, (0, 0, 255), 2) #
Displays text at the upper position of the rectangle on the frame image
```

if len(contours2)>0:

```
    frame = put_english (frame, "Temperature found to exceed
threshold!" ", (10, 10), (255,255,255), 18) # Displays the detected
content in the upper part of the frame image
```

The recognition effect we get in the actual work is shown in Fig. 11.14. It can be seen that under the condition of setting a certain HSV threshold range, the program marks the hot wire end part and achieves the expected effect. The code can be found in the download file "Temperature Thermal Imaging Detection.py."

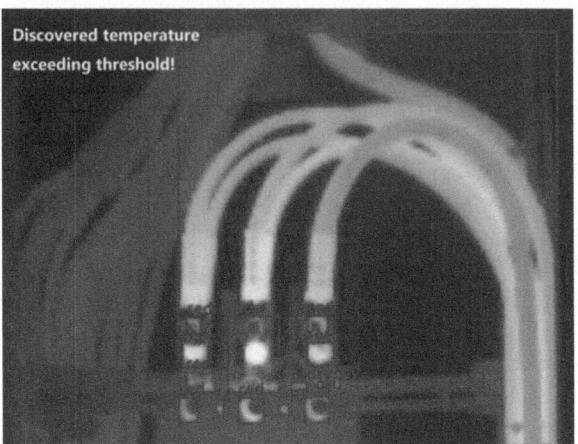

Fig. 11.14 Thermal imaging detection results

11.5 Case Summary

This case study illustrates the application of advanced computer vision technology, underpinned by the robust capabilities of the OpenCV library, to enhance the operational efficiency of a data center. Utilizing slide positioning and edge computing, the vision robot enables meticulous daily examinations, ensuring comprehensive real-time monitoring of power distribution equipment, servers, UPS panel indicators, and detecting abnormal high temperatures. This implementation guarantees continuous, 24-h inspection and instant, precise alarm notifications. Integration with the existing data center monitoring infrastructure significantly heightens the frequency and precision of equipment inspections, thereby considerably mitigating operational and maintenance risks for the bank's data center.

This instance represents a tangible deployment of artificial intelligence within the IT operations and maintenance spectrum of commercial banks, demonstrating substantial utility in expediting fault resolution processes. This, in turn, enhances the quality of operations and maintenance, while simultaneously upholding the political and social responsibilities incumbent upon commercial banks. The implementation of computer vision technology, particularly through edge computing, reveals diverse applications within banking business contexts. For example, it facilitates temperature monitoring of individuals entering and exiting banking premises during pandemics, supports unattended premises and peripheral fire surveillance, and more. Readers are encouraged to adapt the provided code to apply similar technologies to analogous scenarios, illustrating the versatility and practical relevance of these innovations.

The manufacturer's authorised representative in the EU is Springer Nature Customer Service Centre GmbH, Europaplatz 3, 69115 Heidelberg, Germany. If you have any concerns regarding our products, please contact ProductSafety@springernature.com

Printed and bound by CPI Group (UK) Ltd, Croydon, CR0 4YY

26/03/2026

02078939-0007